Advances in Plant Nitrogen Metabolism

Advances in Plant Nitrogen Metabolism is a thoughtful, provocative, and up-to-date volume that presents important physiological, biochemical, and molecular perspectives of the nitrogen metabolism in plants and regulatory networks underlying it. The book is an attempt to team up with global leading research experts working in the field of plant nitrogen metabolism to compile an up-to-date and wide-ranging volume. The main purpose of this book is to present information on the most recent developments including the different modern approaches and methodologies that are being currently employed in the field of plant nitrogen metabolism. We trust that this comprehensive volume will familiarize readers with the detailed mechanisms of nitrogen metabolism and its regulation and the current trends in this field of study.

The book offers comprehensive coverage of the most essential topics, including:

- Role of nitrogen and its assimilation in plants
- Recycling and remobilization of nitrogen during senescence
- Role of phytohormones in nitrogen metabolism
- Biological nitrogen fixation
- Nitrogen biofertilizers: role in sustainable agriculture
- Effect of stress on plant nitrogen metabolism
- Reactive nitrogen species (RNS) in plants
- Nitrogen toxicity in plants, symptoms, and safeguards
- Nitrogen metabolism enzymes: structure, role, and regulation
- Regulatory RNAs and their role in nitrogen metabolism of diazotrophs

As a pivotal contribution to the field, this volume is an invaluable and up-to-date foundation for plant physiologists, plant biochemists, geneticists, molecular biologists, agronomists, environmental researchers, and students of plant science. The book can also be used for the coursework of research and master's students.

Advances in Plant Nitrogen Metabolism

Edited by
Peerzada Yasir Yousuf, Peerzada Arshid Shabir,
and Khalid Rehman Hakeem

CRC Press is an imprint of the
Taylor & Francis Group, an **informa** business

Designed cover image: © Shutterstock

First edition published 2023
by CRC Press
6000 Broken Sound Parkway NW, Suite 300, Boca Raton, FL 33487-2742

and by CRC Press
4 Park Square, Milton Park, Abingdon, Oxon, OX14 4RN

© 2023 selection and editorial matter, Peerzada Yasir Yousuf, Peerzada Arshid Shabir, and Khalid Rehman Hakeem; individual chapters, the contributors

CRC Press is an imprint of Taylor & Francis Group, LLC

Reasonable efforts have been made to publish reliable data and information, but the author and publisher cannot assume responsibility for the validity of all materials or the consequences of their use. The authors and publishers have attempted to trace the copyright holders of all material reproduced in this publication and apologize to copyright holders if permission to publish in this form has not been obtained. If any copyright material has not been acknowledged please write and let us know so we may rectify in any future reprint.

Except as permitted under U.S. Copyright Law, no part of this book may be reprinted, reproduced, transmitted, or utilized in any form by any electronic, mechanical, or other means, now known or hereafter invented, including photocopying, microfilming, and recording, or in any information storage or retrieval system, without written permission from the publishers.

For permission to photocopy or use material electronically from this work, access www.copyright.com or contact the Copyright Clearance Center, Inc. (CCC), 222 Rosewood Drive, Danvers, MA 01923, 978-750-8400. For works that are not available on CCC please contact mpkbookspermissions@tandf.co.uk

Trademark notice: Product or corporate names may be trademarks or registered trademarks and are used only for identification and explanation without intent to infringe.

Library of Congress Cataloging-in-Publication Data
Names: Yousuf, Peerzada Yasir, editor. | Shabir, Peerzada Arshid, editor. | Hakeem, Khalid Rehman, editor.
Title: Advances in plant nitrogen metabolism / edited by Peerzada Yasir Yousuf, Peerzada Arshid Shabir, and Khalid Rehman Hakeem
Description: First edition | Boca Raton, FL : CRC Press, 2023. | Includes bibliographical references and index.
Identifiers: LCCN 2022026530 (print) | LCCN 2022026531 (ebook) | ISBN 9781032163949 (hardback) | ISBN 9781032163956 (paperback) | ISBN 9781003248361 (ebook)
Subjects: LCSH: Nitrogen–Metabolism. | Plants–Metabolism. | Botanical chemistry.
Classification: LCC QK881 .A38 2023 (print) | LCC QK881 (ebook) | DDC 572/.42–dc23/eng/20220608
LC record available at https://lccn.loc.gov/2022026530
LC ebook record available at https://lccn.loc.gov/2022026531

ISBN: 9781032163949 (hbk)
ISBN: 9781032163956 (pbk)
ISBN: 9781003248361 (ebk)

DOI: 10.1201/9781003248361

Typeset in Times
by Newgen Publishing UK

Contents

Foreword vii
Preface ix
About the Editors xi
List of Contributors xiii

Chapter 1   Introduction to Plant Nitrogen Metabolism: An overview 1

                 Sukumar Taria, Ajay Arora, Badre Alam, Sushil Kumar, Ashok Yadav, Sudhir Kumar, Mahesh Kumar, Hirdayesh Anuragi, Rajeev Kumar, Shashi Meena, Rahul Gajghate, and Ayyanadar Arunachalam

Chapter 2   Nitrogen: A Key Macronutrient for the Plant World 19

                 Ruchi Raina and Samina Mazahar

Chapter 3   Nitrogen Deficiency in Plants 29

                 Bilal Ahmad Wani, Sufiya Rashid, Kausar Rashid, Hanan Javid, Junaid Ahmad Magray, Rouf ul Qadir, and Tajamul Islam

Chapter 4   Nitrogen Assimilation in Plants 39

                 Reeta Kumari, Sonal Bhatnagar, and Charu Kalra

Chapter 5   Assimilation of Nitrates in Plants 55

                 Sufiya Rashid, Kausar Rashid, Hanan Javid, Bilal Ahmad Wani, Junaid Ahmad Magray, Rouf ul Qadir, Tajamul Islam, Mubariz Javed, and Irshad Ahmad Nawchoo

Chapter 6   Factors Affecting Nitrogen Uptake, Transport, and Assimilation 69

                 Junaid Ahmad Magray, Shabir Ahmad Zargar, Tajamul Islam, and Hanan Javid

Chapter 7   Role of Nitrogen in Photosynthesis 87

                 Afrozah Hassan, Shabana Gulzar, and Irshad Ahmad Nawchoo

Chapter 8   Recycling and Remobilization of Nitrogen during Senescence 97

                 Aadil Farooq War, Ishfaq Ahmad Sheergojri, Subzar Ahmad Nanda, Mohd Asgar Khan, Ishfaq Ul Rehmaan, Zafar Ahmad Reshi, and Irfan Rashid

Chapter 9   Role of Phytohormones in Nitrogen Metabolism 119

                 Samina Mazahar and Ruchi Raina

Chapter 10	Biological Nitrogen Fixation: An Overview	129

Hanan Javid, Rouf ul Qadir, Sufiya Rashid, Kausar Rashid, Junaid Ahmad Magray, Tajamul Islam, Bilal Ahmad Wani, Shabana Gulzar, and Irshad Ahmad Nawchoo

Chapter 11	Nitrogenase Enzyme Complex: Functions, Regulation, and Biotechnological Applications	143

Anandkumar Naorem, Jyotsana Tilgam, Parichita Priyadarshini, Yamini Tak, Alka Bharati, and Abhishek Patel

Chapter 12	*nod*, *nif*, and *fix* Genes	155

Irfan Iqbal Sofi, Shabir Ahmad Zargar, and Shivali Verma

Chapter 13	Nitrogen Biofertilizers: Role in Sustainable Agriculture	171

Ishfaq Ul Rehman, Bushra Jan, Nafeesa Farooq Khan, Tajamul Islam, Summia Rehman, Ishfaq Ahmad Sheergojri, Irfan Rashid, Aadil Farooq War, Subzar Ahmad Nanda, and Abid Hussain Wani

Chapter 14	Effect of Biotic Stresses on Plant Nitrogen Metabolism	185

Subzar Ahmad Nanda, Ishfaq Ahmad Sheergojri, Aadil Farooq War, Mohd Asgar Khan, and Ishfaq ul Rehmaan

Chapter 15	Salt Stress and Nitrogen Metabolism in Plants	195

Peerzada Yasir Yousuf and Arjumand Frukh

Chapter 16	Reactive Nitrogen Species in Plants	203

Urfi Jahan, Uzma Kafeel, Fareed Ahmad Khan, and Afrin Jahan

Chapter 17	Nitrogen Toxicity in Plants, Symptoms, and Safeguards	213

Summia Rehman, Humara Fayaz, Ishfaq Ul Rehmaan, Kausar Rashid, and Sufiya Rashid

Chapter 18	Nitrogen Metabolism Enzymes: Structure, Role, and Regulation	227

Shafia Liyaqat Nahvi and Afiya Khurshid

Chapter 19	Regulatory RNAs and Their Role in Nitrogen Metabolism of Diazotrophs	237

Azra Quraishi and Mehreen Fatima

Index 253

Foreword

Plant Nitrogen Metabolism is a rapidly advancing field of study where new findings are surfacing in the literature almost daily. The ways of understanding different aspects of this field have undergone a sea change and there is now a far greater emphasis on the elucidation of the molecular mechanisms operative in the regulation of nitrogen metabolism. It has become increasingly difficult for students to keep abreast with the ongoing explosion of knowledge in this field.

This book on nitrogen metabolism – *Advances in Plant Nitrogen Metabolism* – is both authoritative and timely. The book focuses on an in-depth analysis of wide-ranging topics in plant nitrogen metabolism. This up-to-date and wide-ranging volume covers almost all the aspects of nitrogen metabolism that are regularly organized in chapters starting with the overview, followed by the uptake, assimilatory processes, and their regulation. Then the effects of environmental stresses on nitrogen metabolism are discussed. The end chapters deal with biological nitrogen, the interaction of nitrogen fixers with plants, enzymes, and genes involved, and the assimilation of products of nitrogen fixation. This book describes in detail the significant advances that have been made in understanding the physiological, biochemical, and molecular aspects of nitrogen metabolism and its regulation in plants. I trust that this comprehensive volume will familiarize readers with the detailed mechanisms of nitrogen metabolism and its regulation and the current trends in this field of study.

Preface

Nitrogen is one of the primary macronutrients critical for the growth, development, and productivity of plants. It forms an important component of many biomolecules, including proteins, nucleic acids, chlorophyll, cofactors, signaling molecules, secondary metabolites, etc. Although nitrogen is very abundant in the atmosphere (as dinitrogen gas), it is largely inaccessible to plants, making it a scarce resource and often a key limiting nutrient. To maintain the productivity of the burgeoning population there has been dramatic use of nitrogen fertilizers in agriculture. However, the excessive use of these nitrogenous fertilizers is associated with environmental problems. Owing to its importance, nitrogen metabolism has been the focus of research from very earlier times. Significant advances have been made in understanding the physiological, biochemical, and molecular aspects of nitrogen metabolism and its regulation in plants. Large attention has been paid towards finding an alternative solution to the nitrogen conundrum like increasing the nitrogen-use-efficiency of plants and optimizing the use of diazotrophs to fix nitrogen biologically. The development of smart crop plants with improved assimilation and management of nitrogen would reduce the need for intensive nitrogen fertilization and positively influence the environment. The new molecular omics approaches have been utilized to provide new integrated information both about the nitrogen metabolic processes including their regulation under normal and stress conditions and determinants of nitrogen-use-efficiency in plants. Many new genes and proteins involved in the underlying molecular mechanisms have been discovered and used for the development of nitrogen-use-efficient plants.

Advances in Plant Nitrogen Metabolism is a thoughtful, provocative, and up-to-date volume that presents important physiological, biochemical, and molecular perspectives of the regulatory networks underlying nitrogen metabolism in plants. The chapters provide a concise presentation of how the nitrogen is uptaken, transported, assimilated, and remobilized within the plant body. An emphasis has been laid on the recent approaches in understanding the mechanisms underlying nitrogen metabolism and its regulation. Besides, a comprehensive description of recent advances on the influence of phytohormones on nitrogen metabolism, the need, and the strategies of developing nitrogen-use-efficient plants is discussed. The last chapters cover many important facets of biological nitrogen fixation, including the interaction of nitrogen fixers with plants, enzymes, and genes involved, and the assimilation of products of nitrogen fixation.

The book is an attempt to team up with global leading research experts working in the field of plant nitrogen metabolism to compile an up-to-date and wide-ranging volume. The main purpose of this book is to present information on the most recent developments including the different modern approaches and methodologies that are being currently employed in the field of plant nitrogen metabolism. We trust that this comprehensive volume will familiarize readers with the detailed mechanisms of nitrogen metabolism and its regulation and the current trends in this field of study.

About the Editors

Peerzada Yasir Yousuf (PhD., MSc., CSIR-NET, ARS-NET, SET, GATE) earned his B.Sc. from the University of Kashmir in 2008 and M.Sc. and Ph.D. degrees in Botany from Jamia Hamdard, New Delhi in 2010 and 2016, respectively. He worked in the Molecular Ecology Lab as a Senior Research Fellow in the Department of Botany Jamia Hamdard New Delhi. He taught environmental science as guest faculty in Jamia Hamdard in different faculties like the Faculty of Unani Medicine, Faculty of Information Technology, etc. In his Ph.D. program he worked on the biochemical and molecular aspects of stress tolerance in plants. Dr. Yousuf has qualified for many competitive examinations like UGC-CSIR NET (AIR 39), ARS-NET, GATE (AIR 45), and JK-SET. He completed a certification course on Mitochondria from Harvard University, United States. He also pursued NPTEL online certification course on "Bioenergetics of life processes" from Indian Institute of Technology (IIT) Kanpur with Gold + Elite category and was among toppers (Score 91%) besides pursuing NPTEL online certification course on "Introduction to Proteomics" from Indian Institute of Technology (IIT) Bombay with Elite category (Score 71%). He was selected for the SERB National Post-Doctoral fellowship (N-PDF) by the Department of Science and Technology, Government of India in 2016.

Dr. Yousuf has six years of research experience and 11 years of teaching experience. Currently, he holds the position of Assistant Professor of Botany in the Department of Higher Education Jammu and Kashmir. He has published several research articles and book chapters in globally reputed journals like *Frontiers in Plant Science, Genes, Plos One, Protoplasma, Plant Growth Regulation, Environmental Science and Pollution Research*, etc. Dr. Yousuf has presented his work at several national and international conferences. He has also delivered extension lectures in different fields of plant sciences at different institutes. He is a lifetime member of different national and international organizations besides being the recipient of some fellowships. He also serves as a member of the editorial board and is a reviewer of many research journals. He is actively engaged in the research fields of plant physiology and molecular ecology.

Peerzada Arshid Shabir (PhD., MSc., CSIR-NET) received his Ph.D. in the field of Plant Reproductive Ecology and Genetic Diversity and is presently working as an Assistant Professor in the Department of Higher Education, Jammu and Kashmir, India. He is actively engaged in the teaching of Plant Reproductive Ecology, Plant Physiology, Cell and Molecular Biology, and Genetic Engineering of Plants. He has published many research papers, review articles, and book chapters in reputed international journals. He has also been a reviewer of many reputed international journals and has presented many papers in various national and international conferences. His current research interests are to study the genomic bases of phenotypic plasticity and the study of the genomic bases of adaptation and speciation. Dr. Peerzada Arshid Shabir has been conferred with junior and senior research fellowships by Counsel of Scientific and Industrial Research, New Delhi. He has ten years of research experience and four years of teaching experience.

Khalid Rehman Hakeem (Ph.D., FRBS) is Professor at King Abdul-Aziz University, Jeddah, Saudi Arabia. After completing his doctorate (Botany; specialization in Plant Eco-physiology and Molecular Biology) from Jamia Hamdard, New Delhi, India, he worked as a lecturer at the University of Kashmir, Srinagar, India, for a short period. Later, he joined Universiti Putra Malaysia, Selangor, Malaysia, and worked there as a Post-doctorate Fellow and Fellow Researcher (Associate Professor) for several years. Dr. Hakeem has more than ten years of teaching and research experience in plant eco-physiology, biotechnology and molecular biology, medicinal plant research, plant-microbe-soil interactions, as well as in environmental studies. He is the recipient of several fellowships at both national and international levels. He has served as a visiting scientist at Jinan University, Guangzhou, China. Currently, he is involved with several international research projects with different government organizations. To date, Dr. Hakeem has authored and edited more than 80 books with international publishers, including Springer Nature, Academic Press (Elsevier), and CRC Press. He also has to his credit more than 180 research publications in peer-reviewed international journals and 75 book chapters in edited volumes with international publishers. At present, Dr. Hakeem serves as an editorial board member and reviewer for several high-impact international scientific journals from Elsevier, Springer Nature, Taylor and Francis, Cambridge, and John Wiley Publishers. He is included in the advisory board of Cambridge Scholars Publishing, UK. He is also a fellow of the Royal Society of Biology, fellow of Plantae group of the American Society of Plant Biologists, member of the World Academy of Sciences, member of the International Society for Development and Sustainability, Japan, and member of the Asian Federation of Biotechnology, Korea. Currently, Dr. Hakeem is engaged in studying the plant processes at ecophysiological as well as molecular levels.

Contributors

Badre Alam
ICAR-Central Agroforestry Research Institute
Jhansi, India

Hirdayesh Anuragi
ICAR-Central Agroforestry Research Institute
Jhansi, India

Ajay Arora
ICAR-Division of Plant Physiology
Indian Agricultural Research Institute
New Delhi, India

Ayyanadar Arunachalam
ICAR-Central Agroforestry Research Institute
Jhansi, India

Alka Bharati
ICAR-Central Agroforestry Research Institute
U.P., India

Sonal Bhatnagar
Department of Environmental Sciences
SPM College, University of Delhi
New Delhi, India

Mehreen Fatima
Department of Life Sciences
University of Management and
  Technology
Lahore, Pakistan

Humara Fayaz
Department of Botany
University of Kashmir
J&K, India

Arjumand Frukh
Department of Botany
School of Chemical and Life Sciences
Jamia Hamdard
New Delhi, India

Rahul Gajghate
ICAR- Indian Grassland and Fodder
  Research Institute
Jhansi, India

Shabana Gulzar
Govt Degree College for Women Cluster
  University Srinagar
J&K, India

Afrozah Hassan Department of Botany
University of Kashmir
J&K, India

Tajamul Islam
Department of Botany, University of Kashmir
J&K, India

Afrin Jahan
Department of Botany
Aligarh Muslim University
Uttar Pradesh, India

Urfi Jahan
Department of Botany
Aligarh Muslim University
Uttar Pradesh, India

Bushra Jan
Department of Botany
University of Kashmir
J&K, India

Mubariz Javed
Department of Botany
University of Kashmir
J&K, India

Hanan Javid
Department of Botany
University of Kashmir
J&K, India

Uzma Kafeel
Department of Botany
Aligarh Muslim University
Uttar Pradesh, India

Charu Kalra
Department of Botany
Deen Dayal Upadhyaya College,
  University of Delhi
New Delhi, India

Mohd Asgar Khan
Department of Botany
University of Kashmir
J&K, India

Fareed Ahmad Khan
Department of Botany
Aligarh Muslim University
Uttar Pradesh, India

Nafeesa Farooq Khan
Department of Botany
University of Kashmir
J&K, India

Afiya Khurshid
Department of Vegetable Science
Sher-e-Kashmir University of Agricultural
   Science and Technology
Kashmir, India

Rajeev Kumar
ICAR- Indian Institute of Vegetables Research
Varanasi, India

Sushil Kumar
ICAR-Central Agroforestry Research Institute
Jhansi, India

Mahesh Kumar
ICAR-Division of Plant Physiology
Indian Agricultural Research Institute
New Delhi, India

Sudhir Kumar
ICAR-Division of Plant Physiology
Indian Agricultural Research Institute
New Delhi, India

Reeta Kumari
Department of Botany
Deen Dayal Upadhyaya College,
   University of Delhi
New Delhi, India

Junaid Ahmad Magray
Department of Botany
University of Kashmir
J&K, India

Samina Mazahar
Department of Botany
Dyal Singh College, University of Delhi
New Delhi, India

Shashi Meena
ICAR-Division of Plant Physiology
Indian Agricultural Research Institute
New Delhi, India

Shafia Liyaqat Nahvi
Department of Fruit Science
Sher-e-Kashmir University of Agricultural
   Science and Technology
Kashmir, India

Subzar Ahmad Nanda
Department of Botany
University of Kashmir
J&K, India

Anandkumar Naorem
ICAR-Central Arid Zone Research
   Institute
Gujarat, India

Irshad Ahmad Nawchoo
Department of Botany
University of Kashmir
J&K, India

Abhishek Patel
ICAR-Central Arid Zone Research Institute
Gujarat, India

Parichita Priyadarshini
ICAR-Crop Improvement Division
Indian Grassland and Fodder
   Research Institute
U.P., India

Rouf ul Qadir
Department of Botany
University of Kashmir
J&K, India

Azra Quraishi
University of Management and Technology
Lahore, Pakistan

Contributors

Ruchi Raina
Department of Botany
Dyal Singh College, University of Delhi
New Delhi, India

Kausar Rashid
Department of Botany
University of Kashmir
J&K, India

Sufiya Rashid
Department of Botany
University of Kashmir
J&K, India

Irfan Rashid
Department of Botany
University of Kashmir
J&K, India

Ishfaq Ul Rehmaan
Department of Botany
University of Kashmir
J&K, India

Ishfaqul Rehman
Department of Botany
University of Kashmir
J&K, India

Summia Rehman
Department of Botany
University of Kashmir
J&K, India

Sumaiya Rehman
Department of Botany
University of Kashmir
J&K, India

Zafar Ahmad Reshi
Department of Botany
University of Kashmir
J&K, India

Ishfaq Ahmad Sheergojri
Department of Botany
University of Kashmir
J&K, India

Irfan Iqbal Sofi
Department of Botany
University of Kashmir
J&K, India

Yamini Tak
Agriculture University
Rajasthan, India

Sukumar Taria
ICAR-Division of Plant Physiology
Indian Agricultural Research Institute
New Delhi, India

Jyotsana Tilgam
ICAR- National Bureau of Agriculturally
   Important Microorganisms
U.P., India

Shivali Verma
Department of Botany
University of Jammu
J&K, India

Bilal Ahmad Wani
Department of Botany
University of Kashmir
J&K, India

Abid Hussain Wani
Department of Botany
University of Kashmir
J&K, India

Aadil Farooq War
Department of Botany
University of Kashmir
J&K, India

Ashok Yadav
ICAR-Central Agroforestry Research
   Institute
Jhansi, India

Shabir Ahmad Zargar
Department of Botany
University of Kashmir
J&K, India

1 Introduction to Plant Nitrogen Metabolism
An overview

Sukumar Taria[1,2]*, Ajay Arora[1], Badre Alam[2], Sushil Kumar[2], Ashok Yadav[2], Sudhir Kumar[1], Mahesh Kumar[1], Hirdayesh Anuragi[2], Rajeev Kumar[3], Shashi Meena[1], Rahul Gajghate[4], and Ayyanadar Arunachalam[2]

[1]ICAR-Division of Plant Physiology, Indian Agricultural Research Institute, New Delhi, India
[2]ICAR-Central Agroforestry Research Institute, Gwalior Road, Jhansi, India
[3]ICAR- Indian Institute of Vegetables Research, Varanasi, India
[4]ICAR- Indian Grassland and Fodder Research Institute, Jhansi, India
*Corresponding author- physiokunal1@gmail.com

CONTENTS

1.1	Introduction	2
1.2	Nitrogen, Its Occurrence, Importance, and Role in Plants: An Overview	2
1.3	Nitrogen Deficiency in Plants	4
1.4	Nitrogen Assimilation in Plants	4
	1.4.1 Nitrate Reductase Converts Nitrate to Nitrite	4
	1.4.2 Nitrite Reductase Converts Nitrite to Ammonium	5
	1.4.3 Nitrogen Uptake and Transport in Plants: Pathways, Transporters, Etc.	6
1.5	Regulatory RNAs and Their Role in Nitrogen Metabolism	6
1.6	Role Of Nitrogen in Photosynthesis	6
1.7	Recycling and Remobilization of Nitrogen during Senescence	7
1.8	Role of Phytohormones in Nitrogen Metabolism	7
1.9	Regulation of Nitrogen Metabolism under Abiotic Stress	7
	1.9.1 Nitrogen Metabolism under Drought Stress	7
	1.9.2 Nitrogen Metabolism under High-Temperature Stress	8
	1.9.3 Nitrogen Metabolism under Salinity Stress	8
	1.9.4 Nitrogen Metabolism under Heavy Metal Stress	8
1.10	Epigenetic Regulation of Nitrogen Metabolism	9
1.11	Omics Approaches to Nitrogen Metabolism in Plants	9
1.12	Reactive Nitrogen Species (RNS) in Plants	9
1.13	Nitrogen Use Efficiency in Plants: Physiological and Molecular Aspects	10
1.14	Nitrogen Toxicity in Plants, Symptoms and Safeguards	10
1.15	Biological Nitrogen Fixation: An Overview	10
1.16	Nitrogenase Enzyme Complex	11
1.17	*nod*, *nif*, and *fix* Genes	11

DOI: 10.1201/9781003248361-1

1.18 Biochemical and Functional Aspects of Legume Nodule 12
    1.18.1 Synthesis, Transport, and Assimilation of Products of Symbiotic
           Nitrogen Fixation 12
1.19 Nitrogen Biofertilizers: Role in Sustainable Agriculture 12
References 14

1.1 INTRODUCTION

Many biochemical compounds present in plant cells contain nitrogen (e.g., nitrogen is found in the nucleoside phosphates and amino acids that form the building blocks of nucleic acids and proteins, respectively). Only the elements oxygen, carbon, and hydrogen are more abundant in plants than nitrogen. Most natural and agricultural ecosystems show dramatic gains in productivity after fertilization with inorganic nitrogen, attesting to the importance of this element. Since the nitrogen element has been bonded by a triple bond, higher energy is required to break this bond for its assimilation. This element is assimilated into amino acids required for the biosynthesis of organic compounds.

1.2 NITROGEN, ITS OCCURRENCE, IMPORTANCE, AND ROLE IN PLANTS: AN OVERVIEW

Nitrogen is one of the essential elements for the growth, development, and reproduction of plants and animals as well (Leghari *et al.*, 2016; Yousuf *et al.*, 2021). It is also the fourth most ample element in the living biomass after hydrogen, carbon, and oxygen. It naturally occurs in the earth's atmosphere and comprises about 80% of the biosphere (the rest of the biosphere is made up of oxygen and other gases). Hence, its availability in the biosphere is very prominent and omnipresent. Nitrogen is found in amino acids and nucleotides and thus in all proteins and nucleic acids. Furthermore, amino acids are the building blocks of all proteins. Proteins comprise structural components; namely, cells, tissue, and organs besides enzymes and hormones that are of utmost importance for the functioning of all living creatures on the earth. Nitrogen is an element that can combine with other elements to make different compounds. A specialized group of bacteria are the only known organisms that can fix inert compounds of nitrogen into biologically-useful compounds. Nitrogen occurs in both forms (organic and inorganic). The organic forms of nitrogen comprise amino acids, proteins, nucleic acids, nucleotides, and urea. However, inorganic nitrogen is mainly found as N_2 gas, nitrate, nitrite, and ammonium. Most of the nitrogen on the earth is present as N_2 gas. This N_2 gas becomes biologically available only through the different processes (fixation by bacterial, lightning, or volcanic eruptions and human activities). The organic form of nitrogen is converted into inorganic form first by soil microorganisms before it is used by plants, a process known as nitrogen mineralization (Figure 1.1). The conversion of organic nitrogen into inorganic in the soil is mediated by a group of specialized bacteria that normally use different forms of nitrogen to fuel

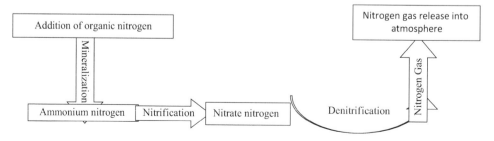

FIGURE 1.1 Nitrogen conversion processes.

some of their metabolic processes. Nitrogen present especially in the form of organic materials, amino acids, proteins, nucleic acids, nucleotides, and urea is transformed into ammonia, (NH_3) or ammonium (NH_4^+) by special types of bacteria during the decomposition process, which is known as ammonification. Besides ammonification, nitrification and denitrification are the other N transformation processes. Nitrification is an aerobic process in which ammonia is converted to nitrites (NO_2^-) and then nitrates (NO_3^-); this process occurs only in the presence of oxygen. However, denitrification is an anaerobic process in which nitrates (NO_3^-) are converted into gaseous nitrogen. Denitrification takes place only when no oxygen or extremely low concentrations of oxygen is available and is known to be harmful to the agricultural system.

Nitrogen is present in many forms in the biosphere. The atmosphere contains a vast 78% by volume of molecular nitrogen (N_2). Acquisition of nitrogen from the atmosphere requires the breaking of a stable triple covalent bond between two nitrogen atoms (N-N) to produce ammonia (NH_3) or nitrate (NO_3^-). These reactions, known as nitrogen fixation, can be performed by both industrial and natural processes.

N_2 combines with hydrogen to form ammonia under elevated temperature (about 200°C) and high pressure (about 200 atmospheres). Extreme temperature and pressure are required to overcome the high activation energy of the reaction. This nitrogen fixation reaction is called the Haber–Bosch process of nitrogen fixation. Lightening also contributes to nitrogen fixation (8% of the nitrogen fixed) and converts water vapor and oxygen into highly reactive hydroxyl free radicals, free hydrogen atoms, and free oxygen atoms that attack molecular nitrogen (N_2) to form nitric acid (HNO_3). This nitric acid then falls to Earth with rain. About 2% of the nitrogen fixed is derived from photochemical reactions between gaseous nitric oxide (NO) and ozone (O_3) that produce nitric acid (HNO_3) (Figure 1.2). The remaining 90% results from biological nitrogen fixation, in which bacteria or blue-green algae (cyanobacteria) fix N_2 into ammonium (NH^{4+}).

Asymbiotic nitrogen fixation: This process includes the conversion of nitrogen into ammonia is done by free-living bacteria (*Chlorobium*, *Rhodospirillum*) and *Cyanobacteria* (*Nostoc*, *Anabaena*).

Symbiotic nitrogen fixation: The conversion of atmospheric nitrogen (unavailable form) into ammonia (available form) is done by microbes [bacteria (*Rhizobium spp.*), algae (blue-green algae), and actinomycetes (*Frankia*)] in symbiotic association with plants.

Associative nitrogen fixation: In this process, the conversion of atmospheric nitrogen (unavailable form) into ammonia (available form) is done by microbes (*Azospirillum*) in casual association with plants.

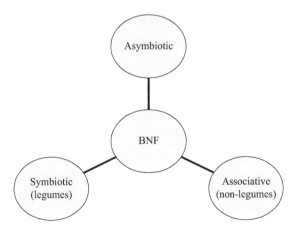

FIGURE 1.2 Diagrammatic view of biological nitrogen fixation.

Nitrogen is known to be an essential macronutrient for the plant to function well and is the main component of amino acids, which form the building blocks of plant proteins and enzymes (Leghari et al., 2016). Proteins build up the structural materials of all living things and enzymes help the myriad of biochemical reactions within a plant. Nitrogen is also a major component of chlorophyll, which is essential for the photosynthetic process in the plant and indirectly influences plant growth and yield. Nitrogen also produces rapid early growth, improves fruit quality, enhances the growth of leafy vegetables, increases the protein content of fodder crops besides encouraging the uptake and utilization of other nutrients including potassium and phosphorous, and controls the overall growth of the plant (Bloom, 2015; Hemerly, 2016; Leghari et al., 2016). Nitrogen plays a vital role in increasing the yield and quality of crops (Leghari et al., 2016). The optimum rate of nitrogen application not only improves photosynthetic processes, leaf area production, leaf area duration, and the net assimilation rate but also overcomes its losses. Optimum nitrogen fertilization significantly influences production and yield attributes (i.e., plant height, number of tillers, number of spikelets, and test weight of crops) (Ali et al., 2011) and also improves sugar quality parameters such as Brix, TSS, reducing sugars, and sucrose content of sweet sorghum (Kumar et al., 2016).

1.3 NITROGEN DEFICIENCY IN PLANTS

Nitrogen is the mineral element that plants require in the greatest amounts. It acts as a constituent of different plant components, including amino acids, proteins, and nucleotides (purines and pyrimidines). Therefore, nitrogen deficiency rapidly inhibits plant growth. Deficiency of this nutrient causes chlorosis, which is the yellowing of the leaves in older leaves due to the ease in movement of nitrogen from older to younger parts. Under severe deficiency, these leaves become completely yellow and fall off the plant. Younger leaves may not show these symptoms initially because nitrogen can be mobilized from older leaves. Thus, a nitrogen-deficient plant may have light green upper leaves and yellow or tan lower leaves.

When nitrogen deficiency develops slowly, plants may have markedly slender and often woody stems. This woodiness may be due to a buildup of excess carbohydrates that cannot be used in the synthesis of amino acids or other nitrogen-containing compounds. Carbohydrates not used in nitrogen metabolism may also be used in anthocyanin synthesis, leading to the accumulation of that pigment. This condition is revealed as a purple coloration in leaves, petioles, and stems of nitrogen-deficient plants of some species, such as tomato and certain varieties of corn. Nitrate reductase is the main molybdenum-containing protein in vegetative tissues, and one symptom of molybdenum deficiency is the accumulation of nitrate that results from diminished nitrate reductase activity.

1.4 NITROGEN ASSIMILATION IN PLANTS

1.4.1 Nitrate Reductase Converts Nitrate to Nitrite

Most of the nitrogen that is absorbed by the plant as nitrate is assimilated into organic nitrogen compounds. The reduction of nitrate into nitrite in the cytosol by a two-electron donor reaction through nitrate reductase (NR) enzyme is the first step of the process (Oaks, 1994). The enzyme nitrate reductase catalyzes this reaction as follows:

$$NO_3^- + NAD(P)H + H^+ + 2e^- \rightarrow NO_2^- + NAD(P)^+ + H_2O$$

where NAD(P)H indicates NADH or NADPH. The most common form of nitrate reductase uses only NADH as an electron donor; another form of the enzyme that is found predominantly in non-green tissues such as roots can use either NADH or NADPH (Warner and Kleinhofs, 1992). The nitrate reductases of higher plants are homodimers, each containing three prosthetic groups: FAD (flavin adenine dinucleotide), heme, and molybdenum complexed to an organic molecule called

Introduction to Plant Nitrogen Metabolism

a pterin (Mendel and Stallmeyer, 1995; Campbell, 1999). The FAD-binding domain accepts two electrons from NADH or NADPH. The electrons then pass through the heme domain to the molybdenum complex, where they are transferred to nitrate (Figure 1.3).

1.4.2 NITRITE REDUCTASE CONVERTS NITRITE TO AMMONIUM

Nitrite (NO_2^-) is a highly toxic ion and therefore plant cells immediately transport the nitrite into chloroplasts in leaves and plastids in roots to the harmful effect of toxic ions. In these organelles, the enzyme nitrite reductase reduces nitrite to ammonium according to the following overall reaction through a six-electron reduction process:

$$NO_2^- + 6Fd_{red} + 8H^+ + 6e^- \rightarrow NH_4^+ + 6Fd_{ox} + 2H_2O$$

where Fd is ferredoxin and the subscripts red and ox stand for reduced and oxidized, respectively. Reduced ferredoxin derives from photosynthetic electron transport in the chloroplasts, whereas NADPH is generated by the oxidative pentose phosphate pathway in non-green tissues.

Chloroplasts and root plastids contain different forms of the enzyme, but both forms consist of a single polypeptide containing two prosthetic groups: an iron-sulfur cluster (Fe_4S_4) and a specialized heme (Siegel and Wilkerson, 1989). These groups acting together bind nitrite and reduce it directly to ammonium, without accumulation of nitrogen compounds of intermediate redox states (Figure 1.4). Nitrite reductase is encoded in the nucleus and synthesized in the cytoplasm with an N-terminal transit peptide that targets it to the plastids (Wray, 1993). Whereas $NO3^-$ and light

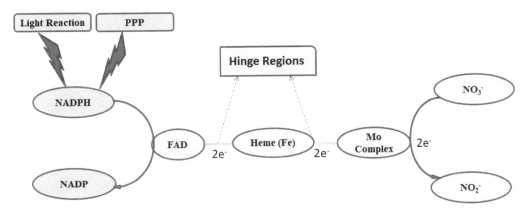

FIGURE 1.3 Conversion of nitrate to nitrite.

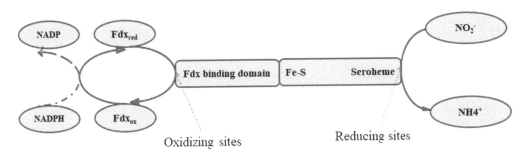

FIGURE 1.4 Conversion of nitrite to ammonium.

induce the transcription of nitrite reductase mRNA, the end products of the process—asparagine and glutamine—repress this induction.

1.4.3 Nitrogen Uptake and Transport in Plants: Pathways, Transporters, Etc.

Nitrate and ammonium are the two important nitrogen forms that are available in soils for plants for assimilation. Ammonium is a predominant form for rice grown in the paddy field, but nitrogen (40% of the total) is also taken up by rice in the form of nitrate due to the nitrification in the rhizosphere zone (Kirk and Kronzucker, 2005). The most important transporter of nitrate in plants is the NITRATE TRANSPORTER 1(NRT1)/PEPTIDE TRANSPORTER (PTR) family (NPF) and NRT2 family. It has been reported that there are 93 NPF members and 4 NRT2 members in the rice genome (Wang et al. 2018c). Various studies inferred that overexpression of various transporters like *OsNPF6.3/OsNRT1.1A*, *OsNRT1.1B*, *OsNPF7.3*, *OsNPF7.7*, *OsPTR9*, and *OsNRT2.3b* in rice enhanced grain yield and nitrogen-use efficiency (Fan et al. 2016b; Fang et al. 2013, 2017; Hu, B. et al. 2015a, b; Hu, J. et al. 2015; Huang et al. 2018; Wang et al. 2018b).

The transport of ammonium is mediated by ammonium transporter/methylammonium permease/Rhesus (AMT/MEP/Rh) family members. There are 12 predicted AMTs identified in the rice genome (Li et al., 2009a; Sonoda et al., 2003). Interestingly analyses of transgenic plants revealed that overexpression of *AMT* genes in rice often causes impaired growth resulting from carbon/nitrogen metabolic imbalance (Bao et al., 2015; Hoque et al., 2006). However, Ranathunge et al. (2014) showed that expression of Os AMT1;1 can enhance the yield in rice.

1.5 REGULATORY RNAS AND THEIR ROLE IN NITROGEN METABOLISM

Several ncRNAs like small non-coding RNAs and long non-coding RNAs are involved in nitrogen response. In *A. thaliana*, AUXIN RESPONSE FACTOR8 (ARF8) is the target of miR167, and nitrate influx inhibited the expression of miR167, thus allowing the accumulation of ARF8 in the pericycle region to stimulate the lateral root growth (Gifford et al., 2008). MiR160 and miR171(Liang et al., 2012) have been ported to regulate root system architecture by regulating the transcription factor genes like ARF10/16/17 and SCARECROW-LIKE PROTEIN 6 (SCL6) respectively (Wang et al., 2005; 2010). miR393 controls the root system architecture in response to both internal and external N concentration by regulating the AUXIN-SIGNALING F-BOX PROTEIN 3 (AFB3) in Arabidopsis (Vidal et al., 2010). It has been demonstrated that miR169 inhibits nitrogen uptake by downregulating the expression of NUCLEAR FACTOR Y SUBUNIT A5 (NFYA5), a positive regulator of nitrate transporter genes AtNRT2.1 and AtNRT1.1 (Zhao et al., 2011). miR826 and miR5090 regulate nitrogen metabolism by regulating key gene ALKENYL HYDROXALKYL PRODUCING 2 (AOP2), which encodes 2-oxoglutarate-dependent dioxygenase involved in glucosinolate biosynthesis (Kliebenstein et al., 2001). lncRNAs like TS120 and TAS3 have been reported to be involved in nitrogen response (Liu et al., 2019; Fukuda et al., 2019). Genome-wide microarray-based analysis identified 14 differentially expressed miRNAs in maize (miR160, miR164, miR167, miR168, miR169, miR172, miR319, miR395, miR397, miR398, miR399, miR408, miR528, and miR827) under chronic (15 days) or transient (2 h) nitrogen deficiency (Xu et al., 2011).

1.6 ROLE OF NITROGEN IN PHOTOSYNTHESIS

Nitrogen (N) is an essential macronutrient that determines growth and development in plants. It serves as a vital constituent of various biomolecules (i.e., proteins, enzymes, nucleic acids, chlorophyll, secondary metabolites, and hormones) that regulate metabolic processes in plants. Its concentration is positively correlated with the photosynthetic rate because a substantial portion of N assimilated by plants goes to the formation of photosynthetic apparatus/chloroplast (Nunes-Nesi et al., 2010). Besides this, N also acts as an important regulator in the manipulation of CO_2 diffusion, apart from taking part in photosynthesis (Xiong et al., 2015). It has been also documented that nitrogenous compounds serve as signaling molecules to regulate the responses of plants to environmental stimuli and provide protection against adverse conditions (Zhong et al., 2017).

Nitrogen is an important element for photosynthesis as it is a component of the tetrapyrrole structure of chlorophyll molecules. It is well known that chlorophyll molecules present in the thylakoid component of chloroplast absorb the light energy and convert it into chemical energy in the form of ATP, which is needed to drive the essential reaction of photosynthetic processes.

1.7 RECYCLING AND REMOBILIZATION OF NITROGEN DURING SENESCENCE

Nutrient remobilization and degradation of macromolecules are the essential physiological processes during plant senescence (Peerzada and Iqbal, 2021) Amongst the plant hormones, cytokinin is the major plant growth regulator involved in the remobilization of the nutrients from the senescent part of the plant to younger parts. It is reported that *OsNAP*, a NAC transcription factor, acts as a checkpoint of age-dependent and ABA-mediated senescence (Liang et al., 2014). *OsNAP* plays a crucial role in regulating the nutrient remobilization from leaves to the filling grains by directly binding to the promoters of nutrient transporters.

1.8 ROLE OF PHYTOHORMONES IN NITROGEN METABOLISM

Different signaling compounds, acting either positively or negatively, control the position of nodule primordia. The nucleoside uridine diffuses from the stele into the cortex in the protoxylem zones of the root and stimulates cell division (Lazarowitz and Bisseling, 1997). Ethylene is synthesized in the region of the pericycle, diffuses into the cortex, and blocks cell division opposite the phloem poles of the root. During ammonia assimilation, it has been shown that OsIDD10 positively regulates the expression of ammonium uptake and assimilation-related genes and is involved in the crosstalk between ammonium and auxin signaling (Xuan et al. 2013, 2018).

1.9 REGULATION OF NITROGEN METABOLISM UNDER ABIOTIC STRESS

1.9.1 Nitrogen Metabolism under Drought Stress

Drought stress-exposed plants have decreased acquisition of essential nutrients such as nitrogen and phosphorous, which leads to alteration in the concentrations of plant metabolites. For example, there is an increase in the synthesis of compatible solutes such as special amino acids (e.g., proline), sugars and sugar alcohols, and glycine betaine (Girousse et al. 1996). Furthermore, drought stress reduces the activities of glutamine synthetase (GS) and glutamate synthase (GOGAT) enzymes, which are key enzymes of nitrogen metabolism (Singh et al.2015). Water stress also leads to reduced ribulose-1, 5-bisphosphate carboxylase/oxygenase (rubisco) and nitrate reductase (NR) activity. In many plants, NR is one of the first enzymes affected by declining water status. Since N serves as a major component in photosynthetic machinery formation, the photosynthetic activity is adversely affected due to declined N metabolism under drought stress (Sinclair et al., 2000; Zhong et al., 2017).

Nitrogen plays a regulatory role in plants for water stress tolerance, but it all depends upon the intensity of stress and N level. Previous studies have demonstrated that plants have better growth and increased resistance of photosynthesis to water stress under a higher dose of N as compared to a low N dose (Wang et al., 2016). Moreover, under higher N levels more N is allocated to soluble organic nitrogenous compounds formation, which acts as osmotica to balance the water status in plant cells under osmotic stress (Singh et al., 2016).

1.9.2 Nitrogen Metabolism under High-Temperature Stress

Similar to drought stress, high-temperature stress also decreases nitrate uptake and nitrogen assimilation, which limit the development, growth, and reproduction of plants (Williams et al., 2013). High-temperature stress-mediated alteration in nitrogen metabolism has been widely reported in several crops like creeping bentgrass, tomato, wheat, and rice. Heat stress alters the assimilation of NO_3^- and NH_4^+ by altering the enzymes involved in N metabolism (Hungria and Kaschuk, 2014; Giri et al., 2017). In this sense, Argentel-Martínez et al. (2019) concluded that heat stress in wheat alters the activity of the enzymes NR and GS involved in nitrogen assimilation.

1.9.3 Nitrogen Metabolism under Salinity Stress

Salinity stress is one of the major abiotic stresses affecting plant growth and development. N metabolism is of great significance for plant survival under salinity stress due to the accumulation of N-containing compatible solutes, such as amino acids and betaines, which provide osmotic adjustment (Teh et al., 2016). Salinity stress limits the uptake of NO_3^- by plant roots and also restricts the process of N assimilation by inhibiting the synthesis and activities of nitrate reductase (NR), nitrite reductase (NiR), glutamine synthetase (GS), and glutamate synthase (GOGAT) (Hossain et al., 2012; Yousuf et al., 2016). NR and NiR activity under salt stress are inhibited by salt ions either by direct inhibition or reduced synthesis (Debouba et al., 2006) whereas GS activity is inhibited mainly due to the low availability of glutamate for the GS enzyme because of the high deviation of glutamate to the synthesis of osmolytes such as proline (Kawakami et al., 2013).

It has been found that under salinity stress there is an elevated level of NH_4^+ content in plants due to rapid proteolysis, which induces the activity of glutamate dehydrogenase (GDH) enzyme that shifts the NH_4^+ assimilation from GS/GOGAT to GDH pathway (Wang et al., 2007). This GDH-mediated $NH4^+$ detoxification and replenishment of glutamate pool are essential for the production of protective metabolites (proline, phytochelatins, etc.). Zhou et al. (2004) reported that the responses of N-metabolizing enzymes greatly vary among species, cultivars, and analyzed tissues under salt stress.

Previous results also suggest that available N source influences N uptake, assimilation, and accumulation (total N, proteins, and amino acids) under salinity stress, and the adverse effect of salinity could be improved by exogenous application of either ammonium or nitrate that depends on the characteristics of the plant species.

1.9.4 Nitrogen Metabolism under Heavy Metal Stress

Heavy metals are a group of metals and metalloids that have relatively high density and are toxic even at ppb levels. Heavy metals are categorized into two groups based on their essentiality criteria such as zinc (Zn), manganese (Mn), copper (Cu), iron (Fe), and nickel (Ni). They are considered as essential heavy metals because they play a pivotal role in physiochemical and metabolic processes of within the plants. On the other hand, cadmium (Cd), mercury (Hg), and lead (Pb)), which are not needed by plants for their growth and development, are called nonessential metals (Fashola et al., 2016). These non-essential heavy metals get absorbed by plant roots because of their resemblance with some essential micronutrients and alter the physiological, biochemical, and molecular processes (Sarwar et al., 2017). Heavy metals cause the production of reactive oxygen species (ROS) that lead to cell membrane damage, which ultimately influences the NO_3^- and $NH4^+$ uptake by plants (Rengel et al., 2015).

Furthermore, heavy metals also alter the activity of various N assimilatory enzymes by binding to the vital sulfhydryl (SH) groups and decreasing the N assimilation process in plants. This decreased N assimilation process further disrupts the general homeostasis of metabolic activities in plants (Riaz et al., 2018b).

1.10 EPIGENETIC REGULATION OF NITROGEN METABOLISM

Epigenetic regulation depicts the changes in the gene's expression without change in DNA sequence such as histone modifications and DNA methylations, which contribute to plant adaptation in different environments (Chang et al., 2020). This modification either changes the chromatin-folding structure or acts as an anchor for other factors. Histone modification occurs as methylation of a specific lysine residue, whereas methylation of cytosine residue occurs in the process of DNA methylation.

Histone modification response refers to the post-translational modification such as phosphorylation, methylation, acetylation, ubiquitination, etc. *AtSDG8* was also identified as a regulator of the N assimilation gene AtASN1 and was required to maintain normal H3K36me3 levels at gene loci encoding nitrate or nitrite reductases (Li et al., 2015). It is reported that *AtHNI9* (High nitrogen insensitive 9) encodes a component of RNA polymerase II complex (IWS1), which was involved in transcription regulation by recruiting chromatin-remodeling factors or depositing H3K36me3 on histone proteins (Girin et al., 2010). Loss of function of AtHNI9 causes decreased level methylated histone, thereby inhibiting high nitrogen repression of AtNRT2.1 (Widiez et al., 2011). It is also reported that AtHNI9 indirectly represses *AtNRT2.1* expression by reducing cellular ROS levels (Bellegarde et al., 2019).

Usually, DNA methylation represses gene expression. It was reported that nitrogen deficiency decreased the genome-wide methylated CG from 26.6% to 12% in the root (Mager et al., 2018), while there were no significant changes in methylation levels in rice has been reported under N deficiency (Kou et al., 2011). Nitrogen assimilation enzyme-like nitrate reductase (NR) is also regulated by DNA methylation. *AtCMT3* can bind to the promoter and gene body of nitrate reductase gene AtNIA2 to methylate the CHG sequences, and this process is inhibited by ammonium treatment, leading to upregulated at *AtNRT2* (Kim et al., 2016).

1.11 OMICS APPROACHES TO NITROGEN METABOLISM IN PLANTS

Omics approaches combine different applications such as genomics, proteomics, transcriptomics, metabolomics, nutrigenomics, and so on to get a complete understanding of various metabolic processes of the cells. Omics approaches are the way to study the complex pattern of nitrogen metabolisms in plants (Diaz et al., 2008). Recent signs of progress in the field of metabolomics have helped to monitor primary metabolites levels including amino acids, sugars, organic acids, fatty acids, and secondary metabolites including carotenoids and phenylpropanoids (Saito and Matsuda, 2010; Lei *et al.*, 2011). This provides basic information about plant responses to physiological or environmental changes triggered by the N status (Tschoep et al., 2009). RNA seq analysis along with the de novo assembly and transcript mapping can be useful for identifying differentially expressed genes (Marguerat and Bahler, 2010; Trapnell et al., 2012). Metabolomics approaches using gas and liquid chromatography with spectroscopic methods such as MS or NMR (Saito and Matsuda, 2010) can help analyze metabolite levels under a particular set of conditions. KEGG (Kanehisa et al., 2014), and KNApSAcK (Afendi et al., 2012), the "metabolite profile"-oriented databases, are a promising source for comprehensive metabolite identification.

1.12 REACTIVE NITROGEN SPECIES (RNS) IN PLANTS

When plants receive excess light energy for the generation of reducing equivalents, surplus electron reduces the oxygen molecule, generating high-energy oxygen molecule that results in the peroxidation of the lipid component of the biological membrane. Reactive oxygen species (ROS) such as H_2O_2 (hydrogen peroxide), O^{2-} (superoxide), and OH^- (hydroxyl) radicals are produced under stress through enhanced leakage of electrons to molecular oxygen (Arora et al. 2002). Plants have evolved both non-enzymatic and enzymatic antioxidant systems to balance the ROS levels in plants to control the damaging effect. Non-enzymatic antioxidants are low molecular weight molecules like β-carotenes, ascorbic acid (AA), α-tocopherol, and reduced glutathione (GSH). Enzymatic components include superoxide dismutase (SOD), guaiacol peroxidase (POD), ascorbate peroxidase (APX), catalase (CAT), polyphenol oxidase (PPO), and glutathione reductase (GR). SOD causes dismutation of superoxide radical into hydrogen peroxide molecule whereas CAT enzyme causes the breakdown of hydrogen peroxide into oxygen and water molecule.

1.13 NITROGEN USE EFFICIENCY IN PLANTS: PHYSIOLOGICAL AND MOLECULAR ASPECTS

In general, nitrogen use efficiency (NUE) is defined as the ratio between the total biomass and N input supplied through a different form of nitrogenous fertilizer applied. NUE is divided into two components such as nitrogen uptake efficiency (NupK), which is defined as the ability of the plant to absorb and uptake N from the soil, whereas nitrogen utilization efficiency (NutE) is defined as the ability to use N to produce biomass and grain yield.

The primary root traits such as root axis number, rooting depth, and root density are the major factors contributing to NUE. Deep root systems are more effective in absorbing nutrients than shallow rooting systems. With regards to nitrogen uptake efficiency, the capture of nitrate at low concentration in the topsoil requires high rooting density. Apart from that, several transcription factors are involved in the regulation of NUE. In nitrate signaling the first identified transcription factor is ANR1 (Zhang and Forde, 1998), and its rice homologs including OsMADS25, OsMADS27, and OsMADS57 are also involved in N signaling and transport of nitrate (Chen et al., 2018; Huang et al., 2019; Yu et al., 2015), thus regulating NUE. In rice the BTB family of transcription factors (BT1 & BT2) negatively regulates nitrate uptake and NUE (Araus et al., 2016).

Several genes encoding key enzymes in nitrogen assimilation have been cloned, including glutamine synthetase (GS) (Funayama et al. 2013; Tabuchi et al. 2005) and glutamate synthase (GOGAT), which have the potential for improving NUE. Concurrent overexpression of genes encoding cytosolic glutamine synthetase (*OsGS1;1*) and chloroplastic glutamine synthetase (*OsGS2*) could increase the grain yield under abiotic stress conditions (James et al., 2018).

1.14 NITROGEN TOXICITY IN PLANTS, SYMPTOMS AND SAFEGUARDS

Plants can store high levels of nitrate, or they can translocate it from tissue to tissue without deleterious effect. However, if livestock and humans consume plant material that is high in nitrate, they may suffer methemoglobinemia, a disease in which the liver reduces nitrate to nitrite, which combines with hemoglobin and renders the hemoglobin unable to bind oxygen. Humans and other animals may also convert nitrate into nitrosamines, which are potent carcinogens. Some countries limit the nitrate content in plant materials sold for human consumption. In contrast to nitrate, high levels of ammonium are toxic to both plants and animals. Ammonium dissipates transmembrane proton gradients that are required for both photosynthetic and respiratory electron transport and for sequestering metabolites in the vacuole. Because high levels of ammonium are dangerous, animals have developed a strong aversion to its smell. The active ingredient in smelling salts, a medicinal vapor released under the nose to revive a person who has fainted, is ammonium carbonate. Plants assimilate ammonium near the site of absorption or generation and rapidly store any excess in their vacuoles, thus avoiding toxic effects on membranes and the cytosol.

1.15 BIOLOGICAL NITROGEN FIXATION: AN OVERVIEW

Initially, a symbiotic association of rhizobia and legume plants is not obligatory. Under the nitrogen-limiting condition, they seek one another through an elaborate exchange of signals. Different host-specific and rhizobial-specific genes are involved in BNF. Two processes occur simultaneously during the root nodule formation process such as the infection process and nodule organogenesis. During the infection process, rhizobia migrate toward the roots in response to chemical signals secreted by the roots such as flavonoids and betaines. Rhizobia become attached to the root hair cell by lectin secreted by the roots and nod factors are released by the rhizobia inducing the root hair curling. The cell wall of the root hair degrades in these regions, also in response to Nod factors, allowing the bacterial cells direct access to the outer surface of the plant plasma membrane (Lazarowitz and Bisseling, 1997). From the root hairs, rhizobia enter into cells of the inner layers of the cortex through infection threads (internal tubular extension of plasma membrane produced by the fusion of Golgi-derived membrane vesicles). The internal tubular extension filled with the proliferating bacteria moves toward the nodule primordium and its membrane fuses with the membrane of the host cells. At first, the bacteria continue to divide, and the surrounding membrane increases in surface area to accommodate this growth by fusing with smaller vesicles. Soon thereafter, upon an undetermined signal from the plant, the bacteria stop dividing and begin to enlarge and to differentiate into nitrogen-fixing endosymbiotic organelles called bacteroids. The membrane surrounding the bacteroids is called the peribacteroid membrane. Biological nitrogen fixation (BNF) occurs in the presence of the nitrogenase enzyme, which causes the eight-electron reduction of dinitrogen into ammonia with concomitant production of hydrogen molecules.

1.16 NITROGENASE ENZYME COMPLEX

The nitrogenase enzymes complex can be separated into two components—the Fe protein and MoFe protein—neither of which has catalytic activity by itself. The Fe protein is the smaller of the two components and has two identical subunits of 30 to 72 kDa each, depending on the organism. Each subunit contains an iron-sulfur cluster that participates in the redox reaction involved in the conversion of N_2 to NH_3. The Fe protein is irreversibly inactivated by O2 with typical half decay times of 30 to 45 seconds (Dixon and wheeler, 1986). Another part of the component, MoFe protein, has four subunits, with a total molecular mass of 180 to 235 kda depending on the species. Each subunit has two Mo-Fe-S clusters. The MoFe protein is also inactivated by oxygen, with a half-decay time of 10 minutes in the air.

1.17 *NOD, NIF,* AND *FIX* GENES

Plant genes specific for nodules are called nodulin (*Nod*) genes, whereas bacterial genes specific for the nodule formations are called nodulation (*nod*) genes (Heidstra and Bisseling, 1996). The nod genes are classified as common nod genes and host-specific nod genes. The common nod genes (*nodA*, *nodB*, and *nodC*) are found in all rhizobial strains, whereas host-specific nod genes (*nodP*, *nodQ*, and *nodH*) differ among the bacterial genes. Only one of the nod genes, the regulatory nod D, is constitutively expressed and its protein product, NodD, regulates the transcription of the other nod genes. The nod genes, which NodD activates, code for nodulation proteins, most of which are involved in the biosynthesis of Nod factors. Nod factors are lipochitin oligosaccharide signal molecules, all of which have chitin β-1-4 linked N-acetyl-D-glucosamine backbone and a fatty acid chain on the C-2 positions of the non-reducing sugar.

Three of the nod genes (nodA, nodB, and nodC) encode enzymes (NodA, NodB, and NodC, respectively) that are required for synthesizing this basic structure (Stokkermans et al., 1995). NodA is an N-acyltransferase that catalyzes the addition of a fatty acyl chain. NodB is a chitin-oligosaccharide deacetylase that removes the acetyl group from the terminal nonreducing sugar. NodC is a chitin-oligosaccharide synthase that links N-acetyl-D-glucosamine monomers.

Host-specific nod genes that vary among rhizobial species are involved in the modification of the fatty acyl chain or the addition of groups important in determining host specificity (Carlson et al.

1995). NodE and NodF determine the length and degree of saturation of the fatty acyl chain; those of Rhizobium leguminosarum bv. viciae and R. meliloti result in the synthesis of an 18:4 and a 16:2 fatty acyl group, respectively. Other enzymes, such as NodL, influence the host specificity of Nod factors through the addition of specific substitutions at the reducing or non-reducing sugar moieties of the chitin backbone.

Dinitrogenase, an enzyme complex (heteromers), is composed of two units such as dinitrogenase reductase (Fe-protein) and dinitrogenase (Mo-Fe protein). Fe protein is encoded by *nifH*, accepts an electron from the ferredoxin, and catalyzes the one-electron reduction of the Mo-Fe protein. The Mo-Fe protein, a heterotetramer, is encoded by the *nifD* and *nifK*, accepts the electron, and binds to hydrogen and N_2, leading to the production of H_2 and ammonia.

A low oxygen concentration environment is required for nitrogenase to function for nitrogen fixation. Under low oxygen concentration, FixL is phosphorylated, which in turn phosphorylates the FixJ. FixJ can bind to DNA and activate transcription of genes needed for nitrogen fixation like nifA and FixK. However, binding of the oxygen to FixL heme protein inhibits the histidine kinase activity repressing the expression of these genes under conditions where nitrogenase would be unstable.

1.18 BIOCHEMICAL AND FUNCTIONAL ASPECTS OF LEGUME NODULE

The symbiosis between legumes and rhizobia is not obligatory. Legume seedlings germinate without any association with rhizobia, and they may remain unassociated throughout their life cycle. Rhizobia also occur as free-living organisms in the soil. Under nitrogen-limited conditions, however, the symbionts seek out one another through an elaborate exchange of signals.

1.18.1 Synthesis, Transport, and Assimilation of Products of Symbiotic Nitrogen Fixation

Ammonia is produced as a result of biological nitrogen fixation. To avoid ammonia toxicity, the plant rapidly converts ammonia into a non-toxic form before transport to the shoot through the xylem. Nitrogen-fixing nodules can be divided into amide transporter or ureides transporter. Amides such as glutamine and asparagines are transported by the temperate region legumes pea, clover, broad bean, and lentil. Glutamine is synthesized through the ATP-dependent reaction of glutamine and ammonia through the glutamine synthetase (GS) enzyme. Asparagine synthesis involves the transfer of the amide nitrogen from glutamine to asparagines by the action of enzymes asparagines synthetase (AS). AS, the enzyme that catalyzes this reaction, is found in the cytosol of leaves and roots and nitrogen-fixing nodules.

Tropical origin legumes export ureides such as citrulline, allantoin, and allantoic acid. Allantoin is synthesized in peroxisomes from uric acid by urate oxidase, and allantoic acid is synthesized from allantoin in the endoplasmic reticulum by allantoinase. All three compounds are ultimately released into the xylem and transported to the shoot, where they are rapidly catabolized to ammonium. Finally, ammonium enters into the assimilation pathways.

1.19 NITROGEN BIOFERTILIZERS: ROLE IN SUSTAINABLE AGRICULTURE

Bio-fertilizers are the products that contain living or latent cells of different microbes that upon application not only increase availability but also boost the supply of primary nutrients to crops (Managalassery *et al.*, 2017). Bio-fertilizers can also be defined as commercial preparations, containing live or dormant cells of efficient strains of nitrogen-fixing, phosphate solubilizing, potassium solubilizing, and cellulitic microorganisms. They are known to increase the nutrient composition of soil through the processes of either nitrogen fixation or solubilizing mineral ions, which ultimately help in stimulating plant growth through the synthesis of growth-promoting substances.

Introduction to Plant Nitrogen Metabolism

Bio-fertilizers help in promoting the growth of crops by increasing the supply of essential nutrients. Most important and commonly used microorganisms as bio-fertilizers are nitrogen-fixing soil bacteria (*Azotobactor*, *Rhizobium*), nitrogen-fixing cyanobacteria (*Anabaena*), phosphate-solubilizing bacteria (*Pseudomonas sp.*), and AM fungi. Likewise, phytohormone (auxin)-producing bacteria and cellulolytic microorganisms are also used as bio-fertilizer formulations. As per suitability and need, the application of bio-fertilizers can be done as a seed treatment, foliar application, root dipping, seedlings dipping, and direct to the soil. Bio-fertilizers are categorized into two main groups: biological nitrogen-fixing bio-fertilizers and phosphate-solubilizing (mobilizing) bio-fertilizers. Bio-fertilizers are considered a low-cost, effective, environmentally friendly, cheap, easy-to-use, and renewable source of plant nutrients. Besides, increasing nutrient availability for plants, bio-fertilizers also secrete growth-promoting substances like hormones, vitamins, etc., which ultimately enhances seed germination, growth, and yield of crops. In the present scenario, when the sustainability of the agricultural systems is of utmost need a variety of commercial bio-fertilizer formulations such as carrier-based bio-fertilizers, liquid bio-fertilizers, and bio-fertilizer consortia are being promoted across the globe to offset the deleterious effect of chemical fertilizers. Based on the type of microorganisms they contain, bio-fertilizers are further classified into different groups as listed in Table 1.1.

Nitrogenous bio-fertilizers include symbiotic nitrogen fixers (rhizobium and Azolla) and associate symbionts (azospirillum). However, non-symbiotic nitrogen-fixing bio-fertilizers include azotobacter. In the symbiotic nitrogen fixation, the organisms exist in the root nodules of the plant and fix nitrogen in the root nodules themselves (Managalassery *et al.*, 2017). Bio-fertilizers are a supplementary component to soil and crop management practices such as crop rotation, organic adjustments, tillage maintenance, recycling of crop residue, soil fertility renovation, and the biocontrol of pathogens and insect pests; these operations can significantly be useful in maintaining the sustainability of various crop productions (Bhardwaj *et al.*, 2014). The use of efficient strains

TABLE 1.1
Different Types of Bio-fertilizers

Groups	Examples
1. Nitrogen-fixing bio-fertilizers	
Free-living	*Azotobacter*, *Bejerinkia*, *Clostridium*, *Klebsiella*, *Anabaena*, and *Nostoc*
Symbiotic	*Rhizobium*, *Frankia*, *Anabaena*, and *Azollae*
Associative symbiotic	*Azospirillum*
2. Phosphate solubilizing bio-fertilizer	
Bacteria	*Bacillus megaterium var*, *Phosphaticum*, *Bacillus subtilis*, and *Bacillus circulans*
Fungi	*Penicillum Spp.* and *Aspergillus awamori*
3. Phosphate mobilizing bio-fertilizers	
Arbuscular Mycorrhiza	*Glomus Spp.*, *Gigaspora Spp.*, *Acaulospora Spp. Scutellospora Spp.*, and *Sclerocystis spp.*
Ectomycorrhiza	*Laccaria spp. Pisolithus Spp*, *Boletus Spp.*, and *Amanita Spp.*
Ericoid Mycorrhiza	*Pezizella ericae*
Orchid Mycorrhiza	*Rhizoctonia solani*
4. Bio-fertilizers for micronutrients	
Bacillus Spp	Silicate and zinc solubilizers
Plant growth-promoting Rhizobacteria	
Pseudomonas	*Pseudomonas fluorescens*

Source: Bhattacherjee and Dey (2014).

of nitrogenous bio-fertilizers not only adds a significant amount of nitrogen in the soil but also improves soil quality and its health besides overcoming soil and water pollution and ultimately providing enhanced quality crop production. Application of *Azotobacter*, *Azospirillum*, and *Rhizobium* in rice enhances its physiological functions and improves the root morphology (Choudhury *et al.*, 2004). It has been found that the inoculation of bio-fertilizers to the soil ecosystem advances soil physicochemical properties, soil microbes' biodiversity, soil health, plant growth and development, and crop productivity (Bhardwaj *et al.*, 2014). Besides improving the grain yield of crops, especially legumes, *Rhizobium* inoculants have also been found to strengthen defense mechanisms in plants against pathogens and herbivores.

REFERENCES

Afendi, F. M., Okada, T., Yamazaki, M., Hirai-Morita, A., Nakamura, Y., Nakamura, K. et al. (2012). KNApSAcK family databases: integrated metabolite–plant species databases for multifaceted plant research. *Plant and Cell Physiology*, *53*(2), e1–e1.

Ali, A., Syed, A. A. W., Khaliq, T., Asif, M., Aziz, M., & Mubeen, M. (2011). Effects of nitrogen on growth and yield components of wheat. (Report). *Biological Sciences*, *3*(6), 1004–1005.

Araus, V., Vidal, E. A., Puelma, T., Alamos, S., Mieulet, D., Guiderdoni, E., & Gutiérrez, R. A. (2016). Members of BTB gene family of scaffold proteins suppress nitrate uptake and nitrogen use efficiency. *Plant Physiology*, *171*(2), 1523–1532.

Argentel Martínez, L., Garatuza Payán, J., Arredondo Moreno, J. T., & Yepez González, E. A. (2019). Effects of experimental warming on peroxidase, nitrate reductase and glutamine synthetase activities in wheat. *Agronomy Research*, 17, 22.

Arora, A., Sairam, R. K., & Srivastava, G. C. (2002). Oxidative stress and antioxidative system in plants. *Current Science*, 1227–1238.

Bao, A., Liang, Z., Zhao, Z., & Cai, H. (2015). Overexpressing of OsAMT1-3, a high affinity ammonium transporter gene, modifies rice growth and carbon-nitrogen metabolic status. *International Journal of Molecular Sciences*, *16*(5), 9037–9063.

Bellegarde, F., Maghiaoui, A., Boucherez, J., Krouk, G., Lejay, L., Bach, L et al. (2019). The chromatin factor HNI9 and ELONGATED HYPOCOTYL5 maintain ROS homeostasis under high nitrogen provision. *Plant Physiology*, *180*(1), 582–592.

Bhardwaj, D., Ansari, M. W., Sahoo, R. K., & Tuteja, N. (2014). Biofertilizers function as key player in sustainable agriculture by improving soil fertility, plant tolerance and crop productivity. *Microbial Cell Factories*, *13*(1), 1–10.

Bhattacherjee, R., Dey, U. (2014). A way towards organic farming; A review. *African Journal of Microbiology Research*, 8(24), 2332–234.

Bloom, A. J. (2015). The increasing importance of distinguishing among plant nitrogen sources. *Current Opinion in Plant Biology*, 25, 10–16.

Campbell, W. H. (1999). Nitrate reductase structure, function and regulation: bridging the gap between biochemistry and physiology. *Annual Review of Plant Biology*, *50*(1), 277–303.

Carlson, R. W., Forsberg, L. S., Price, N. P., Bhat, U. R., Kelly, T. M., & Raetz, C. R. (1995). The structure and biosynthesis of Rhizobium leguminosarum lipid A. *Progress in Clinical and Biological Research*, *392*, 25–31.

Chang, Y. N., Zhu, C., Jiang, J., Zhang, H., Zhu, J. K., & Duan, C. G. (2020). Epigenetic regulation in plant abiotic stress responses. *Journal of Integrative Plant Biology*, *62*(5), 563–580.

Chen, H., Xu, N., Wu, Q., Yu, B., Chu, Y., Li, X. et al. (2018). OsMADS27 regulates the root development in a NO3⁻—Dependent manner and modulates the salt tolerance in rice (*Oryza sativa* L.). *Plant Science*, *277*, 20–32.

Choudhury, A. T. M. A., & Kennedy, I. R. (2004). Prospects and potentials for systems of biological nitrogen fixation in sustainable rice production. *Biology and Fertility of Soils*, *39*(4), 219–227.

Debouba, M., Gouia, H., Suzuki, A., & Ghorbel, M. H. (2006). NaCl stress effects on enzymes involved in nitrogen assimilation pathway in tomato "Lycopersicon esculentum" seedlings. *Journal of Plant Physiology*, *163*(12), 1247–1258.

Diaz, C., Lemaître, T., Christ, A., Azzopardi, M., Kato, Y., Sato, F. et al., (2008). Nitrogen recycling and remobilization are differentially controlled by leaf senescence and development stage in Arabidopsis under low nitrogen nutrition. *Plant Physiology*, *147*(3), 1437–1449.

Dixon, R. O. D., & Wheeler, C. T. (1986). *Nitrogen Fixation in Plants*. Chapman and Hall, New York.

Fan, X., Tang, Z., Tan, Y., Zhang, Y., Luo, B., Yang, M. et al. (2016). Overexpression of a pH-sensitive nitrate transporter in rice increases crop yields. *Proceedings of the National Academy of Sciences*, *113*(26), 7118–7123.

Fang, Z., Bai, G., Huang, W., Wang, Z., Wang, X., & Zhang, M. (2017). The rice peptide transporter OsNPF7. 3 is induced by organic nitrogen, and contributes to nitrogen allocation and grain yield. *Frontiers in Plant Science*, *8*, 1338.

Fang, Z., Xia, K., Yang, X., Grotemeyer, M. S., Meier, S., Rentsch, D. et al. (2013). Altered expression of the PTR/NRT 1 homologue Os PTR 9 affects nitrogen utilization efficiency, growth and grain yield in rice. *Plant Biotechnology Journal*, *11*(4), 446–458.

Fashola, M. O., Ngole-Jeme, V. M., & Babalola, O. O. (2016). Heavy metal pollution from gold mines: environmental effects and bacterial strategies for resistance. *International Journal of Environmental Research and Public Health*, *13*(11), 1047.

Fukuda, M., Nishida, S., Kakei, Y., Shimada, Y., & Fujiwara, T. (2019). Genome-wide analysis of long intergenic noncoding RNAs responding to low-nutrient conditions in Arabidopsis thaliana: possible involvement of trans-acting siRNA3 in response to low nitrogen. *Plant and Cell Physiology*, *60*(9), 1961–1973.

Funayama, K., Kojima, S., Tabuchi-Kobayashi, M., Sawa, Y., Nakayama, Y., Hayakawa, T., & Yamaya, T. (2013). Cytosolic glutamine synthetase1; 2 is responsible for the primary assimilation of ammonium in rice roots. *Plant and Cell Physiology*, *54*(6), 934–943.

Gifford, M. L., Dean, A., Gutierrez, R. A., Coruzzi, G. M., & Birnbaum, K. D. (2008). Cell-specific nitrogen responses mediate developmental plasticity. *Proceedings of the National Academy of Sciences*, *105*(2), 803–808.

Giri, A., Heckathorn, S., Mishra, S., & Krause, C. (2017). Heat stress decreases levels of nutrient-uptake and-assimilation proteins in tomato roots. *Plants*, *6*(1), 6.

Girin, T., El-Kafafi, E. S., Widiez, T., Erban, A., Hubberten, H. M., Kopka, J. et al. (2010). Identification of Arabidopsis mutants impaired in the systemic regulation of root nitrate uptake by the nitrogen status of the plant. *Plant Physiology*, *153*(3), 1250–1260.

Girousse, C., Bournoville, R., & Bonnemain, J. L. (1996). Water deficit-induced changes in concentrations in proline and some other amino acids in the phloem sap of alfalfa. *Plant Physiology*, *111*(1), 109–113.

Heidstra, R., and Bisseling, T. (1996) Nod factor-induced host responses and mechanisms of Nod factor perception. *New Phytologist*, *133*, 25–43.

Hemerly, A. (2016, January). Genetic controls of biomass increase in sugarcane by association with beneficial nitrogen-fixing bacteria. In: *Plant and Animal Genome XXIV Conference. Plant and Animal Genome, during month of January*.

Hoque, M. S., Masle, J., Udvardi, M. K., Ryan, P. R., & Upadhyaya, N. M. (2006). Over-expression of the rice OsAMT1-1 gene increases ammonium uptake and content, but impairs growth and development of plants under high ammonium nutrition. *Functional Plant Biology*, *33*(2), 153–163.

Hossain, M. A., Uddin, M. K., Ismail, M. R., & Ashrafuzzaman, M. (2012). Responses of glutamine synthetase-glutamate synthase cycle enzymes in tomato leaves under salinity stress. *International Journal of Agriculture and Biology*, *14*(4), 509–515.

Hu, B., Wang, W., Deng, K., Li, H., Zhang, Z., Zhang, L., & Chu, C. (2015a). MicroRNA399 is involved in multiple nutrient starvation responses in rice. *Frontiers in Plant Science*, *6*, 188.

Hu, B., Wang, W., Ou, S., Tang, J., Li, H., Che, R. et al. (2015b). Variation in NRT1. 1B contributes to nitrate-use divergence between rice subspecies. *Nature Genetics*, *47*(7), 834–838.

Hu, J., Wang, Y., Fang, Y., Zeng, L., Xu, J., Yu, H. et al. (2015). A rare allele of GS2 enhances grain size and grain yield in rice. *Molecular Plant*, *8*(10), 1455–1465.

Huang, S., Liang, Z., Chen, S., Sun, H., Fan, X., Wang, C. et al. (2019). A transcription factor, OsMADS57, regulates long-distance nitrate transport and root elongation. *Plant Physiology*, *180*(2), 882–895.

Huang, W., Bai, G., Wang, J., Zhu, W., Zeng, Q., Lu, K. et al. (2018). Two splicing variants of OsNPF7. 7 regulate shoot branching and nitrogen utilization efficiency in rice. *Frontiers in Plant Science*, *9*, 300.

Hungria, M., & Kaschuk, G. (2014). Regulation of N2 fixation and NO_3^-/NH_4^+ assimilation in nodulated and N-fertilized *Phaseolus vulgaris* L. exposed to high temperature stress. *Environmental and Experimental Botany*, 98, 32–39.

James, D., Borphukan, B., Fartyal, D., Ram, B., Singh, J., Manna, M. et al. (2018). Concurrent overexpression of OsGS1; 1 and OsGS2 genes in transgenic rice (Oryza sativa L.): impact on tolerance to abiotic stresses. *Frontiers in Plant Science*, 9, 786.

Kanehisa, M., Goto, S., Sato, Y., Kawashima, M., Furumichi, M., & Tanabe, M. (2014). Data, information, knowledge and principle: back to metabolism in KEGG. *Nucleic Acids Research*, 42(D1), D199-D205.

Kawakami, E. M., Oosterhuis, D. M., & Snider, J. L. (2013). Nitrogen Assimilation and Growth of Cotton Seedlings under NaCl Salinity and in Response to Urea Application with NBPT and DCD. *Journal of Agronomy and Crop Science*, 199(2), 106–117.

Kim, J. Y., Kwon, Y. J., Kim, S. I., Kim, D. Y., Song, J. T., & Seo, H. S. (2016). Ammonium inhibits chromomethylase 3-mediated methylation of the Arabidopsis nitrate reductase gene NIA2. *Frontiers in Plant Science*, 6, 1161.

Kirk, G. J. D., & Kronzucker, H. J. (2005). The potential for nitrification and nitrate uptake in the rhizosphere of wetland plants: a modelling study. *Annals of Botany*, 96(4), 639.

Kliebenstein, D. J., Lambrix, V. M., Reichelt, M., Gershenzon, J., & Mitchell-Olds, T. (2001). Gene duplication in the diversification of secondary metabolism: tandem 2-oxoglutarate–dependent dioxygenases control glucosinolate biosynthesis in Arabidopsis. *The Plant Cell*, 13(3), 681–693.

Kou, H. P., Li, Y., Song, X. X., Ou, X. F., Xing, S. C., Ma, J. et al. (2011). Heritable alteration in DNA methylation induced by nitrogen-deficiency stress accompanies enhanced tolerance by progenies to the stress in rice (Oryza sativa L.). *Journal of Plant Physiology*, 168(14), 1685–1693.

Kumar, S., Rao, S. S., & Yakadri, M. (2016). Influence of Nitrogen Levels and Planting Geometry on Sweet Sorghum (Sorghum bicolor) Juice Sugar Quality Traits Under Semi-arid Tropical Environment. *National Academy Science Letters*, 39(5), 317–321.

Lazarowitz, S. G., & Bisseling, T. (1997). Plant development from the cellular perspective: integrating the signals. *The Plant Cell*, 9(11), 1884.

Leghari, S. J., Wahocho, N. A., Laghari, G. M., HafeezLaghari, A., MustafaBhabhan, G., HussainTalpur, K. et al. (2016). Role of nitrogen for plant growth and development: A review. *Advances in Environmental Biology*, 10(9), 209–219.

Lei, Z., Huhman, D. V., & Sumner, L. W. (2011). Mass spectrometry strategies in metabolomics. *Journal of Biological Chemistry*, 286(29), 25435–25442.

Li, B. Z., Merrick, M., Li, S. M., Li, H. Y., Zhu, S. W., Shi, W. M., & Su, Y. H. (2009). Molecular basis and regulation of ammonium transporter in rice. *Rice Science*, 16(4), 314–322.

Li, Y., Meng, X., Zheng, B., & Ding, Y. (2015). Parameter identification of fractional order linear system based on Haar wavelet operational matrix. *ISA Transactions*, 59, 79–84.

Liang, C., Wang, Y., Zhu, Y., Tang, J., Hu, B., Liu, L. et al. (2014). OsNAP connects abscisic acid and leaf senescence by fine-tuning abscisic acid biosynthesis and directly targeting senescence-associated genes in rice. *Proceedings of the National Academy of Sciences*, 111(27), 10013–10018.

Liang, G., He, H., & Yu, D. (2012). Identification of nitrogen starvation-responsive microRNAs in Arabidopsis thaliana. *PloS One*, 7(11), e48951.

Liu, F., Xu, Y., Chang, K., Li, S., Liu, Z., Qi, S. et al. (2019). The long noncoding RNA T5120 regulates nitrate response and assimilation in Arabidopsis. *New Phytologist*, 224(1), 117–131.

Mager, S., & Ludewig, U. (2018). Massive loss of DNA methylation in nitrogen-, but not in phosphorus-deficient Zea mays roots is poorly correlated with gene expression differences. *Frontiers in Plant Science*, 9, 497.

Mangalassery, S., Kumar, M., Dayal, D., & Kumar, S. (2017). Improving productivity of drylands through integrated soil fertility management. In: Mangalassery, M., Dayal, D., & Machiwal, D. (Eds.), Soil and water management strategies for drylands, Kalyani Publisher, Ludhiana, pp. 1–11.

Marguerat, S., & Bähler, J. (2010). RNA-seq: from technology to biology. *Cellular and Molecular Life Sciences*, 67(4), 569–579.

Mendel, R. R., & Stallmeyer, B. (1995). Molybdenum cofactor (nitrate reductase) biosynthesis in plants: First molecular analysis. In: *Current Issues in Plant Molecular and Cellular Biology*. Springer, Dordrecht, pp. 577–582.

Nunes-Nesi, A., Fernie, A. R., & Stitt, M. (2010). Metabolic and signaling aspects underpinning the regulation of plant carbon nitrogen interactions. *Molecular Plant*, 3(6), 973–996.

Oaks, A. (1994). Primary nitrogen assimilation in higher plants and its regulation. *Canadian Journal of Botany*, 72(6), 739–750.

Peerzada, Y.Y., & Iqbal, M. (2021). Leaf senescence and ethylene signaling. In: Aftab, T., & Hakeem, K.R. (Eds.), *Plant Growth Regulators*. Springer, Cham. https://doi.org/10.1007/978-3-030-61153-8_7

Ranathunge, K., El-Kereamy, A., Gidda, S., Bi, Y. M., & Rothstein, S. J. (2014). AMT1; 1 transgenic rice plants with enhanced NH4+ permeability show superior growth and higher yield under optimal and suboptimal NH4+ conditions. *Journal of Experimental Botany*, 65(4), 965–979.

Rengel, Z., Bose, J., Chen, Q., & Tripathi, B. N. (2015). Magnesium alleviates plant toxicity of aluminium and heavy metals. *Crop and Pasture Science*, 66(12), 1298–1307.

Riaz, M., Yan, L., Wu, X., Hussain, S., Aziz, O., Imran, M. et al. (2018). Boron reduces aluminum-induced growth inhibition, oxidative damage and alterations in the cell wall components in the roots of trifoliate orange. *Ecotoxicology and Environmental Safety*, 153, 107–115.

Saito, K., & Matsuda, F. (2010). Metabolomics for functional genomics, systems biology, and biotechnology. *Annual Review of Plant Biology*, 61, 463–489.

Sarwar, N., Imran, M., Shaheen, M. R., Ishaque, W., Kamran, M. A., Matloob, A. et al. (2017). Phytoremediation strategies for soils contaminated with heavy metals: modifications and future perspectives. *Chemosphere*, 171, 710–721.

Siegel, L. M., & Wilkerson, J. O. (1989). Structure and function of spinach ferredoxin-nitrite reductase. In: Wray, J. L. & Kinghorn, J. R. (Eds.), *Molecular and Genetic Aspects of Nitrate Assimilation*. Oxford Science, Oxford, pp. 263–283.

Sinclair, T. R., Pinter Jr, P. J., Kimball, B. A., Adamsen, F. J., LaMorte, R. L., Wall et al. (2000). Leaf nitrogen concentration of wheat subjected to elevated [CO_2] and either water or N deficits. *Agriculture, Ecosystems & Environment*, 79(1), 53–60.

Singh, M., Singh, V. P., & Prasad, S. M. (2016). Responses of photosynthesis, nitrogen and proline metabolism to salinity stress in Solanum lycopersicum under different levels of nitrogen supplementation. *Plant Physiology and Biochemistry*, 109, 72–83.

Singh, N. B., Singh, D., & Singh, A. (2015). Biological seed priming mitigates the effects of water stress in sunflower seedlings. *Physiology and Molecular Biology of Plants*, 21(2), 207–214.

Sonoda, Y., Ikeda, A., Saiki, S., Wirén, N. V., Yamaya, T., & Yamaguchi, J. (2003). Distinct expression and function of three ammonium transporter genes (OsAMT1; 1–1; 3) in rice. *Plant and Cell Physiology*, 44(7), 726–734.

Stokkermans, T. J. W., Ikeshita, S., Cohn, J., Carlson, R. W., Stacey, G., Ogawa, T., & Peters, N. K. (1995) Structural requirements of synthetic and natural product lipo-chitin oligosaccharides for induction of nodule primordia on Glycine soja. *Plant Physiology*, 108, 1587–1595.

Tabuchi, M., Sugiyama, K., Ishiyama, K., Inoue, E., Sato, T., Takahashi, H., & Yamaya, T. (2005). Severe reduction in growth rate and grain filling of rice mutants lacking OsGS1; 1, a cytosolic glutamine synthetase1; 1. *The Plant Journal*, 42(5), 641–651.

Teh, C. Y., Shaharuddin, N. A., Ho, C. L., & Mahmood, M. (2016). Exogenous proline significantly affects the plant growth and nitrogen assimilation enzymes activities in rice (*Oryza sativa*) under salt stress. *Acta Physiologiae Plantarum*, 38(6), 151.

Trapnell, C., Roberts, A., Goff, L., Pertea, G., Kim, D., Kelley, D. R. et al. (2012). Differential gene and transcript expression analysis of RNA-seq experiments with TopHat and Cufflinks. *Nature Protocols*, 7(3), 562–578.

Tschoep, H., Gibon, Y., Carillo, P., Armengaud, P., Szecowka, M. et al. (2009). Adjustment of growth and central metabolism to a mild but sustained nitrogen-limitation in Arabidopsis. *Plant, Cell & Environment*, 32(3), 300–318.

Vidal, E. A., Araus, V., Lu, C., Parry, G., Green, P. J., Coruzzi, G. M., & Gutiérrez, R. A. (2010). Nitrate-responsive miR393/AFB3 regulatory module controls root system architecture in Arabidopsis thaliana. *Proceedings of the National Academy of Sciences*, 107(9), 4477–4482.

Wang, J. W., Wang, L. J., Mao, Y. B., Cai, W. J., Xue, H. W., & Chen, X. Y. (2005). Control of root cap formation by microRNA-targeted auxin response factors in Arabidopsis. *The Plant Cell*, 17(8), 2204–2216.

Wang, L., Mai, Y. X., Zhang, Y. C., Luo, Q., & Yang, H. Q. (2010). MicroRNA171c-targeted SCL6-II, SCL6-III, and SCL6-IV genes regulate shoot branching in Arabidopsis. *Molecular Plant*, 3(5), 794–806.

Wang, R., Xing, X., & Crawford, N. (2007). Nitrite acts as a transcriptome signal at micromolar concentrations in Arabidopsis roots. *Plant Physiology*, *145*(4), 1735–1745.

Wang, W., Hu, B., Yuan, D., Liu, Y., Che, R., Hu, Y. et al. (2018). Expression of the nitrate transporter gene OsNRT1.1A/OsNPF6.3 confers high yield and early maturation in rice. *The Plant Cell*, *30*(3), 638–651.

Wang, Z., Zhang, W., Beebout, S. S., Zhang, H., Liu, L., Yang, J., & Zhang, J. (2016). Grain yield, water and nitrogen use efficiencies of rice as influenced by irrigation regimes and their interaction with nitrogen rates. *Field Crops Research*, *193*, 54–69.

Warner, R. L., & Kleinhofs, A. (1992). Genetics and molecular biology of nitrate metabolism in higher plants. *Physiologia Plantarum*, *85*(2), 245–252.

Widiez, T., Girin, T., Berr, A., Ruffel, S., Krouk, G., Vayssières, A. et al. (2011). High nitrogen insensitive 9 (HNI9)-mediated systemic repression of root NO3– uptake is associated with changes in histone methylation. *Proceedings of the National Academy of Sciences*, *108*(32), 13329–13334.

Williams, A. P., Allen, C. D., Macalady, A. K., Griffin, D., Woodhouse, C. A., Meko, D. M. et al. (2013). Temperature as a potent driver of regional forest drought stress and tree mortality. *Nature Climate Change*, *3*(3), 292–297.

Wray, J. L. (1993). Molecular biology, genetics and regulation of nitrite reduction in higher plants. *Physiologia Plantarum*, *89*(3), 607–612.

Xiong, D., Yu, T., Liu, X., Li, Y., Peng, S., & Huang, J. (2015). Heterogeneity of photosynthesis within leaves is associated with alteration of leaf structural features and leaf N content per leaf area in rice. *Functional Plant Biology*, *42*(7), 687–696.

Xu, Z., Zhong, S., Li, X., Li, W., Rothstein, S. J., Zhang, S. et al. (2011). Genome-wide identification of microRNAs in response to low nitrate availability in maize leaves and roots. *PloS One*, *6*(11), e28009.

Xuan, Y. H., Kumar, V., Zhu, X. F., Je, B. I., Kim, C. M., Huang, J. et al. (2018). IDD10 is Involved in the Interaction between NH 4+ and Auxin Signaling in Rice Roots. *Journal of Plant Biology*, *61*(2), 72–79.

Xuan, Y. H., Priatama, R. A., Huang, J., Je, B. I., Liu, J. M., Park, S. J. et al. (2013). Indeterminate domain 10 regulates ammonium-mediated gene expression in rice roots. *New Phytologist*, *197*(3), 791–804.

Yousuf, P. Y., Ahmad, A., Ganie, A. H., Iqbal, M. (2016) Salt stress-induced modulations in the shoot proteome of Brassica juncea genotypes. *Environmental Science and Pollution Research*, *23*(3), 2391–2401

Yousuf, P. Y., Shabir, P. A., Hakeem, K. R. (2021). miRNAomic Approach to Plant Nitrogen Starvation. *International Journal of Genomics*, 2021, 8560323.

Yu, C., Liu, Y., Zhang, A., Su, S., Yan, A., Huang, L. et al. (2015). MADS-box transcription factor OsMADS25 regulates root development through affection of nitrate accumulation in rice. *PLoS One*, *10*(8), e0135196.

Zhang, H., & Forde, B. G. (1998). An Arabidopsis MADS box gene that controls nutrient-induced changes in root architecture. *Science*, *279*(5349), 407–409.

Zhao, M., Ding, H., Zhu, J. K., Zhang, F., & Li, W. X. (2011). Involvement of miR169 in the nitrogen-starvation responses in Arabidopsis. *New Phytologist*, *190*(4), 906–915.

Zhong, C., Cao, X., Hu, J., Zhu, L., Zhang, J., Huang, J., & Jin, Q. (2017). Nitrogen metabolism in adaptation of photosynthesis to water stress in rice grown under different nitrogen levels. *Frontiers in Plant Science*, *8*, 1079.

Zhou, W., Sun, Q. J., Zhang, C. F., Yuan, Y. Z., Zhang, J., & Lu, B. B. (2004). Effect of salt stress on ammonium assimilation enzymes of the roots of rice (*Oryza sativa*) cultivars differing in salinity resistance. *Acta Botanica Sinica-English Edition-*, *46*(8), 921–927.

2 Nitrogen
A Key Macronutrient for the Plant World

Ruchi Raina[1] and Samina Mazahar[1]*
[1]Assistant Professor, Department of Botany, Dyal Singh College,
University of Delhi, New Delhi, India
*Corresponding author: E-mail: rjain20@gmail.com

CONTENTS

2.1 Introduction 19
2.2 Discovery 20
2.3 Chemistry and Structure of Nitrogen 20
    2.3.1 Electronic Configuration 20
    2.3.2 Lewis Structure 21
    2.3.3 Physical And Chemical Properties 21
        2.3.3.1 Physical properties 21
        2.3.3.2 Chemical properties 22
    2.3.4 Isotopes of Nitrogen 22
2.4 Occurrence and Sources of Nitrogen 22
2.5 Role of Nitrogen in Plants 24
2.6 Factors Affecting Nitrogen Nutrition 24
    2.6.1 Soil pH 24
    2.6.2 Soil Fertility 24
    2.6.3 Soil Texture 25
    2.6.4 Soil Moisture 25
2.7 Deficiency Symptoms of Nitrogen 25
2.8 Effects of Excessive Nitrogen 25
    2.8.1 Increased Foliage Growth 25
    2.8.2 Stunted Root Growth 25
    2.8.3 Underground Water Pollution 26
References 26

2.1 INTRODUCTION

Can life exist without atmosphere? In our opinion, every living organism requires one or the other gases in the air to support the life system. Earth is like a lifeless rock without an atmosphere. Earth's atmosphere is composed of about 78% nitrogen, 21% oxygen, and 0.93% argon. The remainder, less than 0.1%, contains such trace gases as water vapor, carbon dioxide, and ozone. Of these, nitrogen plays a significant role in the life cycle of plants. It is an imperative macronutrient for plants as it builds essential constituents such as proteins and nucleic acids and regulates biological functions.

DOI: 10.1201/9781003248361-2

2.2 DISCOVERY

Early investigations of air by Carl Wilhelm Scheele, a Swedish chemist in 1772, showed that air consists of a mixture of two gases. He named these gases "fire air", as it helps in combustion and "foul air" because it was left after the first one was used up. So, based on the properties of these two gases, fire air was oxygen and the other one was nitrogen. At the same time a Scottish botanist, Daniel Rutherford, was also working on nitrogen. Although Carl Wilhelm Scheele and Henry Cavendish were also working independently in the same field the credit for the discovery and isolation of nitrogen goes to Daniel Rutherford in 1772, since his work was published first. The name nitrogène was suggested by French chemist Jean-Antoine-Claude Chaptal in 1790 when it was revealed that nitrogen is the constituent of nitric acid and nitrates. Nitrogen is an asphyxiant gas and its inability to support life (Greek: zoe) led Lavoisier to name it azote, still the French equivalent of nitrogen.

2.3 CHEMISTRY AND STRUCTURE OF NITROGEN

Nitrogen (N) is a non-metallic element that belongs to Group 15 of the periodic table (Figure 2.1). It is a colorless, odorless, tasteless gas that is the most abundant element in Earth's atmosphere and a major constituent of all living organisms. This Group 15 contains five elements: Nitrogen (N), Phosphorus (P), Arsenic (As), Antimony (Sb), and Bismuth (Bi). This group is also called the nitrogen family. The whole Group 15 family is known as pnicogens and compounds called pniconides. Nitrogen in its molecular form exists as a diatomic molecule (N_2) and has a triple bond between atoms, and is therefore called dinitrogen. It has very high electronegativity next to fluorine and oxygen.

2.3.1 Electronic Configuration

Nitrogen has five electrons in its outermost shell. Two electrons are found in 2s and the rest three are present in the 2p subshell. These three p subshell electrons are unpaired. So the general electronic configuration of nitrogen is represented as $1s^2\ 2s^2\ 2p^3$ (Figure 2.2). It has oxidation states between −3 and +5.

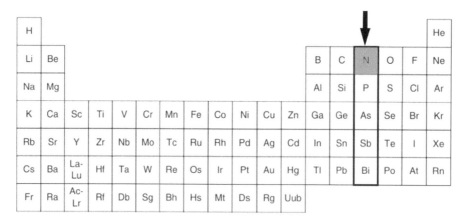

FIGURE 2.1 Periodic table showing position of nitrogen.

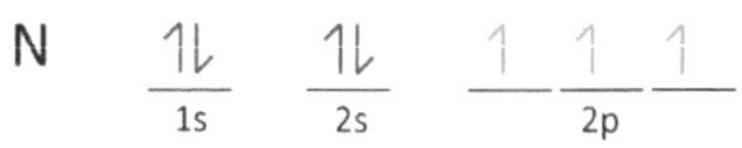

FIGURE 2.2 Electron configuration of nitrogen.

2.3.2 Lewis Structure

Lewis structure is also known as Lewis dot structure, used to represent bonding as well as lone pair of electrons between atoms of a molecule. In the case of nitrogen, three lines (Figure 2.3) represent three bonds between nitrogen atoms. This is called a triple bond. Each bond in a triple bond is a pair of electrons, so a triple bond represents a total of six electrons. As shown in the figure, each N is surrounded by two dots and three lines mark the presence of a total of eight valence electrons, giving it a structure of an octet and making it more stable.

2.3.3 Physical And Chemical Properties

Nitrogen is a chemical element with many physical and chemical properties (Table 2.1).

2.3.3.1 Physical properties
- It is a non-metallic, colorless, odorless, tasteless gas.
- It is non-toxic.
- It has an atomic number 7 and an atomic weight of 14.0067

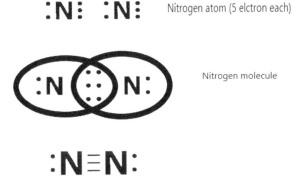

FIGURE 2.3 Lewis dot structure of nitrogen.

TABLE 2.1
Some Properties of Nitrogen

S.No.	Property	value
1.	Symbol	N
2.	Group	15
3.	Discovered	In 1772 by Daniel Rutherford
4.	Atomic number	7
5.	Atomic weight	14.0067
6.	Belting point	−209.86 °C (−345.8 °F)
7.	Boiling point	−195.8 °C (−320.4 °F)
8.	Density (1 atm, 0° C)	1.2506 grams/litre
9.	Usual oxidation states	−3, +3, +5
10.	Electron configuration	$1s^2 2s^2 2p^3$
11.	Atomic radius	1.55 A°
12.	Physical state at 293 K	gas
13.	Natural isotopes	^{14}N, ^{15}N

- It is slightly lighter than air.
- Its water solubility is low (around 23.2 cm^3 per litre of water at 1 atmospheric pressure and 273 K).
- It has a low melting and boiling point (i.e., 63.2 K (-209.86°C) and 77.2 K (-195.79°C), respectively).

2.3.3.2 Chemical properties
- It is chemically unreactive at room temperature.
- The bond distance of nitrogen is 109.8 pm and bond dissociation enthalpy is 941.4 KJ mol^{-1}.
- The low reactivity of nitrogen is due to high bond dissociation enthalpy.
- Reactivity increases with an increase in temperature.
- At higher temperatures, nitrogen reacts with some metals to form ionic nitrides (e.g., lithium nitride and magnesium nitride) and non-metals form covalent nitrides (e.g., ammonia and nitric oxide).

2.3.4 ISOTOPES OF NITROGEN

Nitrogen-14 and Nitrogen-15 are two naturally occurring isotopes of nitrogen. Isotopes with the range from 10 to 25 masses are also found but these are radioactive.

Nitrogen 14 makes up more than 99% of all nitrogen found on Earth. It is the most abundant nitrogen in the atmosphere and living organisms. It is very stable and non-radioactive. It has various applications in food preservation, agricultural practices, and biomedical research. It is believed that Nitrogen-14 is the source of naturally occurring, radioactive carbon-14. Carbon-14 is formed by some nuclear reaction of cosmic radiations with N-14 in the upper atmosphere of Earth. This carbon-14 decays back to nitrogen-14 with a half-life of 5,730 ± 40 years (Godwin, 1962).

Nitrogen-15 is a rare stable and non-radioactive isotope of nitrogen. Two sources of nitrogen-15 are the positron emission of oxygen-15 (CRC) and the beta decay of carbon-15. It is used in the field of medical science like brain research, specifically nuclear magnetic resonance spectroscopy (NMR). It is also sometimes used in agricultural practices and as labels in protein biology. Nitrogen-15 tracing is a technique used to study the nitrogen cycle.

2.4 OCCURRENCE AND SOURCES OF NITROGEN

The element of Group 15 nitrogen does not occur very abundantly in nature. It is about 78% by volume of the total earth's atmosphere, and its abundance in the earth crust is very low. The estimated nitrogen in living biomass (mainly terrestrial plants) is equivalent to about three parts per million of the atmospheric nitrogen. Except for living biomass, nitrogen pools are also found in soil organic matter, rocks (in fact, the largest single pool), coal deposit sediments, organic matter, and nitrate in ocean water. The most common gaseous form of nitrogen in the atmosphere after molecular nitrogen is dinitrogen oxide (Tamm, 1991). Nitrate is the form of nitrogen most used by plants for growth and development. The most common nitrate minerals are sodium nitrate and potassium nitrate.

Nitrogen is a key element of nucleic acid, which determines the genetics of plants and animals. It is also a constituent of proteins and amino acids. There is always a continuous interchange of nitrogen compounds between the atmosphere and the biota of earth. This interchange between the atmosphere and living organisms is known as the nitrogen cycle (Figure 2.4). However, nitrogen has to convert first into usable or fixed form like ammonia. This conversion can be done by lightning strikes or with the help of diazotrophs using the enzyme nitrogenase, a process called nitrogen fixation. Nitrogen fixation is done in different ways like atmospheric, industrial, and biological. In the atmosphere, the conversion of nitrogen gas into nitrogen oxide (NO) takes place with the help of the lightning process. In this process, nitrogen reacts with oxygen gas to form oxides. Moreover, this

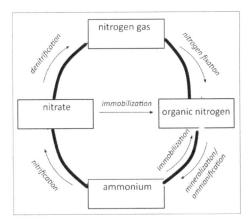

FIGURE 2.4 Nitrogen cycle.

NO also reacts with water to make nitrous or nitric acid. This nitric acid seeps into the soil where it becomes the source of nitrate and is finally used by the plants. Though nitrogen in the atmosphere is quite stable due to the triple bond, the energy and heat produced by the lightning process break this bond and allow nitrogen atoms to react (Tuck, 1976). An industrial process of nitrogen fixation also plays an important role. The most common industrial process to produce ammonia is the Haber process. It is the process of reduction of nitrogen to produce nitrogen fertilizer, revolutionizing modern-day technology (Smil, 2004). This process requires high pressure and temperature of about 200 atm and 400°C, respectively. Fertilizer production is the largest source of fixed nitrogen by humans in the terrestrial ecosystem. However, along with fertilizers, this fixed nitrogen is also used to make explosives and other products. The use of nitrogen fertilizers has intensified globally. Expansion of the human population has also been marked due to the usage of these fertilizers (Erisman et al., 2008). Biological nitrogen fixation (BNF) is an important alternative to nitrogen fertilizer (de Bruijn, 2016). It is a process in which atmospheric nitrogen gets converted into ammonia by a few prokaryotes like N-fixing bacteria, *Rhizobium* and *Azotobacter*, using an enzyme complex known as nitrogenase. With the help of this process, an N pool is maintained in the ecosystem and N loss can be replenished. Prokaryotes involved in BNF can be symbiotic or non-symbiotic ones, but the putative role of BNF in the agroecosystem is played by a symbiotic microorganism such as *Rhizobium*, which infects the roots of leguminous plants through root nodule formation and helps the crop by providing nitrogen. The total N2 fixation in the world (terrestrial + oceanic) is estimated to be about 170 Tg of which legumes account for almost half of the total estimated value (i.e., 80 Tg) and industrial N fixation (production of N fertilizer) for the other half (i.e., approx. 88 Tg). However, the potential contributions by free-living N2-fixing bacteria are difficult to quantify, even under the most favorable environments (Subbha Rao, 2017). The fixed nitrogen produced by the above-entioned processes is passed to the different living organisms by one or the other method and returned to the soil after the death of these organisms through the activities of soil microorganisms, which provide the nitrogen in its ready form again to plants and then to animals and humans. Excess of fixed nitrogen is poisonous to plants, animals, and humans, and is known as nitrogen pollution. Nitrogen pollution is caused due to the overproduction of some forms of nitrogen (bonded with carbon, oxygen, or hydrogen) also known as reactive nitrogen (Driscoll et al., 2013). The presence of this reactive form of nitrogen in soil and water has been shown to enhance vegetative growth beyond the favorable limit. In the current situation, the The whole nitrogen cycle depends upon the nitrogen produced naturally or with the help of anthropogenic activities. However, with the increase in human population demand for food is also increasing and is leading to mass production of synthetic nitrogen fertilizers for agricultural activities (Ghaly et al., 2015). Human activities such as the production and use of commercial fertilizer and fossil fuels in industrial processes, energy generation, and transportation has altered the nitrogen cycle and caused disturbances to other natural cycles such as carbon, sulfur, and oxygen.

2.5 ROLE OF NITROGEN IN PLANTS

Nitrogen is an essential nutrient for the growth, development, and reproduction of plants to complete their life cycle (Anas et al., 2020, Tafteh and Sepaskhah, 2012; Yousuf et al., 2021). Despite nitrogen being one of the most abundant elements on Earth, nitrogen deficiency is probably the most common nutritional problem affecting plants worldwide. Nitrogen from the atmosphere and Earth's crust is not directly available to plants. Around 3–4% nitrogen is present in healthy plants specifically in its upper ground part. This concentration is significantly higher as compared to other nutrients. Nitrogen is an essential element of all the amino acids in plant structures, which are the building blocks of plant proteins, important in the growth and development of vital plant tissues and cells like the cell membranes and chlorophyll. It is a component of nucleic acid that forms DNA, a genetic material that plays a significant role in the transfer of crop traits. Plant metabolism is incomplete without nitrogen. The role of nitrogen is very important in different physiological processes. It provides darker green color to plants. It stimulates root growth as well. Nitrogen stimulates the productivity and yield of crops (Messignam et al., 2009). Nitrogen also helps to increase photosynthetic rate, leaf area, and leaf biomass (Rafiq et al., 2010). It also enhances the productivity of seed crops in terms of higher seed number and yield. Studies have shown that seed mass, quality, and vigor are also increased (Amjed et al., 2013, Khaliq et al., 2008, Khaliq et al., 2009, Nasim et al., 2011). Nitrogen also plays a pivotal role in horticulture plants. It increases the pulp content as well as fruit yield (Satapathy and Banik, 2002) in horticulture crops. Nitrogen also encourages the uptake of other important nutrients like potassium and phosphorus for the overall growth and development of plants (Bloom, 2015, Hemmerley, 2016). Increased nitrogen also helps to increase CO_2 assimilation rate and increase plant resistivity for different environmental stresses like limited water supply or saline soil conditions Bondada et al., 1996, Chen et al., 2010). However, some inherent factors like soil moisture, soil aeration, and electrical conductivity (salt content) affect nitrogen cycling and nitrogen losses through leaching, runoff, or denitrification.

2.6 FACTORS AFFECTING NITROGEN NUTRITION

There are many factors that influence the magnitude of nitrogen uptake in plant communities. These include soil air and moisture, soil temperature, soil fertility, soil pH, age of the plant, climatic factors, etc. (Dong et al., 2015). Some of these factors are discussed below.

2.6.1 SOIL PH

Soil pH affects the growth of plants as well as microflora responsible for uptake of nutrients. In general, a soil pH between the range of 6.0 to 7.0 has been shown to be the best for optimum uptake. The bacteria of the genus *Rhizobium* are often negatively affected by soil pH values less than 6.0. If soil pH is less than 6.0, and where it is economical to do so, limestone may be added to increase soil pH and thus increase BNF. However, there are some exceptions (e.g., symbiosis between red clover and its microsymbiont function well at pH less than 6.0).

2.6.2 SOIL FERTILITY

Reduction in soil fertility can be related to limited soil nitrates or deficiency of essential nutrients limiting plant growth and development. Excessive soil nitrate affects the level of microflora in the soil.

2.6.3 Soil Texture

Soil texture is an important factor in nitrogen nutrition. Sandy and coarse soils do not hold much nitrogen whereas clay loam and clay and loamy soil have the maximum capacity to hold N for plants. Organic manures can be added to enhance the texture of soil (Leghari et al., 2016).

2.6.4 Soil Moisture

Moisture present in the soil helps the intake of N by plants, whereas the availability and uptake process is adversely affected by drought conditions, which also result in loss of N due to the process of volatilization (nitrogen losses into the atmosphere in the form of ammonia gas due to lack of moisture). Therefore, the application of optimum water is always advisable (Leghari et al., 2016).

2.7 DEFICIENCY SYMPTOMS OF NITROGEN

Nitrogen is a highly mobile element in plants. The important deficiency symptoms include:

1. Chlorosis of the leaves and yellowing of older leaves takes place due to inhibition of chloroplast and chlorophyll synthesis.
2. Necrosis of plant leaves takes place and sometimes complete death of plant in case of severe deficiency.
3. Deficiency leads to stunted or dwarf growth.
4. Yield is reduced to a great extent in terms of fruit size and number.
5. Reduction in the fertilization process, early dropping of flower, and advancement in flower bud formation.
6. Leaves become more susceptible to disease and pest attacks.
7. Reduction in sugar content of crops and prolonged growth period and maturity is observed in many crops.

2.8 EFFECTS OF EXCESSIVE NITROGEN

Nitrogen plays a very important role as a macronutrient in the life cycle of plants. However, excessive nitrogen causes serious effects in plants (Elhanafi et al., 2019). Some of the effects are discussed below.

2.8.1 Increased Foliage Growth

As discussed earlier, nitrogen in plants increases the chlorophyll content of plants, increases the size of leaves, and increased surface area. Thus, excess nitrogen increases the foliage growth, which leads to deprivation of this macronutrient to other plants and inhibits their growth. Moreover, all energy flow is redirected to foliage as compared to the reproductive organs of plants, which causes reduced growth even in the growing season.

2.8.2 Stunted Root Growth

As all the energy is directed toward the foliage part of plants, the plant may be destabilized in its soil position due to poor growth of the root. These roots succumb to nitrogen-induced stresses that damage the plant throughout its length. Moreover, if a plant becomes very tall due to foliage growth it could be damaged or uprooted by heavy wind.

2.8.3 Underground Water Pollution

All the nitrogen present in the soil cannot be absorbed by the plants. This excess nitrogen is leached down in the groundwater through water runoff. Moreover, nitrate, which is formed by microbial activity from nitrogen, is also leached in the underground water, leading to an excess of nitrate in water supplies causing serious effects in our ecosystem.

REFERENCES

Amjed A., Ahmad A., Khaliq T., Anser A., and Ahmad M., Nitrogen nutrition and planting density effects on sunflower growth and yield: A review. Pakistan Journal of Nutrition. 2013, 12 (12), 1024–1035.

Anas M., Liao F., Verma K. K., Sarwar M. A., Mahmood A., Chen Z. L., Li, Q., Zeng X.P., and Li1 Y. L. Y-R. Fate of nitrogen in agriculture and environment: agronomic, eco-physiological and molecular approaches to improve nitrogen use efficiency. Anas et al. Biological Research. 2020, 53, 47 https://doi.org/10.1186/s40659-020-00312-4.

Bloom, A.J., The increasing importance of distinguishing among plant nitrogen sources. Current Opinion in Plant Biology. 2015, 25, 10–16.

Bondada B., Oosterhuis D., Norman R., and Baker W. Canopy photosynthesis, growth, yield, and boll 15 N accumulation under nitrogen stress in cotton. Crop Science. 1996, 36, 127–33.

Chen W., Hou Z., Wu L., Liang Y., and Wei C. Effects of salinity and nitrogen on cotton growth in arid environment. Plant Soil. 2010, 326, 61–73.

de Bruijn F.J. Biological nitrogen fixation. Book Summary. Advances in Microbiology, 2016, 6, 407–411. http://dx.doi.org/10.4236/aim.2016.66040

Dong, T., J. Li, Y. Zhang, H. Korpelainen, U. Niinemets and C. Li. Partial shading of lateral branches affects growth, and foliage nitrogen-and water-use efficiencies in the conifer *Cunninghamia lanceolata* growing in a warm monsoon climate. Tree Physiology. 2015, 35 (6), 632–643.

Driscoll C, Whitall D, Aber J, Boyer E, Castro M, et al. Nitrogen pollution in the Northeasterns United States: Sources, effects and management options. Bioscience. 2003, 53, 357–374.

Elhanafi L., Houhou M., Rais C., Mansouri I., Elghadraoui L., and Greche H. Impact of excessive nitrogen fertilization on the biochemical quality, phenolic compounds, and antioxidant power of *Sesamum indicum* L seeds. Journal of Food Quality. 2019, Article ID 9428092 | https://doi.org/10.1155/2019/9428092

Erisman J. W., Sutton M. A., Galloway J, Klimont Z., and Winiwarter W. How a century of ammonia synthesis changed the world. Nature Geoscience. 2008, 1 (10), 636–639.

Ghaly A. E., and Ramakrishnan V. V., Nitrogen sources and cycling in the ecosystem and its role in air, water and soil pollution: A critical review. Journal of Pollution Effects and Control. 2015, 3 (2), 1–26.

Godwin, H. "Half-life of radiocarbon." Nature. 1962, 195 (4845), 984.

Hemerly, A., Genetic controls of biomass increase in sugarcane by association with beneficial nitrogen-fixing bacteria. In: Plant and Animal Genome XXIV Conference. Plant and Animal Genome, 2016 during month of January.

Khaliq T. A., Ahmad A., Hussain and Ali M. A. Impact of nitrogen rate on growth, yield and radiation use efficiency of maize under varying environments. Pakistani Journal of Agricultural Science. 2008, 45,1–7.

Khaliq, T. A., Ahmad A., Hussain and Ali M. A., Maize hybrid response to nitrogen rates at multiple locations in semi arid environment. Pakistan Journal of Botany, 2009, 41(1), 207–224.

Leghari S.J., Wahocho N. A., Laghari G. M., and Laghari A. H. Role of nitrogen for plant growth and development: A review. Advances in Environmental Biology. 2016, 10(9), 209–218.

Massignam, A. M., Chapman S. C., Hammer G. L., and Fukai S., Physiological determinants of maize and sunflower achene yield as affected by nitrogen supply. Field Crops Research, 2009, 113: 256–267.

Nasim W., Ahmad A.,Wajid J., Akhtar J., and Muhammad D. Nitrogen effects on growthland development pf sunflower hybrids under agro-climatic conditions of Multan. Pakistan Journal of Botany. 2011, 43, 2083–2092.

Rafiq M. A., Ali A., Malik M. A., and Hussain. Effect of fertilizer levels and plant densities on yield and protein contents of autumn planted maize. Pakistan Journal of Agriculture Science. 2010, 47, 201–208.

Satapathy S.K., and Banik B.C., Studies on nutritional requirement of mango cv. Amrapali. Orissa Journal of Horticulture 2002, 30(1), 59–63.

Smil V. 2004. *Enriching the Earth: Fritz Haber, Carl Bosch, and the Transformation of World Food Production*, MIT Press.

Subba Rao A., Jha P., Meena B.P., Biswas A. K., Lakaria B. L., and Patra A. K. Nitrogen processes in agroecosystems of India. The Indian nitrogen assessment sources of reactive nitrogen. Environmental and Climate Effects, Management Options, and Policies. 2017, 59–76.

Tafteh A., and Sepaskhah A. R. Yield and nitrogen leaching in maize field under different nitrogen rates and partial root drying irrigation. International Journal of Plant Production. 2011, 6, 93–113.

Tamm C.O. Introduction: Geochemical occurrence of nitrogen. Natural nitrogen cycling and anthropogenic nitrogen emissions. Nitrogen in Terrestrial Ecosystems, 1991, 81, ISBN: 978-3-642-75170-7.

Tuck A. F. Production of nitrogen oxides by lightning discharges. Quarterly Journal of the Royal Meteorological Society. 1976, 102 (434), 749–755.

Yousuf P. Y., Shabir P. A., and Hakeem K. R. (2021). miRNAomic approach to plant nitrogen starvation. International Journal of Genomics, 2021, 8560323.

3 Nitrogen Deficiency in Plants

Bilal Ahmad Wani, Sufiya Rashid, Kausar Rashid, Hanan Javid, Junaid Ahmad Magray, Rouf ul Qadir, and Tajamul Islam

Plant Reproductive Biology Genetic Diversity and Phytochemistry Research Laboratory, Department of Botany, University of Kashmir, Hazratbal Srinagar-190006, J and K, India
Email; biobilal968@gmail.com

CONTENTS

3.1 Introduction 29
3.2 Diagnosis of Nitrogen Deficiency in Plants 30
3.3 Effect of Nitrogen Deficiency on Plant Growth and Development 31
3.4 Efficiency of Plants to Use Nitrogen 32
3.5 Nitrogen Management Practices 34
    3.5.1 Method of Broadcasting 34
    3.5.2 Placement Method 34
    3.5.3 Starter Approach Application 35
    3.5.4 Foliar Application 35
    3.5.5 Fertigation Technique 35
3.6 Conclusion 35
References 35

3.1 INTRODUCTION

Nitrogen plays a prominent role in the metabolism of plants. All critical processes in plants are related to proteins, in which nitrogen is an important component. As a consequence, the use of N_2 is obligatory to achieve good crop outputs. Nitrogen plays a key role not only in agriculture by improving yields, but also by enhancing food qualities (Ullah et al., 2010). Each type of plant is special and has an optimal array of nutrients as well as a minimum required level. Below this minimal, plants begin to display signs of nutrient deficiency. The use of N at optimum rates improved photosynthesis, development of leaf area, and also net assimilation rates (Yousuf et al., 2016; Ahmad et al., 2009). The higher leaf area and overall leaf biomass of plants are directly correlated with crop yields (Rafiq et al., 2010). The adequate use of nitrogen through efficient management methods has improved crop yields globally (Smil. 2001). The nutrient concentrations above the maximum level can also lead to improper growth of plants due to their toxicity. Therefore, application and placement of the proper amount of nutrients are critical. Nitrogen-deficient plants tend to show signs of chlorosis and stunted appearance. Nitrogen is quite mobile and is easily transported from one part to another part within the plant body. When nitrogen uptake by a plant is inadequate, nitrogenous substances in the older leaves are hydrolyzed to yield nitrogen, and nitrogen is dislocated to the younger leaves resulting in older leaves displaying marked chlorosis. Stem tissue is likely to be the most appropriate plant portion to be sampled and analyzed for N status assessment of actively developing annual plants. Symptoms of deficiency are the simplest diagnostic strategies for testing an N fertilization program. There are various reports that suggest that deficiency of nitrogen influences plant growth, carbon fixation, leaf gas exchange and chlorophyll fluorescence parameters, thylakoid membrane organization, functions of enzyme RUBP, carbohydrate levels, metabolites, and respiratory enzymes of seedlings, etc.

3.2 DIAGNOSIS OF NITROGEN DEFICIENCY IN PLANTS

Testing of the soil, as well as plant tissues, is critical to determine soil and plant nutrient quality. By analyzing this data, plant and soil scientists may find out the nutrient requirements of a specific plant in a specific soil (Gardner et al., 1967). Plants possess nitrogen not only in inorganic forms but also in functional organic compounds as well. It is therefore appropriate to find the N status of plants by analyzing chlorophyll content and other organic and inorganic protein reserves of nitrogen. Nitrogen is an inorganic form within the plant body that indicates the active supply obtainable for metabolism for some time. Organic reserves reveal what supply of nitrogen the plant has before that period (Macy. 1936). The N used in chlorophyll synthesis helps to visually diagnose the N deficiency by affecting the color of the plant. The deficiency symptoms of N may range from mild to severe, which are sometimes accompanied by less visible symptoms to extreme changes in plant morphology. Acute deficiency is a strong indication that plant productive potential has been impaired (English et al., 1978). Despite the absence of visual signs, some decrease in productivity is always evident. Commonly N deficiency results in stunted growth of the plant with underdeveloped leaves, soft stems, and fewer lateral roots and tillers. Due to less chlorophyll development after impaired N supply leaves show a pale to yellowish-green appearance. Symptoms are mainly noticeable to the oldest leaves as the nitrogen is readily moved from the older leaves. Both inorganic forms and hydrolyzed products of organic forms are transferred from the oldest tissue for re-use in younger leaves (Hambidge. 1947). The leaves can produce yellow, red, or purple colors at advanced phases of growth as these colored pigments predominate chlorophyll. As the N shortness becomes more prominent, the oldest leaves grow orange, beginning from the tip and spreading along with the whole leaf and leaving it dead. The youngest leaves retain greenness until the deficiency becomes acute (Fageria et al., 2010). A thorough explanation of the basic symptoms of N deficiency on a crop-to-crop basis is still unknown. There are a variety of reviews, textbooks, and basic sources that provide this sort of detail on some crops (Hambidge, 1941; Sprague, 1964; Black, 1968; English & Maynard, 1978). In Poaceae, the signs of nitrogen deficiency at higher stages are quiet. The older leaves of the plant start to turn yellow from their tips, followed by the death of these leaves. In the case of corn, the dying of the tissue follows a V-shaped pattern and continues upwards reaching the midrib. In other types of grasses, these symptoms are not as definite, usually associated with yellowing and burning of lower leaves (Jones et al., 1990). In the case of cotton (*Gossypium hirsutum* L.), the symptoms of nitrogen deficiency are less prominent. However, it is associated with small leaf size, less intermodal growth, and a lack of vigor of the plant. When nitrogen deficiency intensifies it results in turning the color of leaves from green to pale yellow. Finally, these leaves turn brown, die, and shed off from the plant in a similar fashion as natural leaf drop at maturity (Gardner et al., 1967). The symptoms of N deficiency in fruit and nut tree crops differ from plant to plant but are usually associated with a pale green to yellowish-green, sparse and stunted foliage, and also twigs may show dieback. However, in some species, colors other than yellow appear, such as brown, orange, red, and purple. The oldest leaves fall and fruits are less and underdeveloped (Tucker, 1984). To examine the nitrogen status of the plant, which is beneficial in the diagnosis of nitrogen deficiency, a reference point must be created first. This point is different from plant to plant and also varies in the case of plant components, location, and growth stage of the plant. This point of reference is referred to as a critical concentration. Macy (1936) first suggested the idea of critical concentration. It has been defined as the concentration of a nutrient at a point that separates the deficiency zone from the optimum zone as shown in Figure 3.1. According to Ulrich (1952), the critical

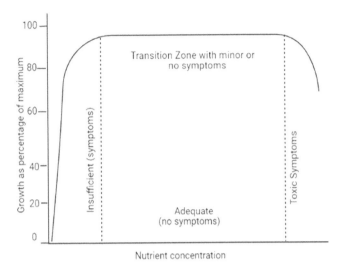

FIGURE 3.1 Relation between plant growth and nutrient (N) concentration.

concentration is slightly below the maximum yield concentration and in the transition zone between deficiency and adequacy. These two points of view vary just marginally and are indistinguishable in practice.

3.3 EFFECT OF NITROGEN DEFICIENCY ON PLANT GROWTH AND DEVELOPMENT

Nitrogen is a key constituent of chlorophyll, which is the site of photosynthesis and therefore critical for biomass accumulation and productivity. Nitrogen plays a crucial function in many physiological processes. It imparts a dark green color to plants, enhances the growth and development of the leaf and stem, and also stimulates root development. N produces rapid early development, improves fruit qualities, encourages the growth of leafy vegetables, improves the protein amount of food crops, helps in the absorption and use of other nutrients such as potassium and phosphorus, and regulates overall plant growth (Bloom. 2015; Hemerly. 2016). The deficiency of nitrogen disrupts normal development, results in the occurrence of chlorosis and the appearance of red or purple marks on the leaf, and restricts the growth of the lateral buds. Generally, the signs of deficiency first occur on older leaves followed by leaf senescence. However, the extra use of nitrogen has also undesirable effects on plant growth, as it imparts extra dark-green color to the leaf and affects plant growth, fruit number, and quality (Bianco et al., 2015). Plants uptake nitrogen mostly in nitrate and ammonia forms. However, in a few soils such as submerged soils, NH_4^+ is considered to be more beneficial than in the case of rice (King et al., 1992). For good growth and production, plants need an optimal concentration of nitrogen. Doses below or above optimum directly decrease crop yields, whereas higher N doses also cause detrimental effects on plants, and this problem continues to focus on crop production (Magistad et al., 1945). All crops such as cereals, oilseeds, fiber, and sugarcane need a reasonable quantity of nitrogen for rapid growth and production of higher harvests of higher quality. Nitrogen has a vital role to play in improving the production of wheat, rice, sugarcane, and cotton. Nitrogen fertilization in the case of wheat influences growth, plant height (cm), number of tillers per meter square, number of spikelets per spike, grains per spike, spike length, and yield parameters (Ali et al. 2000). Increased nitrogen levels have a direct effect on crop grain yield. The maximum wheat yield of 3,848 tonnes per hectare was achieved by applying 180 kg N (Amjed et al., 2011). N_2 at a rate of 120 kg per hectare proves good for rice plant length, number of tillers, dry weight, length of panicle, number of grains filled/rubber, the yield of straw, organic yield, harvest index, gain cost ratio, and yield (4.66 tons per hectare) (Malik et al., 2014). Similarly, the judicious use of nitrogenous fertilizer is more important in the cotton crop. Nitrogen is a key nutrient that influences the growth of the plant, fruit development, and cotton yield (Boquet et al., 1914). Nitrogen levels have major variations in yield components and seed yield of cotton. Application of nitrogen such as 100 kg per hectare increases cottonseed yield and weight of bolls per plant by promoting more monopodial and sympodial branches (Nadeem et al., 2010). Nitrogen is also the cornerstone of the sugarcane crop as it strengthens the vegetative parts and thus raises the weight of the cane. China and India achieved a double yield after the application of about 300 kg nitrogen per hectare. In other countries, N is typically used at a rate of between 150 and 200 kg per hectare. The yield and quality of sugarcane have increased significantly by increasing the rate of N at 195.5 kg per hectare. Poor adverse effects on the quality of the sugarcane were also reported at 299 kg of nitrogen per hectare (Hemalatha, 2015). The yield of oilseed crops such as mustard and sunflower cultivated worldwide is greatly affected by nitrogen fertilization (180kg N per hectare) in semi-arid conditions (Nasim et al., 2012. Nitrogen requirements for horticultural crops are valuable, but their role is not yet well explained for some plants particularly in ornamental and large fruit tree crops, chiefly due to lack of understanding in developing countries. An adequate N rate is important for the improvement of horticulture crops. The main crops in this field are citrus, mango, onion, and potatoes. Nitrogen holds a good place to increase the yield and qualities of mango fruit. Applying N fertilizer to mango trees significantly improved fruit qualities pulp content and fruit number per tree (Satapathy et al., 2002). The regular application of N fertilizer increased the yield of fruits whereas the recommended N rate for younger plants is generally based on the analysis of leaf N or the examination of N in harvested fruit as in the case of citrus (Sharma et al., 2000). The suggested rates start at 0.2–0.5kg of N per tree per year. In Texas (USA), N levels for six to eight-year-old citrus plants range from 0.5 to 0.6 kg per tree per year (Sauls, 2002). Nitrogen is used in the collection of biomass from roots and leaves, as it plays a prominent role in the vegetative growth of vegetables. It enhances the size and color of the spinach leaf (Wang et al., 2008). It is an important component of proteins, nucleic acids, chlorophyll, and growth hormones (Barker et al., 1974). It contributes a great deal to agriculture and enhances protein-based metabolic processes that lead to increased vegetative, reproductive growths and hence crop yields.

3.4 EFFICIENCY OF PLANTS TO USE NITROGEN

Various variables affect the efficiency of N use such as plant age, plant characteristics, soil behavior, and climatic variables. The general principle is that the productivity of nitrogen usage becomes higher when soil and climate are more favorable for plant growth and development. Application of nitrogen at the initial growth phases (germination and seedling development) when plants do not have a well-developed root system and have lower leaf number by foliar means is more effective than through soil. In these situations, applying a high dose of N to the crops is just a waste of money because it would not yield positive results. N-use efficiency of plants increases from seedling stage to fully mature plants as shown in Figure 3.2. When the roots of the plant are fully established and left, then N usage increases in size as well. The well-developed root system increases the uptake of N and large size enhances the photosynthesis process, stimulating plant physiological action that assists in the efficiency of N use. At this stage of the plant, the foliar application of N is useful compared to that stage of the plant where the leaf area is smaller. Plant N consumption effectiveness typically decreases as the plant moves to the completion of the life cycle process at the maximum maturity stage of the plant and usually stops vegetative growth.

The quality of use of plant N is significantly affected by soil and climate conditions (Tottingham, 1919; Dong et al., 2015). These are classified as soil pH, texture, removal of crops, leaching, oxidation and reduction, volatilization, combustion, erosion, and performance efficiency of N use. Soil

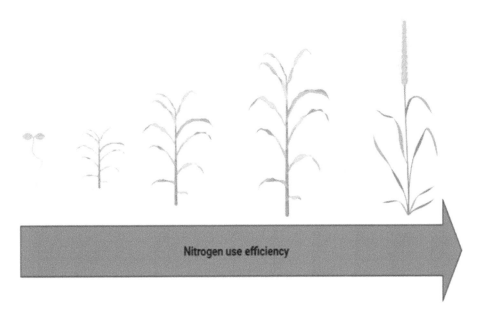

FIGURE 3.2 Nitrogen use efficiency increase from seedling stage to fully mature plant.

pH interferes with N use and good N use performance was found at standard pH range (6.5–7.0). On the other hand, in the nitrogen management system, soil texture is theoretically considered; sandy and coarse soils hold less N. The clay loam and loamy soils have the highest potential for plants to retain N. Therefore, crops grown in clay and loamy soils have greater N usage efficiency. With the addition of organic manures, soil texture can be improved. With the addition of organic manures, soil texture can be improved. For the proper intake via plants, optimum availability of N is required. Among the nitrogen-limiting factors, one of the big problems is leaching. N is added to the seed, dissolved in irrigation water, and leaches down to the downward portion from the topsoil surface. In a simple sense, a lower nitrogen supply decreases plant growth and production.

Soil compaction: The compacted soil has problems with aeration as well as water movement. For the process of breaking N (nutrients) and its dissemination in the root zone, which intensifies the N uptake efficiency, optimum quantities of water and air are necessarily required. Organic manure like farmyard manure (FYM), compost, green manure (GM), poultry, and others enhance soil productivity and soil texture and foster microbial activity by encouraging the use of N and eventually nitrogen efficacy. Moist soils contribute to plant uptake of N, although soil drought conditions adversely affect its availability and process of absorption, which also contributes to soil loss of N due to volatilization. The efficiency of better use of N can be obtained by adding optimal water to plants. In certain soils where other nutrients are excessively available, the abundance and utilization effectiveness of N is decreased. Many high-yielding and good-response varieties of different crops such as wheat, cotton, etc., have been developed in plant breeding and genetic engineering. Such varieties have the potential for more effective absorption and use of N.

Oxidation and reduction: where electrons(s) are either lost or gained, which may result in transformation of nutrients from useful to toxic forms. Due to the high volatilization process, nitrogen begins to be absent in the soil. It can occur due to deficit of wetness on the surface of the soil, particularly in arid regions where water is insufficient and temperatures are high. The rate of transpiration increases by high temperatures and thereby increases the demand for water in the rhizosphere, leading to drought around the plant. N loss can also take the form of ammonia gas in the atmosphere.

Rapid volatilization decreases the effectiveness of using N in plants. To withstand such conditions, a small amount should be applied to maintain the plant's continuous nitrogen response. Further burning and erosion can affect the availability of nitrogen to plants: both terms are defined, though differently, as the loss of nitrogen from the soil. Burning of field crops or straw leads to killing or suppressing the growth of soil microorganisms that are needed for soil health. It also affects soil structure and texture thereby decreases the availability of nutrients, particularly nitrogen. About 98–100% of N is lost by burning straw of wheat, flax, and oats (Heard et al., 2006). Erosion as a result of a wind storm and fast water flows of rain also result in big losses of nitrogen as the upper surface of the soil is removed. Thus, the importance of the content of N and utilization efficiency by plants is considerable.

3.5 NITROGEN MANAGEMENT PRACTICES

The nitrogen application method and its timing are very essential in the development of crops. The broadcasting application of N was popular in traditional cultivation systems in the history of agriculture and is still used because of the simple and rapid distribution of nitrogen fertilizers, but higher losses and low use efficiency of N is still a problem. Along with the foliar application and so on, the traditional N application practice has been substituted with the aid of new techniques such as fertigation and flooded nitrogen fertilizer application technology. The new N application methods greatly reduce the chance of N losses and increase the plant's uptake. Some key items must be considered before adding nitrogen, such as

A. Plant/Crop varieties and their root characteristics
B. Plant N needs to reach its stage of growth
C. Physio-chemical properties of soil
D. Status of soil moisture
E. Source and form of water for irrigation
F. The irrigation rate and the frequency of the N fertilizer to be used and the formulation type

The method of application of fertilizer also plays a key part in nutrient management. Below are some fertilizer application strategies.

3.5.1 Method of Broadcasting

The fertilizer broadcasting application technique is described as the application of fertilizer into the soil and crop. Fertilizer delivery in the whole field is uniform. The maximum fertilizer amount of is required for broadcasting. This technique is ideal for growing dense crops. The granular type of nitrogen is often used. In addition, the broadcast technique is subdivided into a basal application and top dressing.

Basal application: In this process, the fertilizer is applied at the time of sowing and the fertilizer is evenly spread over the whole region of the field and mixed with the soil.

Topdressing: This technique is useful for densely grown crops. It is also best used for the production of paddy and wheat. In addition to benefits, when applied directly to the soil surface of the standing crop, the disadvantages include N loss is higher from the broadcasting process and weeds easily absorb N as compared to crop of interest, thereby reducing its efficiency of use.

3.5.2 Placement Method

The placement method involves the application of N fertilizer into the soil on a regular or recommended basis. This practice is effective when N fertilizer is available in limited amounts and the roots of the plant perform slowly and feebly due to soil characteristics. Fertilizer placement method is also grouped into near placement or localizing addition and deep placement, etc. Localizing placement is when N fertilizers are directly added to developing plants or seeds in the soil. The goal is to provide N to the root zone of the plant adequately and increase its mobilization as much as possible. The technique where ammonium (NH4+) is often used to maximize the plant uptake process is another hand-deep adding or placement of N. The primary purpose is to prevent run-off and leaching loss of N and also prevent it from consumption by weeds.

3.5.3 Starter Approach Application

This is one of the latest application techniques for N fertilization. During rice transplanting, N is mixed with P and K in the ratio 1:2:1 and applied to plants. The primary benefit of this technique is the promotion of vigorous seedling growth.

3.5.4 Foliar Application

Nitrogen fertilizer is dissolved in water and sprayed on the plant leaves. The foliar application of N is found to be important for optimizing crop yield by improving plant growth and development. Sprayers are widely used in this technique.

3.5.5 Fertigation Technique

This technique involves the application of N solution through the flow of irrigation. Tanks are usually used in this technique. This N application method is considered to be the most cost-effective and suitable for nitrogen fertilizers since it minimizes the risk of N losses and ensures maximum accessibility and improved uptake performance of N for plants.

3.6 CONCLUSION

In-plant metabolism, nitrogen plays a critical role and the yield penalty is often associated with its deficiency. Improper and non-judicious use of nitrogen is a major factor for limited yields with the lowest crop quality worldwide. The present chapter showed that different plant physiological processes are affected by nitrogen deficiency and how these deficiencies can be diagnosed. The simplest diagnostic method for assessing the N status of the plant, which can be used for a fertilizer program, is looking at deficiency symptoms. The most definitive signs of nutrient deficiency are possibly the visual symptoms of N deficiency. However, these can rarely be used without a decrease in growth as well as final yield as the sole basis for corrective action; some lasting damage has occurred by the time the impact is apparent. Chemical determinations act as the prime basis for fertilizer application decisions or assessment of fertilizer activities in a practical program using diagnostic techniques as a guide for N fertilization. Suggestions from agricultural experts and soil scientists should be taken to apply the required amount of fertilizers to crops. The application of N by foliar means is more effective as compared to broadcasting methods to avoid N losses and to improve the N-use efficiency of plants.

REFERENCES

Ahmad, S., Ahmad, R., Ashraf, M. Y., Ashraf, M., & Waraich, E. A. (2009). Sunflower (*Helianthus annuus* L.) response to drought stress at germination and seedling growth stages. *Pakistan Journal of Botany*, *41*(2), 647–654.

Alagappan, S., & Venkitaswamy, R. (2016). Impact of different sources of organic manures in comparison with TRRI practice, RDF, and INM on growth, yield, and soil enzymatic activities of rice-green gram cropping system under the site-specific organic farming situation. *American-Eurasian Journal of Sustainable Agriculture, 10*(2), 1–9.

Ali, A., Choudhry, M. A., Malik, M. A., & Ahmad, R. (2000). Effect of various doses of nitrogen on the growth and yield of two wheat (*Triticum aestivum* L.) cultivars. *Pakistan Journal of Biological Sciences (Pakistan), 3*(6), 1004–1005.

Ali, A., Syed, A. A. W., Khaliq, T., Asif, M., Aziz, M., & Mubeen, M. (2011, October). Effects of nitrogen on growth and yield components of wheat.(Report). *International Journal of Biological Sciences, 3*(6), 1004–1005.

Amjed, A., Ahmed, A., Syed, W. H., Khalid, T., Asif, M., Aziz M., & Miubeen M. (2011). Effect of nitrogen on growth and yield components of wheat (report). *Sci. Ini.* (Lahore), 23(4): 331–332.

Barker, A. V., Maynard, D. N., & Mills, H. A. (1974). Variations in nitrate accumulation among spinach cultivars. *Journal of the American Society for Horticultural Science, 99*, 32–134.

Bianco, M. S., Cecílio Filho, A. B., & de Carvalho, L. B. (2015). Nutritional status of the cauliflower cultivar 'Verona' grown with omission of out added macronutrients. *PloS One, 10*(4), e0123500.

Black, C. A. (1968). *Soil-Plant Relations*. 2nd Ed., John Willey and Sons Inc.

Bloom, A. J. (2015). The increasing importance of distinguishing among plant nitrogen sources. *Current Opinion in Plant Biology, 25*, 10–16.

Boquet, D. J., Moser, E. B., & Breitenbeck, G. A. (1994). Boll weight and within-plant yield distribution in field-grown cotton given different levels of nitrogen. *Agronomy Journal, 86*(1), 20–26.

Dong, T., Li, J., Zhang, Y., Korpelainen, H., Niinemets, Ü., & Li, C. (2015). Partial shading of lateral branches affects growth, and foliage nitrogen-and water-use efficiencies in the conifer *Cunninghamia lanceolata* growing in a warm monsoon climate. *Tree Physiology, 35*(6), 632–643.

English, J. E., & D. N. Maynard. (1978). A key to nutrient disorders of vegetable plants. *Horticultural Science, 13*, 28–29.

Fageria, N. K., Baligar, V. C., & Jones, C. A. (2010). *Growth and Mineral Nutrition of Field Crops*. CRC Press.

Gardner, B. R., & Tucker, T. C. (1967). Nitrogen effects on cotton: II. Soil and petiole analyses. *Soil Science Society of America Journal, 31*(6), 785–791.

Hambidge, G. (1941). *Hunger signs in crops*. American Society of Agronomy and National Fertiizer.

Heard, J., Cavers, C., & Adrian, G. (2006). Up in smoke-nutrient loss with straw burning. *Better Crops, 90*(3), 10–11.

Hemalatha, S. (2015). Impact of nitrogen fertilization on quality of sugarcane under fertigation. *IJRSI, 2*(3), 37–39.

Hemerly, A. (2016, January). Genetic controls of biomass increase in sugarcane by association with beneficial nitrogen-fixing bacteria. In *Plant and Animal Genome XXIV Conference. Plant and Animal Genome, during month of January*.

Jones Jr, J. B., Eck, H. V., & Voss, R. (1990). Plant analysis as an aid in fertilizing corn and grain sorghum. *Soil Testing and Plant Analysis, 3*, 521–547.

King, B. J., Siddiqi, M. Y., & Glass, A. D. (1992). Studies of the uptake of nitrate in Barley: V. Estimation of root cytoplasmic nitrate concentration using nitrate reductase activity—Implications for nitrate influx. *Plant Physiology, 99*(4), 1582–1589.

Macy, P. (1936). The quantitative mineral nutrient requirements of plants. *Plant Physiology, 11*(4), 749.

Magistad, O. C., Reitemeier, R. F., & Wilcox, L. V. (1945). Determination of soluble salts in soils. *Soil Science, 59*(1), 65–76.

Malik, T. H., Lal, S. B., Wani, N. R., Amin, D., & Wani, R. A. (2014). Effect of different levels of nitrogen on growth and yield attributes of different varieties of basmati rice (Oryzasativa L.). *International Journal of Scientific & Technology Research, 3*(3), 444–448.

Massignam, A. M., Chapman, S. C., Hammer, G. L., & Fukai, S. (2009). Physiological determinants of maize and sunflower grain yield as affected by nitrogen supply. *Field Crops Research, 113*(3), 256–267.

Miller, A. J. (July 2014) Plant Mineral Nutrition. In: eLS. John Wiley & Sons, Ltd. DOI: 10.1002/9780470015902.a0023717

Nadeem, M. A., Ali, A., Tahir, M., Naeem, M., Chadhar, A. R., & Ahmad, S. (2010). Effect of nitrogen levels and plant spacing on growth and yield of cotton. Pakistan Journal *of Life and Social Sciences, 8*(2), 121–124.

Nasim, W., Ahmad, A., Hammad, H. M., Chaudhary, H. J., & Munis, M. F. H. (2012). Effect of nitrogen on growth and yield of sunflower under semi-arid conditions of Pakistan. *Pakistan Journal of Botany*, *44*(2), 639–648.

Rafiq, M. A., Ali, A., Malik, M. A., & Hussain, M. (2010). Effect of fertilizer levels and plant densities on yield and protein contents of autumn planted maize. *Pakistan Journal of Agricultural Sciences*, *47*(3), 201–208.

Satapathy, S. K., & Banik, B. C. (2002). Studies on nutritional requirement of mango cv. Amrapali. *Orissa Journal of orticulture*, 30(1), 59–63.

Sauls, J. W. (2002). Texas citrus and subtropical fruits: Nutrition and fertilization. *Internet: posted at* http://aggie-horticulture. tamu. edu/citrus/nutrition L, *2288*.

Sharma, R. C., Mahajan, B. V. C., Dhillon, B. S., & Azad, A. S. (2000). Studies on the fertilizer requirements of mango cv. Dashehari in sub-montaneous region of Punjab. *Indian Journal of Agricultural Research*, *34*(3), 209–210.

Smil, V. (2001). *Enriching the earth: Fritz Haber, Carl Bosch, and the transformation of world food production*. MIT press.

Sprague, H. B. (1964). *Hunger Signs in Crops: A Symposium*. Madison American Society of Agronomy Inc..

Tottingham, W. E. (1919). A preliminary study of the influence of chlorides on the growth of certain agricultural Plants 1. *Agronomy Journal*, *11*(1), 1–32.

Tucker, T. C. (1984). Diagnosis of nitrogen deficiency in plants. *Nitrogen in Crop Production*, 247–262.

Ullah, M. A., Anwar, M., & Rana, A. S. (2010). Effect of nitrogen fertilization and harvesting intervals on the yield and forage quality of elephant grass (Pennisetum purpureum) under mesic climate of Pothowar plateau. *Pakistan Journal of Agricultural Sciences*, *47*(3), 231–234.

Ulrich, A. (1952). Physiological bases for assessing the nutritional requirements of plants. *Annual Review of Plant Physiology*, *3*(1), 207–228.

Wang, Z. H., Li, S. X., & Malhi, S. (2008). Effects of fertilization and other agronomic measures on nutritional quality of crops. *Journal of the Science of Food and Agriculture*, *88*(1), 7–23.

Woon, C. K. (1977). Seasonal changes in stem carbohydrate and petiole nitrate in cotton. PhD dissertation. Universtiy of Arizona, Tuscon (Diss. Abstr. 78-05776).

Yousuf, P. Y, Ganie, A. H., Khan, I., Qureshi, M. I., Ibrahim, M. M., Sarwat, M., Iqbal, M., Ahmad, A. (2016). Nitrogen-efficient and nitrogen-inefficient Indian mustard showed differential expression pattern of Proteins in response to elevated CO and low nitrogen. *Frontiers in Plant Science*, *7*, 1074.

4 Nitrogen Assimilation in Plants

Reeta Kumari[1], Sonal Bhatnagar[2], and Charu Kalra[1]
[1]Department of Botany, Deen Dayal Upadhyaya College, University of Delhi, New Delhi, India
[2]Department of Environmental Sciences, SPM College, University of Delhi, New Delhi, India

CONTENTS

4.1 Nitrogen in the Environment 39
4.2 Nitrogen – An Important Element for Plant Growth 40
    4.2.1 Nitrogen Fixation 40
    4.2.2 Ammonification 41
    4.2.3 Nitrification 42
    4.2.4 Denitrification 42
    4.2.5 Nitrogen Assimilation in Plants 42
4.3 Nitrogen Uptake in Plants 44
    4.3.1 Role of Phloem in Amino Acid Translocation 46
    4.3.2 Reduction of Nitrate to NH_3 46
    4.3.3 Fixation of NH_4^+ 47
4.4 Enzymes Responsible for Nitrogen Assimilation 47
    4.4.1 Nitrate Reductase (NR) 47
    4.4.2 Nitrite Reductase (NIR) 48
4.5 Factors Affecting Nitrogen Absorption and Assimilation 48
    4.5.1 Soil Parameters 49
        4.5.1.1 Soil pH 49
        4.5.1.2 Soil Salinity and Fertility 49
    4.5.2 Climate 49
    4.5.3 Metals 49
    4.5.4 Ultraviolet B Radiation 50
    4.5.5 Gaseous Factors 50
    4.5.6 Nutritional Factors 50
    4.5.7 Environmental Factors 50
    4.5.8 Other Factors 51
4.6 Conclusion 51
References 51

4.1 NITROGEN IN THE ENVIRONMENT

Nitrogen is a key macronutrient for plant growth, development, and tolerance to biotic and abiotic stresses. The delivery of mineral nutrients, particularly nitrogen, has a significant impact on the output of an agricultural crop (Sawan, 2006). It has been proven to increase global food production while reducing world hunger when used as fertilizer. This is the most important part of achieving optimal plant growth and development with enhanced food yield and productivity. N has a function in practically all plant metabolic activities since it is a component of amino acids, which are necessary to create proteins and other related chemicals (Tucker, 1999). Plants are completely dependent on inorganic nitrogen, which exists in cellular components such as phospholipids, chlorophyll, alkaloids, enzymes, hormones, nucleic acids, and proteins. The atmosphere contains 78% nitrogen, but plants do not immediately absorb a large amount of nitrogen. Nitrogen is absorbed by plants in the form of nitrate, ammonia, nitrite, urea, and their mixtures. Nitrogen can be obtained by a variety of methods, including atmospheric nitrogen fixation, biological resources, and commercial nitrogen fixation.

Most plants use the accessible inorganic forms of N as NH_4^+ and NO_3^- taken by the roots (Lasa et al., 2002; Wickert et al., 2007; Oh et al., 2008) from the soil solution, which then undergoes intricate systems of assimilation, transformation, and mobilization inside plants (Shah, 2008; Oh et al., 2008). Whereas most inorganic N is supplied to plants as nitrate in most soils, ammonia can be the primary N ion in some soils and hydroponic cultural practices (Lasa et al., 2002). Agricultural and natural plants, on the other hand, can absorb amino acids from the soil (Godlewski and Adamczyk, 2007). When the amount of nitrogen application is optimal, it will increase the photosynthetic process, the production of leaf area, the duration of leaf area, and the net assimilation rate. For the integration of alternative agriculture, efficiency in the use of nitrogen (NUE) is essential. Inorganic nitrogen meets the requirement of cell components for nitrogen. This requires two approaches: i) molecular nitrogen fixation in the air; or ii) assimilation of nitrate or ammonia contained in water or soil (Fageria and Baligar, 2005).

4.2 NITROGEN – AN IMPORTANT ELEMENT FOR PLANT GROWTH

Nitrogen is also the most important component of plant cytoplasm. The transparent substance that makes up protoplasm is the live matter in cells. It is required for floral differentiation, rapid shoot growth, flower bud health, and improved fruit set quality. It also catalyzes the minerals around it. It occurs in healthy soils and provides energy to plants to grow and produce fruit or vegetables. It is also a significant part of amino acids, which are the structural units of proteins. Plants wither and perish when they are deprived of proteins. In plant cells some proteins serve as structural units, while others serve as enzymes, enabling many of the biochemical processes that allow life to exist. Nitrogen is found in energy-transfer molecules like ATP (adenosine triphosphate). Adenosine triphosphate (ATP) permits cells to preserve and utilize the energy released during metabolism. Ultimately, nitrogen is found in nucleic acids like DNA, which is the genetic material that permits cells to develop and multiply. Limited root growth, changes in root architecture, reduced plant biomass, and decreased photosynthesis are all symptoms of nitrogen deficit. As a result, it's critical to understand how N metabolism works and how it's regulated (Wen et al., 2019).

The nitrogen cycle is the transition between distinct chemical forms of nitrogen that occurs on Earth as a result of biological, physical, and geologic activities. Only a few bacteria can transform nitrogen into biologic molecules, which are found primarily inside living cells as shown in Figure 4.1. The four principal nitrogen-transformation (biologic) processes found in nature are nitrogen fixation, ammonification, nitrification, and denitrification (Pidwirny, 2006).

4.2.1 Nitrogen Fixation

The loss of nitrogen from the soil through denitrification is offset by the addition to the pool through conversion of atmospheric dinitrogen to a combined or fixed form. This process of reducing dinitrogen to ammonia is known as nitrogen fixation/dinitrogen fixation by nitrogenase enzyme. Nitrogen fixation is the conversion of N_2 to NH_3 (ammonia), which is a multistep process by certain bacteria. Some of these bacteria (*Rhizobium*) live in symbiosis with legume plants,

Nitrogen Assimilation in Plants

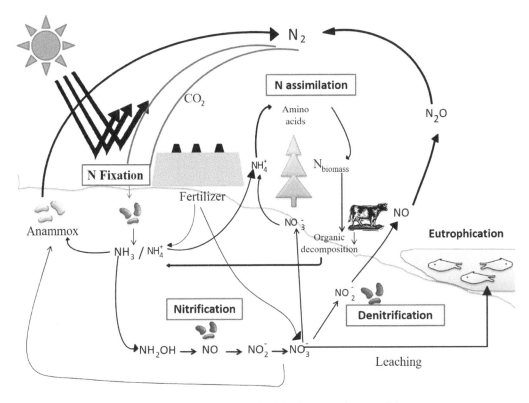

FIGURE 4.1 The nitrogen cycle illustrates the relationships between the essential processes.

while others (cyanobacteria and Azotobacter) are free-living bacteria. Soybeans, alfalfa, beans, and pears are examples of legume plants. Lighting and UV radiations provide sufficient energy to convert dinitrogen to atmospheric nitrogen oxides (NO, N_2O) (Figure 4.1). Ammonia is converted to proteins, nucleic acids (DNA), and other nitrogen-containing organic molecules (NH_3, nitrogenous organic molecules: proteins, nucleic acids, and so on) by this process (Olivera et al., 2004).

$$N_2 + 8e^- + 8H^+ + 16ATP \rightarrow 2NH_3 + H_2 + 16ADP + 16P$$

4.2.2 Ammonification

The conversion of organic nitrogen to ammonia by various microorganisms where excretory materials are degraded by a group of bacteria such as *Bacillus*, *Proteus*, *Pseudomonas*, *Clostridium*, and some fungi. This process is known as ammonification. They secrete enzyme proteinases, which degrade proteins to peptides. These peptides are broken into their amino acid units by the enzyme peptidases. Amino acids formed by the breakdown of proteins are further degraded by many microorganisms. Ammonia produced in the above steps may be lost to the atmosphere and reenters the atmosphere. However, most of the ammonia is absorbed by the plants or can be converted to nitrate by nitrification (Rysgaard et al., 1996).

Amino acid
Urea   Ammonification   Ammonium
Nucleotide   ────────────────────────→ NH_4^+
Amino sugar   *Bacillus, Pseudomonas, Fungi*

4.2.3 Nitrification

Nitrification is a respiratory process to obtain energy where conversion of ammonia to nitrate occurs. This takes place by a group of nitrifying bacteria, *Nitrosomonas/Nitrococcus* and *Nitrobacter*. In the first step, ammonia is oxidized to nitrite by the bacteria *Nitrosomonas*. In the next step, nitrite is oxidized by *Nitrobacter*, converting nitrite to nitrate. These are obligate chemoautotrophs that use O_2 as the final electron acceptor. Nitrification produces nitrate, which is a vital nitrogen source for plants. Nitrate is quickly consumed by microorganisms during denitrification (Rysgaard et al., 1996).

$$2NH_4^+ + 3O_2 \rightarrow 2NO_2^- + 4H^+ + 2H_2O$$

$$2NO_2^- + O_2 \rightarrow 2NO_3^-$$

4.2.4 Denitrification

Denitrification, or nitrate reduction, is the process of bacteria in soils converting nitrate to gaseous nitrogen molecules such as N_2O, NO, and N_2. It occurs under anaerobic conditions and generates NADH. During respiration, bacteria use nitrate as a substitute for oxygen and convert it to various nitrogenous compounds. The reduction of nitrate is a respiratory process known as denitrification. The nitrate is converted into N_2 by denitrifiers (*Thiobacillus denitrificans*), which are eventually discharged into the atmosphere as N_2O, NO, and N_2. Approximately 150 million metric tons of nitrogen are lost in the atmosphere annually (Rysgaard et al., 1996).

$$NO_3^- \longrightarrow NO_2^- \longrightarrow NO \longrightarrow N_2O \longrightarrow N_2$$
Nitrate   Nitrite   Nitric oxide   Nitrous oxide   Nitrogen gas

4.2.5 Nitrogen Assimilation in Plants

Once the process of nitrogen fixation and nitrification have been carried out, the plants can absorb NO_3^- by active transport. The absorbed nitrates are used to synthesize amino acids and nucleotides (Raven and Smith, 1976). The uptake and incorporation of nitrogen into plant tissue is known as assimilation. The synthesis of organic nitrogen molecules such as amino acids from inorganic nitrogen compounds found in the environment is known as nitrogen assimilation. Plants, fungi, and certain bacteria that can't fix nitrogen rely on their capacity to absorb nitrates or ammonium to satisfy their nutritional needs. Nitrate (NO_3^-) and ammonium (NH_4^+) are two forms of nitrogen that plants take from the soil. Nitrate is generally the most easily absorbed form of available nitrogen in aerobic soils when nitrification is feasible. Plants absorb ammonium ions by ammonia transporters. Nitrate is assimilated in the roots as well as the leaves. In developed herbaceous plants, nitrate assimilation takes place predominantly in the leaves, however, nitrate assimilation in the roots plays a significant role in the early stages of growth. Many woody plants (such as trees and shrubs), as well as legumes like soybeans, digest nitrate mostly through their roots. The nitrogen in NO_3^- is transformed to a higher-energy form in nitrite (NO_2^-), then to an even higher-energy form in ammonium (NH_4^+), and lastly to the amide nitrogen of glutamine in nitrate (NO_3^-) assimilation.

Nitrate absorbed by the plants gets converted to amino acids and amides before incorporating into proteins and other macromolecules. A particular mechanism containing at least a nitrate transporter, a nitrate reductase, and a nitrite reductase is necessary to convert nitrate to ammonium. The assimilation of nitrogen involves several enzymes and precursors. Major among the former are nitrate reductase, nitrite reductase, glutamine synthase (GS), glutamine oxoglutarate aminotransferase (GOGAT), glutamate dehydrogenase, and alanine aminotransferase. In non-nitrogen-fixing systems, both GS and GOGAT are commonly found in root and leaf cells. GS is found in the cytosol of root cells and both the cytosol and chloroplasts of leaf cells. GOGAT is a

Nitrogen Assimilation in Plants

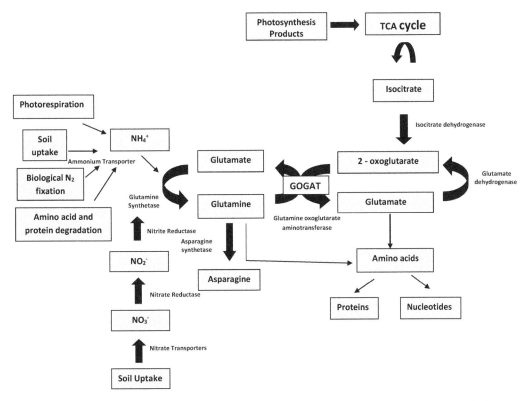

FIGURE 4.2 Nitrogen-assimilation pathway in higher plants. Inorganic nitrogen in the form of nitrate or ammonia becomes incorporated into amino acids and other organic molecules.

plastid enzyme, localized in the chloroplasts of leaves and plastids in roots. Depending on its location, GOGAT may use ferredoxin, NADH, or NADPH as electron donors (Torre and Ávila, 2021). The combined activity of the enzymes GS and GOGAT facilitate the integration of ammonium into organic molecules in the form of glutamine and glutamate. The ketoglutarate, which is essentially required in the assimilation reactions catalyzed by the enzyme GOGAT, is produced as a result of the tricarboxylic acid cycle (TCA cycle) by utilizing carbon from photosynthesis. Ultimately amino acids lead to the formation of proteins, nucleotides, and other N-containing metabolites (Lu et. al., 2016) (Figure. 4.2).

The overall summary equation of reduction of nitrate is as follows:

$$NO^{3-} + 8\ e^- + 10\ H^+ \rightarrow NH_4 + 3H_2O$$

This is a two-step reaction, which can be explained as follows: In the first step, nitrate is converted to nitrite. The reaction is catalyzed by the enzyme nitrate reductase. This requires NADH as an electron donor. In some species, it is NADPH as a coenzyme. It also requires FADH as a prosthetic group, cytochrome as an electron carrier, and molybdenum (Mo) as an activator of an enzyme.

$$NADH + H^+ + FAD \rightarrow NAD^+ + FADH_2$$

$$FADH_2 + Mo(ox) \rightarrow Mo(Red) + 2H^+ + FAD$$

$$Mo(Red) + 2H^+ + NO_3 \rightarrow Mo(Ox) + NO_2$$

In the second step, nitrite is reduced to ammonium ion. This reaction is catalyzed by an enzyme nitrite reductase. In the reaction ferredoxin acts as electron donor. The overall reaction is as follows:

$$NO_2 + 6 e^- + 4H^+ \rightarrow NH_4^+ + 2H_2O$$

This procedure uses the same amount of ATP as 12 ATPs per nitrogen (Bloom et al., 1992). To convert molecular nitrogen (N_2) into ammonia (NH_3), plants such as legumes create symbiotic interactions with nitrogen-fixing bacteria. The first stable result of fixation is ammonia (NH_3); however, at physiological pH, ammonia is protonated to create the ammonium ion (NH_4^+). It takes about 16 ATPs per nitrogen molecule to fix nitrogen biologically and then assimilate NH_3 into an amino acid. (Pate and Layzell, 1990; Vande Broek et al., 1996).

4.3 NITROGEN UPTAKE IN PLANTS

It has been well established that the availability of N often limits plant productivity, therefore research on its acquisition until its assimilation into plant metabolites has gained substantial attention. Many plants increase their N contents when exposed to a mixture of NO_3^- and NH_4^+ (Miller and Cramer, 2004; Hachiya et al., 2012). According to Glass et al. (2002), the net N influx via roots depends on total N influx and total N efflux. However, both NO_3^- and NH_4^+, responsible for the major nitrogen pool of the plant, not only have some common metabolic pathways. At low external concentrations both ions tend to be absorbed actively by root cells, with the help of two high-affinity transport systems (HATS) for NO_3^- (one constitutive and the other inducible) and one HATS for NH_4^+, where uptake of both ions depends on N status of the plant concerning diurnal regulation. In *Arabidopsis thaliana*, molecular studies have indicated the presence of seven HATS for NO_3^- and five for NH_4^+. Despite these similarities, their uptake and extent of utilization vary with plant species, which is further governed by several specific factors. Recent reports also demonstrated that plants can take up N directly from organic molecules (Näsholm et al., 2009) such as amino acids and peptides. In *Arabidopsis thaliana* the role of many transporters and their associated genes has been elucidated (Rentsch et al., 2007; Tegeder and Rentsch, 2010). The capability of the plant to use organic N increases its accessibility for the plethora of N sources (Paungfoo-Lonhienne et al., 2012; Warren, 2014).

Nitrate nitrogen is utilized in numerous processes including absorption, vacuole storage, xylem transport, reduction, and incorporation into organic forms (Wickert et al., 2007). Marquez et al. in 2007 reported that the majority of nitrate assimilation takes place in the roots depending on the age of the plant and space limitation for root growth. Nitrate taken up by plants is reduced to nitrite by nitrate reductase (NR) (Kuoadio et al., 2007; Cao et al., 2008; Rosales et al., 2011). This enzyme catalyzes the reduction of nitrate to nitrite with pyridine nucleotide in N assimilation in higher plants (Ahmad and Abdin, 1999). A high-reactive form of nitrite, chloroplasts, and plastids in leaves and roots rapidly transfer nitrite from the cytosol into chloroplasts. There are two enzymes in these organelles that are responsible for the final reduction of nitrogen to nitrite (Rosales et al., 2011).

Plant leaves act as a sink for nitrogen in the vegetative stage. From here it is remobilized for later usage. The procedure for calculating total N is time-intensive and potentially dangerous as it includes the usage of sulfuric acid. Therefore, leaf tissue and sap nitrate content are often considered by farmers as markers of the plant's nitrogen status before fertilizer application. Fan et al. in 2007 demonstrated that leaf tissue nitrate levels can be used to determine the nitrate stored in the vacuole. This happens after nitrate allocated to the leaves is briefly held in vacuoles and subsequently remobilized, particularly when N supply is insufficient to satisfy demand (von der Fecht-Bartenbach et al., 2010) or during senescence; N is delivered primarily via amino acids. Up to 80% of the nitrogen in cereals such as rice and wheat grains comes from the leaves (Kant et al., 2011). Because most plants store nitrate in their vacuoles and can handle high ion concentrations, it's fair to suppose that nitrate serves as an osmotic agent (Wickert et al., 2007). According to the findings of Cookson et al. (2005) in leaf cells of *A. thaliana* and Fan et al., 2006 in barley root cells, there exists a close link between cytosolic nitrate activity and nitrate reductase activity (NRA), as cytosolic nitrate activity is important for determining the thermodynamic gradients for transport to and from the vacuole (Miller and Smith, 1992; De Angeli et al., 2006).

Nitrogen is taken up by roots in the form of NO_3^- or NH_4^+. N_2 fixation is the formation of NH_4 as a result of nitrate reduction. The glutamine synthetase (GS) reaction is the first step in NH_4 incorporation into amino acids. In roots and leaves, NH_4^+ is converted into amino acids, which are then integrated into proteins. Some proteins' primary, if not exclusive, purpose is to serve as a storage facility for amino acids (Figure 4.3). Two nitrate transport systems have been identified in crops that work in consonance to absorb nitrate from the soil and transport it all over the plant (Tsay et al., 2007; Masclaux-Daubresse et al., 2010). Nitrate reduction is known to happen in both shoots and roots, in two different sites. Initially, reduction occurs in the cytosol forming nitrite, and later on, in plastids/chloroplasts nitrite reduction takes place. Nitrate reductase (NR) catalyzes the reduction of nitrate to nitrite in the cytosol (Meyer and Stitt, 2001) Nitrite then moves to chloroplast to be further reduced to ammonium by nitrite reductase (NiR). Nitrite is toxic and is rarely found at high concentrations in plants. This is strengthened by the fact that the activity of NiR (per gram dry weight of tissue) is usually many times higher than the activity of NR. The resulting ammonia is then rapidly assimilated into organic compounds via the GS/GOGAT system already described. The Nii genes coding for the NiR enzyme has been successfully cloned from several species (Meyer and Stitt, 2001). Though the major source of ammonium in plastid is reduced nitrite it might also be provided by processes such as photorespiration or amino acid recycling. According to Keys and his coresearchers (1978) the photorespiration pathway in the leaves of C_3 plants is responsible for a five- to ten-fold higher ammonium flux than the process of nitrate reduction. The GS/GOGAT cycle

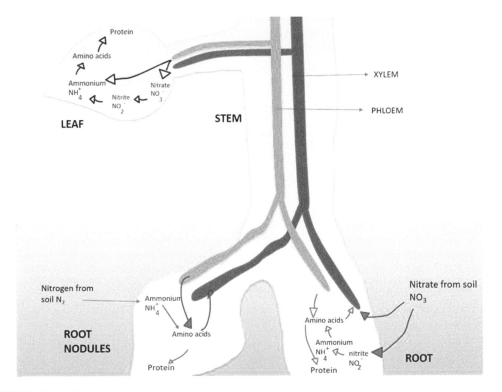

FIGURE 4.3 Nitrogen uptake in plants.

plays a significant role in ammonium assimilation (Lea and Forde, 1994). GS converts glutamate to glutamine by fixing ammonium on the molecule. GOGAT catalyzes the reaction between glutamine and 2-oxoglutarate, resulting in the formation of two molecules of glutamate (or glutamate synthase, GOGAT). NADH or ferredoxin (Fdx) serves as the major source of reducing power for NR, NiR, and GOGAT whereas ATP is vital for glutamine synthetase and asparagine synthetase. Finally, assimilation of nitrogen in organic forms as amino acids takes place.

4.3.1 Role of Phloem in Amino Acid Translocation

Phloem acts as a significant element in amino acid transfer from sources to sinks. Proteins decompose to form amino acids that are to be loaded in the phloem. In this process before phloem loading, the central vacuole of mesophyll cells may serve as a provisional storage site for amino acids. In tobacco, the total amount of amino acids exported from leaf blades increased five-fold during leaf aging (Masclaux-Daubresse et al., 2006). In pea, asparagine; in cereals, tobacco, and tomato glutamine is the most translocated amino acid. In *Arabidopsis* phloem sap chiefly contains asparagine, glutamate, and glutamine (Masclaux-Daubresse et al., 2010).

Both asparagine and glutamine levels spike in the phloem sap during aging and senescence plays a significant role in making nitrogen available for remobilization from the senescing leaf. In a review by Rentsch and colleagues in 2007, it was clearly stated that the *Arabidopsis* genome encodes 67 (putative) amino acid transporters belonging to 11 gene families. However, the nature of the amino acid transporter involved in phloem loading during senescence is not clear yet (van der Graaff et al., 2006; Masclaux-Daubresse et al., 2008).

4.3.2 Reduction of Nitrate to NH_3

Nitrate is converted to nitrite in the mesophyll cells first by nitrate reductase in the cytoplasm, and subsequently to NH_4^+ by nitrite reductase in the chloroplasts. Nitrate is transported into root cells via a two-proton symport mechanism. An H^+-P-ATPase generates a proton gradient across the plasma membrane, which drives nitrate uptake against a concentration gradient. Mitochondrial respiration provides the majority of the ATP required for the production of the proton gradient. Nitrate uptake usually stops when respiration inhibitors or uncouplers cease mitochondrial ATP generation in the roots (Miflin and Habash, 2002). The plasma membrane of root cells contains several nitrate transporters, including one with a relatively low affinity and another with a very high affinity, the latter of which is induced only when metabolism requires it. The nitrate absorption capability of the roots is thus adjusted to the ambient conditions. Plants can grow when the external nitrate content is as low as 10×10^{-6} mol/L due to the effectiveness of the nitrate absorption systems. The nitrate absorbed by the root cells might be temporarily retained in the vacuole. Nitrate is converted to NH_4^+ in the epidermal and cortical cells of the root. This NH_4^+ is mostly employed in the production of glutamine and asparagine. The xylem vessels can deliver these two amino acids to the leaves. However, when the root's capacity for nitrate assimilation reaches its maximum, nitrate is released into the xylem vessels and transferred to the leaves by the transpiration stream. A proton symport is most likely involved in the uptake into the mesophyll cells. The vacuole can store large amounts of nitrate. During the day, nitrate absorption can deplete this vacuolar storage, which can then be refilled at night. Thus, the highest nitrate level is found in spinach leaves in the early morning.

N is transferred into and out of proteins in various organs and transported throughout tissues in a restricted number of transport molecules during plant growth and development. Some organic nitrogen is transferred amongst molecules through transaminases and glutamine-amide transferases, but a large percentage is liberated as NH_3 and reassimilated by glutamine synthetase (GS) (Miflin and Habash, 2002). In most plant species, there are two GS isoenzymes (cytosolic

GS1 and plastidic GS2), each with a separate sub-cellular compartmental location (Tobin and Yamaya, 2001; Cao et al., 2008).

4.3.3 Fixation of NH_4^+

Glutamine synthetase in chloroplasts converts freshly generated NH_4^+ to glutamate at the expense of ATP, creating glutamine. Because glutamine synthetase has such a high activity and affinity for NH_4^+, the NH_4^+ generated by nitrite reductase is entirely absorbed into glutamine (Tsay et al., 2007; Hans-Walter et al., 2011). The NH_4^+ emitted during photorespiration is also fixed by glutamine synthetase. The amount of NH_4^+ produced by the oxidation of glycine is around 5 to 10 times larger than that produced by nitrate assimilation due to the high rate of photorespiration. As a result, nitrate absorption accounts for just a small part of glutamine synthesis in the leaves. The cytoplasm of leaves also contains an isoenzyme of GS. Glutamine and α-ketoglutarate are converted to two molecules of glutamate by GS (also known as glutamine-oxoglutarate aminotransferase, abbreviated as GOGAT). In this reaction, ferredoxin is utilized as a reductant. NADPH-dependent GS is found in some chloroplasts and leucoplasts. The substrate analog azaserine (Fi), which is poisonous to plants, inhibits GSs.

α-Ketoglutarate is transported into the chloroplasts by a specialized translocator in counter-exchange for malate, and the glutamate produced is transported out of the chloroplasts into the cytosol by another translocator, likewise in trade for malate. Another translocator in the chloroplast envelope transfers glutamine in the opposite direction of glutamate, allowing glutamine to be exported from the chloroplasts (Hans-Walter et al., 2011).

4.4 ENZYMES RESPONSIBLE FOR NITROGEN ASSIMILATION

4.4.1 Nitrate Reductase (NR)

Nitrate reductase is a homodimeric multi-component enzyme complex with a molecular weight of 200,000–300,000 daltons and three electron-transferring prosthetic groups per subunit: flavin (FAD), heme b557, and Mo cofactor (Moco) in addition to NAD(P)H-binding and dimerization domains. The enzyme is very flexible and shows two hinge regions i) between Mo cofactor and heme domains, and ii) between heme and FAD domains. Moco of the first type is used in nitrate reductase and sulfite oxidase. The second type is found in xanthine dehydrogenase and aldehyde oxidase. Moco consists of Mo covalently bound to two S atoms in the tricyclic molecule. Molybdenum in Moco is bound to a third S-ligand either of the cysteine residue (below, top molecule) or a terminal S (below, bottom molecule). NR catalyzes nitrate (NO_3^-) reduction to nitrite, the first step of NO_3^- assimilation pathway, which is the main one for nitrogen acquisition by crop plants and for handling signaling role in different plant processes (Alvarez et al., 2012; Krapp et al., 2014). They are found in plants and fungi, use NADH or NADPH as reducing equivalents, and harbor a heme and a FAD cofactor in addition to the molybdenum active site. Barbier and coworkers first reported the reduction of nitrate from *Pichia angusta* by nitrate reductase (Chamizo-Ampudia et al., 2017).

Structure of nitrate reductase enzyme

TABLE 4.1
Nitrogen Assimilation Enzymes and Biochemical Reactions

Enzymes	Reactions
NR	$NO_3^- + NAD(P)H + H^+ + 2e^- = NO_2^- + NAD(P) + H_2O$
NiR	$NO_2^- + Fd\ (red) + H^+ + 2e^- = NH_4^+ + Fd\ (ox) + 2\ H_2O$
GDH	glutamate + H_2O + NAD/NADP = NH_4^+ + 2-oxoglutarate + NADH/NADPH
AS	glutamine + aspartate + ATP = asparagine + glutamate + AMP + PPi
GS1/GS2	glutamate + NH_4^+ + ATP = glutamine + ADP + Pi
NADH-GOGAT	glutamine + 2-oxoglutarate + 2 NADH = 2 glutamate + NAD
AspAT	glutamate + oxaloacetate = aspartate + 2-oxoglutarate
Fd-GOGAT	glutamine + 2-oxoglutarate + 2 Fd (red) = 2 glutamate + 2 Fd (ox)

Source: (Lam et al., 1996).

Abbreviations: GS1- cytoplasmic glutamine synthetase; GS2-chloroplastic glutamine synthetase; Fd-GOGAT – ferredoxin dependent glutamate synthase; NADH-GOGAT – NADH dependent glutamate synthase; GDH- Glutamate dehydrogenase; AspAt- aspartate aminotransferase; AS- asparagine synthetase; Fd, ferredoxin, pi- inorganic phosphate, ppi- pyrophosphate.

In addition to catalyzing the reduction of nitrate by reduced pyridine nucleotides, the NAD(P)H: nitrate reductase exhibits two other activities that can be assayed separately and involve only part of the overall electron-transport capacity of the enzyme. Diaphorase activity results in the reduction by NAD(P)H of various 1- and 2-electron acceptors (cytochrome c, ferricyanide, and other oxidants). The terminal nitrate reductase catalyzes the NAD(P)H-independent reduction of nitrate by reduced flavin nucleotides or viologens. Both moieties participate jointly in the sequential transfer of electrons from NAD(P)H to nitrate (Table 4.1).

4.4.2 NITRITE REDUCTASE (NIR)

Nitrite reductase (NIR) is a key enzyme in the dissimilatory denitrification chain, catalyzing the reduction of NO_2^- to NO. Although this has been a matter of debate for a long time, NO is now accepted as a product of NO_2^- reduction and an obligatory intermediate in most denitrifiers; it is further reduced to N_2O by NO reductase. NiR catalyzes the reduction of nitrite to nitric oxide (NO). The enzyme is a homotrimer; each monomer contains two mononuclear copper sites: one so-called type-1 copper site that acts as an electron transfer center and the other so-called type-2 copper site where the catalysis occurs (Quesada et al., 1994). These two copper sites have distinct absorption properties. The type-1 site belongs to the large family of blue copper sites in biology and has a cysteine ligand. At the oxidized form (Cu^{2+}), this cysteine ligand gives rise to an intense sulfur-to-Cu^{2+} charge transfer absorption near 600 nm ($\varepsilon > 4000\ mol^{-1}\ l\ cm^{-1}$); at the reduced form ($Cu^{1+}$), this site has no absorption in the visible region because of the d10 electron configuration of Cu^{1+} ions. On the contrary, the type-2 copper site has histidine and water-based ligands and shows little or no absorption in the visible region regardless of its oxidation state (Navarro et al., 2000).

4.5 FACTORS AFFECTING NITROGEN ABSORPTION AND ASSIMILATION

The expression and activity of transport systems and enzymes in response to environmental conditions or stresses, and also the overall health of the species, is influenced by a variety of regulatory processes. The absorption of NO_3^- has been demonstrated to be sensitive to (1) low temperature, (2) inhibitors of both respiration and protein synthesis, and (3) anaerobic conditions in several investigations. Several factors influence nitrogen assimilation, which is discussed in detail below (Renseigné, et al., 2007).

4.5.1 Soil Parameters

Various parameters of soil such as soil drainage, the texture of the soil, and slope steepness impact the transportation of nitrogen. Other factors such as rainfall, temperature, and conditions of the area such as soil moisture, aeration (oxygen levels), and its salt content affect the rate of mineralization of N. Decomposition of organic matter releases N faster in warm and humid climatic conditions and slower in cool dry climates. The release of N is quicker in well-aerated soils and quite slower in well-saturated soils. Nitrogen easily leaches out of the root zone in the form of nitrate. The potential of leaching depends upon soil texture and soil water content. Soils having poor drainage or saturated with water causes denitrification to occur resulting in loss of N as gas and thereby resulting in the emission of potent greenhouse gases, yield reduction, and the increased expense of N fertilizer.

4.5.1.1 Soil pH

pH of soil affects both the host plant and the bacteria involved thereby in the symbiotic relationship. Soil pH affects the uptake of nutrients from the soil. A pH value between 6.0 and 7.0 is considered the best environment for optimum uptake by the forage plant. The bacteria most often involved is *Rhizobium*, which is negatively affected by soil pH (< 6.0). If soil pH is less than 6.0 limestone can be added to increase soil pH. Few exceptions are there as in the case of the symbiosis between red clover and its microsymbiont *Rhizobium leguminosarum biovar trifolii* where the association function well at pH values less than 6.0.

4.5.1.2 Soil Salinity and Fertility

Soil salinity is one of the important factors rendering soil unfit for agriculture. Cordovilla et al. (1999) found that N-fixing plants were more sensitive to salinity than were N-fertilized plants. Although nitrate reductase activity in leaves is significantly impacted by osmotic stress produced by NaCl (Silveira et al., 2001), the presence of salt in irrigation water may interfere with the quantity of N utilized by plants, resulting in reduced N availability (Debouba et al., 2006). NO_3 levels were highly decreased in both the leaves and roots in tomato (*Lycopersicon esculentum*) seedlings under increasing salinity. Plants accumulate ammonium, nitrate, and free amino acids in response to salt stress, whereas ammonium incorporation into amino acid complexes diminishes. Soil fertility is linked to either an excess of soil nitrates or a nitrogen deficit that limits plant growth and development. Excess soil nitrate levels, in general, lower BNF.

4.5.2 Climate

Various factors such as low air/soil temperature, lack of sunshine, or drought will likely reduce the assimilation and fixation of nitrogen. Although farm owners have little influence over climate change, certain measures may be made to reduce the impact of climate change. For example, date plants can be chosen to relieve difficulties with overly cold or warm soils. Drought-resistant crops can be irrigated on time to reduce the impact of dry spells.

4.5.3 Metals

Recently contamination of soil by heavy metals has become a serious issue of environmental concern. Heavy metals are considered soil pollutants because of their wide availability and their toxic effect on plants (Yadav, 2010). Besides being toxic, few metals have positive and negative side effects on the metabolic processes affecting growth and development. As a result of its favorable influence on the absorption and transportation of water and mineral nutrients as well as transpiration of seedlings, Lanthanum (La) may enhance the transport of nitrate in plants. It was considerably worse in plants treated with 100 mM Cd (12 percent of the net nitrate absorption) (Gouia et al., 2000). High Al concentrations restrict NO_3 uptake in plant species studied and therefore cause a deficit in N metabolism.

4.5.4 Ultraviolet B Radiation

Increased penetration of UV-B on Earth has raised concerns regarding its damaging impact on crop plants. UV-B radiation affects all the major processes of photosynthesis, including photochemical reactions (thylakoid membranes), enzymatic processes (Calvin cycle), stomatal limitations to CO_2 diffusion, destruction of amino acid residues, and oxygen-mediated damage of unsaturated fatty acids in plant cell membranes. Under UV-B light, Cao et al. (2007) discovered that mineral uptake and assimilation were considerably reduced.

4.5.5 Gaseous Factors

A low concentration of NO_2 did not affect the organic N content of the plants, or the concentration of organic N in leaves and roots, with the exception that it slightly affects the concentration of organic N in the leaves of plants grown at low nitrate levels. Nitrogen dioxide, which the plant absorbs, can be converted into nitrate and nitrite. Exposure to high NO_2 causes an increase in nitrate concentration (Qiao and Murray, 1998). In the case of cucumber and sunflower leaves, high assimilation of CO_2 enhanced nitrate reduction by increasing the synthesis and activity of NR, and sugars resulting from CO_2 assimilation probably act as positive regulatory metabolites (Aguera et al., 2006). Guo et al. (2007) observed the effect of increased CO_2 (or low O_2) decrease rates of photorespiration and thereby enhance photosynthesis and growth ~35% in most plants (C_3 plants). Bloom et al. (2010) reported that in wheat plants receiving NO_3 as a sole N source atmospheric CO_2 enrichment does not stimulate the growth of the plant as per those receiving NO.

4.5.6 Nutritional Factors

Nitrate accumulation in crops and vegetables depends on the amount and type of nutrients present in the soil and also depends upon the time of application, the amount, and chemical composition of the fertilizers (Zhou et al., 2000, Nazaryuk et al., 2002). Application of nitrogen at the beginning of the vegetative cycle is effective in regulating nitrate accumulation, as the plant and soil nitrate concentrations decrease with plants utilizing the reserve food nutrients with its growth (Vieira et al., 1998). Artificial fertilizers with a base of ammonia or a mixture of nitrate and ammonium when applied to the soil can affect the nitrate content in plants (Inal and Tarakcioglu, 2001; Santamaria et al., 2001) Nutrient balance plays an important role in affecting the nutrient status of plants (Ahmed et al., 2000). Increase rate of potassium when applied allows the uptake and transportation of nitrate towards leaves and apex of the plant facilitates the metabolism and utilization of nitrate thereby reducing nitrate accumulation in some vegetable crops (Ahmed et al., 2000; Zhou et al., 2000; Ruiz and Romero, 2002). Application of other nutrients such as salicylic acid, molybdenum, and calcium through leaves can reduce the nitrate content of plants (Xu et al., 2005; Tzung et al., 1995).

4.5.7 Environmental Factors

A large variety of environmental factors affects nitrate accumulation in plants. Santamaria et al. (2001) observed an interaction between light intensity, nitrogen availability, and temperature on nitrate accumulation in several species of the Brassicaceae family. Under low light conditions increase in temperature increases the nitrate accumulation on one hand, while on the other hand under high light intensity increase in temperature increases the nitrate content in presence of a high supply of nitrogen supply. The effect of artificial lighting in greenhouses on nitrate accumulation in lettuce was also studied by Chadjaa et al. (2001). Grzebelus and Baranski (2001) studied the effect of climate on nitrate accumulation and observed that nitrate content was less in the year that had high rainfall and during wet and warm years nitrate accumulation increased, and this did not depend on whether the nitrogen originated from organic or mineral sources (Custic et al., 2003). In comparison to the type and application rates of fertilizers, the meteorological conditions of a region had a greater impact on nitrate levels in plants (Custic et al., 2003). Nitrate accumulation also varies with the types of seasons, as it is higher in the autumn-winter season than in spring (Vieira et al., 1998, Santamaria et al., 1999). It was concluded that plants in winter are not able to use all the nitrogen available in the soil due to the low availability of favorable light and temperature conditions.

4.5.8 Other Factors

Various other factors such as the source of N, time and season of application, method of application, irrigation and crop residue management, type of crop, etc., all can affect and alter the number of N losses. Management of nitrogen in sandy soils is important because of the high potential for leaching losses. The selection of an appropriate N rate is one of the key criteria for the management of N loss. A measure that increases the organic content and reduces compaction in the soil is necessary to stabilize crop N supply, thereby increasing the soil aeration and reducing N losses in soil due to denitrification.

4.6 CONCLUSION

The assimilation of nitrate is a complicated and regulated process. Several extracellular and intracellular stimuli, such as nitrate, NH_4, sunlight, and hormone, affect the enzymes involved in this process. The main factors in this mechanism are nitrate and light. Another set of enzymes is responsible for assimilating and reassimilating ammonia. Plant growth requires regulation of nitrogen and carbon metabolism, which is maintained by complex signaling pathways such as protein kinases/phosphatases, calcium, inositol triphosphates, and others. The phases of nitrogen assimilation to the level of amino acids are frequently separated in space in higher plants. As a result, ammonia in the roots, whether through absorption, nitrate reduction, or nitrogen fixation, is primarily converted into amides, which are subsequently distributed throughout the plant. Similarly, glutamine has been discovered to be the most important transport molecule formed from NO_3^- reduction in the leaf. GOGAT levels in the assimilatory tissue must be significant to synthesize enough glutamate to operate as an ammonia acceptor under certain environments. However, in tissues that receive nitrogen in the form of amides from the transport stream, processes must exist to convert this amide to amino acids. GOGAT may convert glutamine to x-amino acids, and NAD-dependent GOGAT has been identified in developing endosperms and cotyledons, confirming this. Various factors play important roles in affecting the nitrogen assimilation pathways. These can be natural or anthropogenic.

REFERENCES

Aguera, E.; Ruano, D.; Cabello, P.; de la Haba, P. Impact of atmospheric CO_2 on growth, photosynthesis and nitrogen metabolism in cucumber (*Cumumis sativus* L.) plants. *J. Plant Physiol.* 2006, 163, 809–817.

Ahmad, A.; Abdin, M. Z. NADH: nitrate reductase and NAD(P)H: nitrate reductase activities in mustard seedlings. *Plant Sci.* 1999, 14, 18.

Ahmed, A.H.H.; Khalil, M.K.; Farrag, A.M. Nitrate Accumulation, Growth, Yield and Chemical Composition of Rocket (*Eruca vesicaria* subsp. sativa) Plant as Affected by NPK Fertilization, Kinetin, and Salicylic Acid, in Proceedings of ICEHM 2000, Cairo University, Egypt, 2000, pp. 495–508.

Alvarez, J. M.; Vidal, E. A.; Gutierrez, R. A. Integration of local and systemic signaling pathways for plant N responses. *Curr. Opin. Plant Biol.* 2012, 15, 185–191.

Binbin, W.; Chen, L.; Xiling, F.; Dongmei, L.; Ling, L.; Xiude, C.; Hongyu, W.; Xiaowen, C.; Xinhao Z.; Hongyan, S.; Wenqian, Z.; Wei, X. Effects of nitrate deficiency on nitrate assimilation and chlorophyll synthesis of detached apple leaves. *Plant Physiol. Biochem.* 2019, 142, 363–371.

Bloom, A. J.; Burger, M.; Asensio, J. S. R.; Cousins, A. B. Carbon dioxide enrichment inhibits nitrate assimilation in wheat and arabidopsis. *Science* 2010, 328, 899–903.

Cao, R.; Huang, X. H.; Zhou, Q.; Cheng, X. Y. Effects of lanthanum (III) on nitrogen metabolism of soybean seedlings under elevated UV-B radiation. *J. Environ. Sci.* 2007, 19, 136, 113–166.

Cao, Y.; Fan, X. R.; Sun, S. B.; Xu, G. H.; Hu, J.; Shen, Q. R. Effect of nitrate on activities and transcript levels of nitrate reductase and glutamine synthetase in rice. *Pedosphere.* 2008, 18, 664–673.

Chadjaa, H.; Vezina, L. P.; Dorais, M.; Gosselin, A. Effects of lighting on the growth, quality, and primary nitrogen assimilation of greenhouse lettuce (Lactuca sativa L.). *Acta Hortic.* 2001, 559, 325–331.

Chamizo-Ampudia, A.; Sanz-Luque, E.; Llamas, A.; Galvan, A.; Fernandez, E. Nitrate reductase regulates plant nitric oxide homeostasis. *Trends Plant Sci.* 2017, 22, 163–174.

Cookson, S. J.; Williams, L. E.; Miller, A. J. Light–dark changes in cytosolic nitrate pools depend on nitrate reductase activity in Arabidopsis leaf cells. *Plant Physiol.* 2017, 138, 1097–1105.

Cordovilla, M. D. P.; Ligero, F.; Lluch, C. Effects of NaCl on growth and nitrogen fixation and assimilation of inoculated and KNO3 fertilized *Vicia faba* L. and *Pisum sativum* L. plants. *Plant Sci.* 1999, 140, 127–136.

Custic, M.; Poljak, M.; Coga, L.; Cosic, T.; Toth, N.; Pecina, M. The influence of organic and mineral fertilization on nutrient status, nitrate accumulation, and yield of head chicory. *Plant Soil Environ.* 2003, 49, 218–222.

De Angeli, A.; Monachello, D.; Ephritkhine, G.; Frachisse, J. M.; Gambale, F.; Barbier-Brygoo, H. The nitrate/proton antiporter AtCLCa mediates nitrate accumulation in plant vacuoles. *Nature* 2006, 442, 939–942.

Debouba, M.; Gouia, H.; Suzuki, A.; Ghorbel, M. H. NaCl stress effects on enzymes involved in nitrogen assimilation pathway in tomato "Lycopersicon esculentum" seedlings. *J. Plant Physiol.* 2006, 163, 1247–1258.

Fageria, N.K.; Baligar, V.C. Enhancing nitrogen use efficiency in crop plants. *Adv. Agron.* 2005, 88: 97.

Fan, X.; Gordon-Weeks, R.; Shen, Q.; Miller, A. J. Glutamine transport and feedback regulation of nitrate reductase activity in barley roots lead to changes in cytosolic nitrate pools. *J. Exp. Bot.* 2006, 57, 1333–1340.

Glass, A. D.; Britto D. T.; Kaiser, B. N.; Kinghorn, J. R.; Kronzucker, H. J.; Kumar, A.; Okamoto, M.; Rawat, S.; Siddiqi, M. Y.; Unkles, S. E.; Vidmar, J. J. The regulation of nitrate and ammonium transport systems in plants Anthony D.M. *J. Exp. Bot.* 2002, 53(370), Inorganic Nitrogen Assimilation Special Issue, 855–864.

Godlewski, M.; Adamczyk, B. The ability of plants to secrete proteases by roots. *Plant Physiol. Biochem.* 2007, 45, 657–664.

Gouia, H.; Ghorbala, M. H.; Meyer, C. Effects of cadmium on the activity of nitrate reductase and other enzymes of the nitrate assimilation pathway in the bean. *Plant Physiol. Biochem.* 2000, 38, 629–638.

Grzebelus, D.; Baranski, R. Identification of accessions showing low nitrate accumulation in a germplasm collection of garden beet. *Acta Hortic.* 2001, 563, 253–255.

Guo, S. W.; Zhou, Y., Gao.; Y. X. Li, Y.; Shen, Q. R. New insights into the nitrogen form effect on photosynthesis and photorespiration. *Pedosphere* 2007, 17, 601–610.

Hachiya, T.; Watanabe, C.K.; Fujimoto, M. Nitrate addition alleviates ammonium toxicity without lessening ammonium accumulation, organic acid depletion, and inorganic cation depletion in Arabidopsis thaliana shoots. *Plant Cell Physiol.* 2012, 53, 577–591.

Hans-Walter H.; Brigit P.; Fiona, H. *Plant Biochemistry*, 4th edition, Amsterdam; Boston; Heidelberg: Elsevier; Academic Press, 2011.

Inal, A.; Tarakcioglu, C. Effects of nitrogen forms on growth, nitrate accumulation, membrane permeability, and nitrogen use efficiency of hydroponically grown bunch onion under boron deficiency and toxicity, *J. Plant Nutr.* 2001, 24, 1521–1534.

Kant, S.; Bi, Y. M.; Rothstein, S. J. Understanding plant response to nitrogen limitation for the improvement of crop nitrogen use efficiency. *J. Exp. Bot.* 2011, 62, 1499–1509.

Keys, A.; Bird, I.; Cornelius, M.; Lea, P.; Wallsgrove, R.; Miflin, B. Photorespiratory nitrogen cycle. *Nature.* 1978, 275, 741–743.

Kouadiao, J. Y.; Kouakou, H. T.; Kone, M.; Zouzou, M.; Anno, P. A. Optimum conditions for cotton nitrate reductase extraction and activity measurement. *Afr. J. Biotechnol.* 2007, 6, 923–928.

Krapp, A.; David, L. C.; Chardin, C.; Girin, T.; Marmagne, A.; Leprince, A. S. Nitrate transport and signaling in Arabidopsis. *J. Exp. Bot.* 2014, 65, 789–798. DOI: 10.1093/jxb/eru001

Lam, H.M..; Coschigano, K.; Oliveira, I.C.; Melo-Oliveira, R.; Coruzzi, G. The molecular-genetics of nitrogen assimilation into amino acids in higher plants. *Ann. Rev. Plant Physiol.and Plant Mol. Biol.* 1996, 472(1), 596–593.

Lasa, B.; Frechilla, S.; Aparicio-Tejo, P.M.; Lamsfus, C. Role of glutamate dehydrogenase and phosphoenolpyruvate carboxylase activity in ammonium nutrition tolerance in roots. *Plant Physiol. Biochem.* 2002, 40, 969–976.

Lea, P. J.; Forde, B. G. The use of mutants and transgenic plants to study amino acid metabolism. *Plant Cell Environ.* 1994, 17, 541–556.

Lu, J.; Zhang, L.; Lewis, R.; Bovet, L.; Goepfert, S.; Jack, A.; Crutchfield, J.; Huihua, J.; Dewey, R.; Expression of a constitutively active nitrate reductase variant in tobacco reduces tobacco-specific nitrosamine accumulation in cured leaves and cigarette smoke. *Plant Biotech J.* 2016, 14. 10.1111/pbi.12510.

Masclaux-Daubresse, C.; Daniel-Vedele, F.; Dechorgnat, J.; Chardon, F.; Gaufichon, L.; Suzuki, A.; Nitrogen uptake, assimilation, and remobilization in plants: challenges for sustainable and productive agriculture. *Annals Bot.* 2010, 105(7), 1141–1157.

Masclaux-Daubresse, C.; Reisdorf-Cren, M.; Orsel, M. Leaf nitrogen remobilisation for plant development and grain filling. *Plant Biol.* 2008, 10 (Suppl. 1), 23–36.

Masclaux-Daubresse, C.; Reisdorf-Cren, M.; Pageau, K. Glutamine synthetase-glutamate synthase pathway and glutamate dehydrogenase play distinct roles in the sink-source nitrogen cycle in tobacco. *Plant Physiol.* 2006,140, 444–456.

Meyer, C.; Stitt, M. Nitrate reductase and signaling. In: P.J. Lea and J.-F. Morot-Gaudry (Eds.). *Plant Nitrogen*. New York: Springer, 2001, 37–59.

Miflin, B. J.; Habash, D. Z. The role of glutamine synthetase and glutamate dehydrogenase in nitrogen assimilation and possibilities for improvement in the nitrogen utilization of crops. *J. Exp. Bot.* 2002, 53 (370), 979–987.

Miller, A. J.; Smith, S. J. The mechanism of nitrate transport across the tonoplast of barley root cells. *Planta* 1992, 187, 554–557.

Miller, A. J.; Cramer, M. D. Root nitrogen acquisition and assimilation. *Plant Soil.* 2004, 274, 1–36.

Näsholm, T.; Kielland. K.; Ganeteg. U. Uptake of organic nitrogen by plants. *The New Phytol.* 2009, 182, 31–48.

Navarro, M.T.; Guerra, E.; Fernández, E.; Galván, A. Nitrite reductase mutants as an approach to understanding nitrate assimilation in *Chlamydomonas reinhardtii*. *Plant Physiol.* 2000, 122(1), 283–290.

Nazaryuk, V.M.; Klenova, M.I.; Kalimullina, F.R. Eco-agrochemical approaches to the problem of nitrate pollution in agroecosystems. *Russ. J. Ecol.* 2002, 33, 392–397.

Non Renseigné, Shahid Umar, Muhammad Iqbal. *Nitrate Accumulation in Plants, Factors Affecting the Process, and Human Health Implications. A Review*. Agronomy for Sustainable Development, Springer Verlag/EDP Sciences/INRA, 2007, 27 (1), pp. 45–57. final-00886336

Oh, K.; Kato, T.; Xu, H.L. Transport of nitrogen assimilation in xylem vessels of green tea plants fed with NH_4^+-N and NO_3^--N. *Pedosphere.* 2008, 18, 222–226.

Olivera, M.; Tejera, N.; Iribarne Carmen & Ocaña, A. & Lluch, Carmen. Growth, nitrogen fixation and ammonium assimilation in common bean (*Phaseolus vulgaris*): Effect of phosphorus. *Physiol Plant.* 2004, 121, 498–505.

Pate, J.S.; Layzell, D. B. Energetics and biological costs of nitrogen assimilation. In *The Biochemistry of Plants: Intermediary Nitrogen Metabolism*. In: B.J. Miflin and P.J. Lea, (Eds.). Academic Press, London, 1990, 16, 402.

Paungfoo-Lonhienne, C.; Visser, J.; Lonhienne, T. G. A.; Schmidt, S. Past, present and future of organic nutrients. *Plant Soil.* 2012, 359, 1–18.

Pidwirny, M.; *The Nitrogen Cycle*. Fundamentals of Physical Geography, 2nd Ed., 2006.

Qiao, Z.; Murray, F. The effects of NO_2 on the uptake and assimilation of nitrate by soybean plants. *Environ. Exp. Bot.* 1998, 10: 33–40.

Quesada, A.; Galván, A.; Fernández, E. Identification of nitrate transporters in *Chlamydomonas reinhardtii*. *Plant J.* 1994, 5, 407–419.

Rave J. A. N.;. Smith F. A. Nitrogen assimilation and transport in vascular land plants in relation to intracellular pH regulation. *New Phytol.* 1976, 76, 415–431.

Rentsch, D.; Schmidt, S.; Tegeder, M. Transporters for uptake and allocation of organic nitrogen compounds in plants. *FEBS Lett.* 2007, 581, 2281–2289. doi: 10.1016/j.febslet.2007.04.013.

Rosales, E. P. M. F.; Iannone, M.; Groppa, D. M.; Benavides, P. Nitric oxide inhibits nitrate reductase activity in wheat leaves. *Plant Physiol. Biochem.* 2011, 49, 124–130.

Ruiz, J.M.; Romero, L. Relationship between potassium fertilization and nitrate assimilation in leaves and fruits of cucumber (*Cucumis sativus*) plants. *Ann. Appl. Biol.* 2002, 140, 241–245.

Rysgaard, S.; Risgaard-Petersen, N.; Sloth, N.P. Nitrification, denitrification, and nitrate ammonification in sediments of two coastal lagoons in Southern France. *Hydrobiologia.* 1996, 329, 133–141.

Santamaria, P. Nitrate in vegetables: toxicity, content, intake and EC regulation. *J. Sci. Food Agr.* 2006, 86, 10–17.

Sawan, Z. M. Egyptian cotton (*Gossypium barbadense* L.) yield as affected by nitrogen fertilization and foliar application of potassium and mepiquat chloride. *Commun. Biom. Crop Sci.* 2006, 1(2), 99–105.

Shah, S. H. Effects of nitrogen fertilization on nitrate reductase activity, protein and oil yields of *Nigella sativa* L. as affected by foliar GA3 application. Turk. J. Bot. 2008, 32, 165–170.

Silveira, J. A. G.; Melo, A. R. B.; Viegas, R. A.; Oliveira, J. T. A. Salinity-induced effects on nitrogen assimilation related to growth in cowpea plants. *Environ. Exp. Bot.* 2001, 46, 171–179.

Tegeder, M.; Rentsch, D. Uptake and partitioning of amino acids and peptides. *Mol. Plant.* 2010, 3, 997–1011. doi: 10.1093/mp/ssq047

Tobin, A. K.; Yamaya, T. Cellular compartmentation of ammonium assimilation in rice and barley. *J. Exp. Bot.* 2001, 52, 591–604.

Torre, F.D.L.; Ávila, C. Special issue editorial: Plant nitrogen assimilation and metabolism. *Plants* 2021, 10, 1–4.

Tsay, Y. F.; Chiu, C. C.; Tsai, C. B.; Ho, C. H.; Hsu, P. K. Nitrate transporters and peptide transporters. *FEBS Lett.* 2007, 581, 2290–2300.

Tucker, M. R. *Essential plant nutrients: Their presence in North Carolina soils and role in plant nutrition.* Agronomic Division. NCDA&CS. 1999 [Online] Available at: www. ncagr.gov/agronomi/ pdffiles/essnutr.pdf (Accessed 2010 Dec. 15).

Tzung, W. J.; Po, W. Y.; Wu, J.T.; Wang, Y. P. Effects of some environmental factors on nitrate content of Chinese cabbage (*Brassica chinensis* L.). *J. Chin. Agr. Chem. Soc.* 1995, 33, 125–133.

van der Graaff, E.; Schwacke, R.; Schneider, A.; Desimone, M.; Flugge, U. I.; Kunze, R. Transcription analysis of Arabidopsis membrane transporters and hormone pathways during developmental and induced leaf senescence. *Plant Physiol.* 2006, 141, 776–792.

Vande Broek, A.; Keijers, V.; Vanderleyden, J. Effect of oxygen on the free-living nitrogen fixation activity and expression of the *Azospirillum brasilense* nifH gene in various plant-associated diazotrophs. *Symbiosis.* 1996, 21, 25–40.

Vieira, I.S.; Vasconselos, E.P.; Monteiro, A.A. Nitrate accumulation, yield and leaf quality of turnip greens in response to nitrogen fertilization. *Nutr. Cycl. Agroecosys.* 1998, 51, 249–258.

von der Fecht-Bartenbach, J.; Bogner, M.; Dynowski, M.; Ludewig, U. CLC-b mediated NO_3^-/H^+ exchange across the tonoplast of *Arabidopsis* vacuoles. *Plant Cell Physiol.* 2010, 51, 960–968.

Warren, C. R. Organic N molecules in the soil solution: what is known, what is unknown and the path forwards. *Plant Soil.* 2014, 375, 1–19. doi: 10.1007/s11104- 013-1939-y.

Wickert, S.; Marcondes, J.; Lemos, M. V.; Lemos, E. G. M. Nitrogen assimilation in citrus-based on CitEST data mining. *Genet. Mol. Biol.* 2007, 30, 810–818.

Xu, C.; Wu, L.H.; Ju, X.T.; Zhang, F.S. Role of nitrification inhibitor DMPP (3,4-dimethylpyrazole phosphate) in nitrate accumulation in greengrocery (Brassica campestris L. ssp. chinensis) and vegetable soil. *J. Environ. Sci.* (China) 2005, 17, 81–83.

Yadav, S. K. Heavy metals toxicity in plants: An overview on the role of glutathione and phytochelatins in heavy metal stress tolerance of plants. *S. Afr. J. Bot.* 2010, 76, 167–179.

Zhou, Z. Y.; Wang, M. J.; Wang, J. S. Nitrate and nitrite contamination in vegetables in China. *Food Rev. Int.* 2000, 16, 61–76.

5 Assimilation of Nitrates in Plants

Sufiya Rashid[1],*, Kausar Rashid[1], Hanan Javid[1], Bilal Ahmad Wani[1], Junaid Ahmad Magray[1], Rouf ul Qadir[1], Tajamul Islam[1], Mubariz Javed[2], and Irshad Ahmad Nawchoo[1]

[1]Plant Reproductive Biology, Genetic Diversity and Phytochemistry Research Laboratory, Department of Botany, University of Kashmir, Srinagar, J & K, India
[2]Plant Tissue Culture Research Laboratory, Department of Botany, University of Kashmir, Srinagar, J & K, India
Corresponding author e-mail: Sufiiayadar@gmail.com

CONTENTS

5.1 Introduction 55
5.2 Nitrate Assimilation 56
5.3 Nitrate Uptake 56
5.4 Enzymes Involved in Nitrate Assimilation 57
    5.4.1 Nitrate Reductase 57
        5.4.1.1 Subunits 57
    5.4.2 Nitrite Reductase 59
5.5 Regulation of Nitrate Transport and Reduction 60
5.6 Ammonium Assimilation 61
    5.6.1 Glutamine Synthetase (GS) 61
    5.6.2 Glutamate Synthase 62
    5.6.3 Glutamate Dehydrogenase 62
5.7 Fixed Nitrogen is Exported as Asparagine and Ureides 63
5.8 Conclusion 64
References 65

5.1 INTRODUCTION

Nitrogen is one of the most crucial elements essential for all forms of organisms because it is a chief component of proteins, nucleic acids, and other biomolecules. In plants, the amount of nitrogen present varies from 1.5 to 2% of plant dry matter and 16% of total protein. Though being present in huge amounts in the atmosphere, molecular nitrogen is used by only a few organisms like bacteria including cyanobacteria for their requirements. Only a few plant species (mainly legumes) can endure a symbiosis with prokaryotes that can fix di-nitrogen gas. Therefore, plants developed mechanisms to cope with the low availability of nitrogen. Ammonia, mineral nitrogen (nitrates and ammonium), nitrogen oxides (Nox), and amino acids and peptides are the nitrogen forms used by most plants (Wang and Macko, 2011). Mineral nitrogen, ammonium, and nitrate are the predominant sources of nitrogen accessible in the soils (Qu et al., 2016; Wang et al., 2018). For most plants and algae and many fungi and bacteria, nitrate is a very significant source of nitrogen. Plants obtain 100 times more nitrogen through nitrate assimilation than other nitrogen acquisition processes like nitrogen fixation (Ali et al., 2007; Nasholm et al., 2009).

5.2 NITRATE ASSIMILATION

The main nitrogen source for higher plants is nitrate. In the biosphere, about 99% of the organic nitrogen is derived from the assimilation of nitrate (Heldt, 1997). Nitrate is transferred into root cells where it is stored in the vacuole, reduced in the cytoplasm, or brought to the leaves, where it can be stored or reduced (Crawford, 1995). The assimilatory nitrate reduction pathway being a very imperative physiological process forms the major pathway by which inorganic nitrogen is assimilated into carbon skeleton in higher plants, algae as well as cyanobacteria and fungi. The nitrate assimilation process involving nitrate uptake and its reduction to ammonium is carried out by two highly regulated enzymes, one cytosolic nitrate reductase (NR) and the other plastidic nitrite reductase (NiR). The reduced ammonium is fixed into the carbon skeleton forming glutamate and glutamine from 2-oxoglutarate supplied by the TCA cycle and amino acid transamination reaction (Ali, 2020).

5.3 NITRATE UPTAKE

The nitrate uptake system in plants must be flexible and robust because, in the face of external nitrate concentrations that can fluctuate by five orders of magnitude, plants must transport enough nitrate to gratify the total need for nitrogen. To function efficiently in the face of such environmental variation, plants have developed an active, regulated, and multiphasic transport system. The energy driving the absorption of nitrate comes from the proton gradient maintained by the

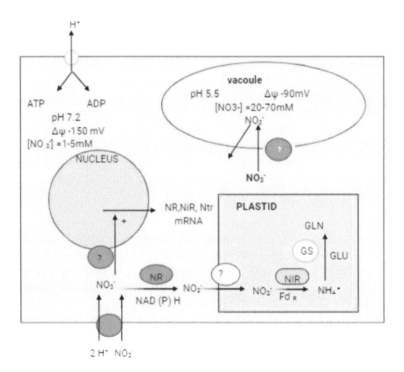

FIGURE 5.1 Diagrammatic representation of Nitrate assimilation pathway.

H+-ATPase across the plasma membrane. Initially, nitrate absorption occurs across the plasma membrane of epidermal and cortical cells of the root (Larsson and Ingemarsson, 1989). Consequentially transport takes place across the tonoplast membrane and the plasma membrane of cells in the vascular system and leaf (Jackson et al., 1986). Between –100 and –250 mV electrical gradient, the transport of anionic nitrate into a root cell must occur (Figure 5.1). The plasma membrane undergoes a rapid and transient depolarization during nitrate uptake, becoming up to 60 mV more positive inside the cell within 1 to 2 min (Glass et al., 1992). Using inhibitors of the plasma membrane H+-ATPase and alkalization of the external medium, the nitrate uptake was also inhibited (McClure et al., 1990a, 1990b). Conversely, the uptake of chlorate into plasma membrane preloaded chlorate vesicles, which is an analog of nitrate driven by acidifying the external medium (Ruiz-Cristin and Briskin, 1991).

Various scientific studies show that electrogenic proton co-transport is the driver of nitrate uptake in plants. No depolarization would be observed if only one proton was co-transported with each nitrate anion. Nitrate absorption, therefore, requires two or more protons co-transported with each nitrate ion. The depolarization is reversible; within 1 to 2 min after exposure to nitrate, the membrane potential starts to recover. The mechanism for this recovery has not yet been revealed, but it is likely to be the result of H+-ATPase activation. Related depolarizations and proton co-transport mechanisms for chloride and phosphate uptake have been studied (Sakano, 1990; Ullrich and Novacky, 1990; Sakano et al., 1992).

5.4 ENZYMES INVOLVED IN NITRATE ASSIMILATION

5.4.1 Nitrate Reductase

The next step in the nitrogen assimilation pathway after the uptake of nitrate into the cell is the reduction of nitrate to nitrite. This stage competes with both nitrate efflux from the cell and nitrate transport into the vacuole (Figure 5.1). The enzyme nitrate reductase (NR) carries out this reduction reaction and is located predominantly in the cytosols of root epidermal and cortical cells and shoot mesophyll cells (Vaughn and Campbell, 1988; Fedorova et al., 1994). NR involves the transfer of two electrons from NAD(P)H to nitrate through three redox centers, which comprise two prosthetic groups (flavin adenine dinucleotide [FAD] and heme) and a MoCo cofactor, which comprises molybdate and pterin complex.

$$NO_3^- + NAD(P)H + H^+ \rightarrow NO_2^- + NAD(P)^+ + H_2O$$

Hence, the nitrate-to-nitrite conversion is an irreversible reaction, and this stage is important for nitrate acquisition in plants, fungi, and algal species (Campbell, 1999). Based on the electron donor they use, different forms of NR are found: an NADH-specific form in higher plants and algae, and a NAD(P)H-bispecific form in higher plants, algae, and fungi. Only birch has been reported to have a bispecific shape (Friemann, 1991).

5.4.1.1 Subunits

NR exists as a homodimer in higher plants, composed of two similar subunits of around 110kD (~900 amino acids). Each subunit contains one FAD, heme-Fe, and Mo-molybdopterin equivalent (Mo-MPT). Dimerization is necessary for the action, and a native homodimer can further dimerize to a tetramer. Tetramers of around 500 kDa can also exist in microorganisms. An internal electron chain powers the conversion of nitrate to nitrite (Figure 5.2) (Campbell, 2002). Five structurally distinct domains are revealed in the 3D structure of NR: (i) Mo-MPT domain; (ii) dimer interface domain; (iii) cytochrome b (Cb) domain; (iv) FAD-binding domain; and (v) NAD(P)H-binding domain. As a result of the mixture of FAD and NADH domains, cytochrome b reductase fragment (CbR) is made. The fusion of the domains of FAD and NADH results in the synthesis of the fragment of cytochrome

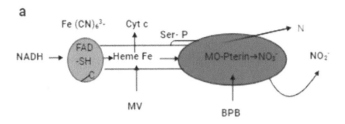
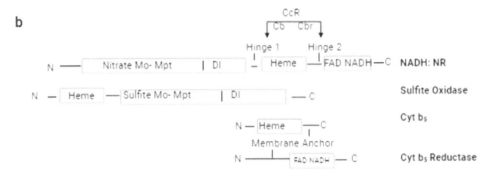

FIGURE 5.2 Models for Nitrate Reductase Structure. (a) Functional model of NR. (b) Sequence model of the enzyme.

b reductase (CbR), which combines with the domain of Cb to create the fragment of cytochrome c reductase (CcR).

Three sequences do not have similarities to any other protein and differ between the various types as well. They are (i) an N-terminal acidic, which may be engaged in NRs regulations; (ii) a Hinge I region, which comprises phosphorylated, regulatory serine, and a trypsin proteolytic site; and (iii) Hinge II, which acts as proteinase site. NR is therefore constructed from subunits that are thought to have formed independently and that may seem to be connected to other configurations of protein units. For instance, the CbR fragment is comparable to cytochrome b5 reductase, while the Cb fragment is closely comparable to cytochrome b_5. It is important to point out that NR Mo-MPT and interface domains are similar to sulfite reductase (SOX), which causes reduction of sulfite to sulfate by reduced cytochrome c (Campbell, 1999).

There are two active sites in each monomeric unit of NR complex, one for NAD(P)H electron acceptance and the other for nitrate to nitrite conversion. The enzyme shows a ping-pong process in which NAD(P)+ and NAD(P)H compete at one active site and the other active site nitrate and nitrite. In between each monomeric end, the domains consist of two active sites: 1) the nitrate-nitrite reduction site in between the molybdopterin and dimer interface domains; and 2) the electron-donor active site in between the FAD and NADH domains (Campbell, 1999; Campbell, 2002). Each monomeric unit of NR comprises three cofactors, namely, FAD, heme, and MPT, and two metal ions, iron, and molybdenum. FAD and the two ions can be found in either a reduced or oxidized state. Consequently, NR can be found to exist in 12–18 possible oxidized or reduced states. The redox potentials specify electron flow within the NR from the electron donor to the active site of the enzyme, which reduces nitrate (Campbell, 2002). Molybdopterin is a peculiar co-factor and can be seen to exist in some other enzymes in plants like xanthine dehydrogenase, aldehyde oxidase, and sulfite oxidase (SOX), although SOX activity in plants is doubtful. A 31-amino acid sequence having an invariant cysteine residue associated with Mo-MPT binding has been found to exist in these enzymes. This is typical of molybdopterin oxidoreductases of eukaryotes (Schwarz and Mendel, 2006).

Prokaryotic NR also comprises a pterin cofactor and molybdenum but varies from eukaryotic molybdopterin in possessing an additional nucleotide. The bacterial enzymes involved in denitrification are membrane-bound terminal electron acceptors or soluble enzymes. An NR activity has been studied in *Desulfo vibrio*, referred to as NAP, and some other prokaryotic NRs and was found as distinctive in having an iron-sulfur redox center (Fe_4S_4) in the same polypeptide along with the Mo-MPT co-factor. Although no eukaryotic nitrate reductase has been studied to possess a Fe-S center in its electron transport chain, the NAP is like eukaryotic NR in having an arginine residue involved in nitrate binding (Campbell, 1999). An active site mutant with invariant cysteine replaced by serine was able to bind NADH and unable to transfer electrons from NADH to FAD. Hence, this cysteine residue (Cys889 in AtNR2) is supposed to be the inhibitor-sensitive thiol of NR, which is associated with electron transfer but not in NADH binding (Campbell, 1999).

5.4.2 Nitrite Reductase

The next step in the nitrate assimilation pathway is the reduction of nitrite to ammonia by the enzyme nitrite reductase (NiR). NiR catalyzes the transfer of six electrons involved in the reduction of NO_2^- to NH_4^+:

$$NO_2^- + 8H^+ + 6e^- \rightarrow NH_4^+ + 2H_2O$$

Unlike NR, which is confined to cytoplasm, NiR is primarily found within chloroplasts in leaf and plastids in root and other non-green tissues. However, the presence of an extrachloroplastic NiR has been reported in cotyledons of mustard (Schuster and Mohr, 1990a; Schuster and Mohr, 1990b). In green leaves, reduced ferredoxin generated by photosystem I act as an electron donor. In roots, a ferredoxin-like protein acts as the electron donor, which attains reducing potential from NADPH generated by the pentose phosphate pathway found in the plastids (Yoneyama and Suzuki, 2019). The addition of nitrate was shown to enhance the levels of the Fd and NADPH-dependent ferredoxin NADP+oxidoreductase in isolated pea root plastids (Bowsher et al., 1993).

The NiR enzymes exist as monomeric metalloproteins of about 60–64 kDa comprising the siroheme prosthetic group, the binding site for nitrite binds, and a 4Fe-4S cluster at its active center (Hase et al., 2006). The NiR apoprotein is synthesized in the nucleus, and an N-terminal transit peptide targets the protein to the chloroplast. Among plant species, NiR proteins depict high conservation (75–80%). Cyanobacterial NiR has more resemblance with the plant than with enterobacterial NiR whereas plant, bacterial, and fungal NiR show slightly less degree of homology (Fariduddin et al., 2018). Two isoforms of NiR have been found in maize and four in *N. tabacum* (Sivasankar et al., 1997). The enzyme possesses two functional domains and co-factors: an fdx-binding domain on the N-terminal and a domain on the C-terminal, which has a binding site for nitrite, two redox centers, a 4Fe-4S center, and a siroheme. Four cysteines found in two clusters provide a link between the domains as well as S ligands for the 4Fe-4S cluster. Cysteine's role in co-factor binding is known by mutating the residues next to them to bulky side chains, which decreases the NiR activity. The siroheme is an iron tetra hydroporphyrin with eight COOH containing side chains (Maia and Moura, 2014). In the literature, there are no reports that suggest that the leaf and root NiRs are separate proteins. NiR mutants are difficult to isolate as plants show non-viability to accumulated nitrite toxicity. Conditional-lethal mutant defective in NiR apoprotein gene, *Nir1*, in barley has been reported (Duncanson et al., 1993). Evidence indicates that the 60 kDa NiR may be an altered form of a larger, native NiR. A putative native NiR of 85 kDa extracted from the spinach has a three- to four-fold higher specific activity with reduced ferredoxin than with reduced methyl viologen and can be divided into two protein components of 61 and 24 kDa. High rates of methyl viologen are retained by the 61-kDa fragment, but not ferredoxin NiR activity. The 24-kDa fragment contains a ferredoxin-binding site but has no enzymatic activity (Kishorekumar et al., 2020).

5.5 REGULATION OF NITRATE TRANSPORT AND REDUCTION

Nitrate assimilation regulation is one of its interesting aspects. Nitrate acts as the primary signal for regulating nitrate assimilation, but light, cytokinin, C02 levels, circadian rhythms, carbon, and nitrogen metabolites such as sucrose and glutamine all are important in regulation. A complicated regulatory pathway involving both transcriptional and post-transcriptional mechanisms helps in linking nitrate assimilation with photosynthesis and carbon metabolism, which consumes energy and carbon skeletons (Crawford, 1995).

In barley seedlings, nitrate reductase mRNA was detected approximately 40 minutes after the addition of nitrate, and maximum levels were attained within 3 hours. In contrast to the rapid mRNA accumulation, there was a gradual linear increase in nitrate reductase activity, reflecting the slower synthesis of the protein. The initiation of gene expression is one of the widespread responses in fungi and plants to nitrate exposure. Nitrate transporters, NR, NiR, glutamine synthetase, and ferredoxin-dependent glutamate synthase are encoded by responding genes (Crawford and Arst, 1993; Hoff et al., 1994). The induction can be seen to take place within minutes in plants and even in response to very low concentrations of nitrate (down to 10 μM). Ammonia blocks the induction in fungi and algae but not higher plants, because of catabolite repression. A nitrate regulation study in fungi and algae has led to the identification of regulatory genes that are important for both nitrate induction and ammonia repression. In fungi, the nitrate induction-mediating genes (NIRA from Neurospora and NIT-4 from Aspergillus) encode positive regulators with zinc finger DNA-binding domains identical to GAL4 from yeast (Burger et al., 1991a; 1991b). Ammonia repression-mediating genes (AREA of Aspergillus and NIT-2 of Neurospora) also code for positive regulators with zinc finger DNA-binding motifs that are different from NIRA and NIT-4 (Kudla et al., 1990). NI72 regulatory gene in algae, Chlamydomonas, was isolated by transposon tagging (Schnell and Lefebvre, 1993), which codes for a positive regulator whose expression is repressed by ammonia. The repression of N/P therefore mediates the catabolite repression of nitrate assimilatory genes in Chlamydomonas. In higher plants, no nitrate regulatory genes have been identified. However, in the promoters of the spinach NiR gene and the Arabidopsis NR gene, cis-acting regulatory regions have been detected, conferring nitrate-inducible expression to reporter genes in transgenic tobacco (Lin et al., 1994).

Light is a significant regulatory signal for NR and NiR expression in addition to nitrate. A phytochrome-mediated pathway increases NR and NiR mRNA levels in etiolated plants on transferring to the light (Becker et al., 1992; Mohr et al., 1992). Red light induction of NR and NiR genes is decreased in the phytochrome-deficient *ama* mutant of tomato (Becker et al., 1992). NR mRNA levels rise in the light in green tissues, but when nitrate is present a constitutive low level exists in the dark. When green plants are placed in a diurnal cycle, the levels of NR and NiR mRNAs fluctuate even if the plants are reverted to constant light conditions (Becker et al., 1992; Pilgrim et al., 1993; Lillo, 1994). Interestingly, NR mRNA levels do not fluctuate but remain raised throughout the diurnal period, when NR is inactivated by inhibitors or by mutation even though mRNA levels of other light-regulated genes, such as cab, appear to oscillate (Pouteau et al., 1989). Perhaps in NR-deficient plants oscillations in the concentration of a metabolite such as glutamine that affects NR mRNA levels can cease (Deng et al., 1991; Vincentz et al., 1993). Carbon metabolites signal show fluctuation in nitrate assimilation and can surpass the light signal. For nitrate reduction, carbon metabolites are very essential because carbon skeletons are important for ammonia fixation and the energy obtained from reduced carbon is required for nitrate reduction in non-green tissues. NR activity and mRNA tend to increase when excised leaves or cells in culture are subjected to sucrose treatment in the dark (Cheng et al., 1992; Vincentz et al., 1993; Lillo, 1994). In addition, NR promoter elements confer induction of sucrose in the dark on reporter genes (Cheng et al., 1992; Vincentz et al., 1993). It is not currently clear how sugars trigger the expression of the NR gene, or which metabolite is the proximal inducer. Interestingly, the levels of NiR mRNA do not react to sugar (Vincentz et al., 1993). Glutamine has a suppressive effect on levels of NR mRNA.

Assimilation of Nitrates in Plants

Another post-transcriptional mechanism, phosphorylation, which is both fast and reversible, also regulates NR. When plants are subjected to the dark treatment or low CO_2 environment, NR activity decreases 3- to 10-fold (Huber et al., 1992; Riens and Heldt, 1992). NR activity is restored under light conditions and optimum CO_2 levels. Phosphorylation mediates the inactivation of NR (Huber et al., 1992, 1994; S.C. Huber et al., 1994). It has been shown that NRs of spinach, arabidopsis, and maize are phosphorylated on serines, producing multiply phosphorylated peptides when digested with trypsin or cyanogen bromide (Huber et al., 1992, 1994; LaBrie and Crawford, 1994). When NR is subjected to too much dark treatment, it is inactivated, which leads to an increase of phosphorylation of two (in spinach) or three (in maize) of these peptides (Huber et al., 1992,1994). NR reactivation is consistent with the dephosphorylation of these peptides and can be blocked with phosphatase inhibitors (Huber et al., 1992; MacKintosh, 1992; Kaiser and Huber, 1994a). Thus, phosphorylation plays an essential role in modulating NR activity hastily and reversibly in reaction to carbon and light signals. To avoid the accumulation of toxic nitrite, this NR control can be crucial when levels of reduced ferredoxin become too low, especially in the dark, to limit the reduction of nitrite in the plastid.

5.6 AMMONIUM ASSIMILATION

To avoid ammonium toxicity, plant cells rapidly convert the ammonium generated from nitrate assimilation or photorespiration into amino acids. The chief pathway for this conversion involves the sequential actions of two enzymes, namely glutamine synthetase and glutamate synthase (Lea et al., 1992). These enzymes together form the GS/GOGAT cycle, which is the major path of ammonia assimilation in plants. This GS/GOGAT cycle is also essential for the reassimilation of ammonia, which is formed from many different metabolic pathways like photorespiration, the degradation of amino acids, and the metabolism of phenylpropanoids. Aminotransferases led to the transfer of nitrogen into other amino acids such as aspartate and alanine after the incorporation of ammonia into glutamate by GS AND GOGAT. Thus, the cycle of GS/GOGAT lies at the boundary of the C and N metabolism (Ali, 2020).

5.6.1 Glutamine Synthetase (GS)

Conversion of glutamate to glutamine is ATP-dependent catalyzed by glutamine synthetase using ammonium as substrate and a divalent cation such as Mn^{2+}, Mg^{2+}, or Co^{2+} as a co-factor.

$$Glutamate + NH_4 + ATP \rightarrow Glutamine + ADP + Pi + H_2O$$

In higher plants, GS protein is formed from eight identical or almost identical subunits and has a molecular mass of ~320–380 kDa (Sechley et al., 1992). A small gene family encodes these subunits. GS exists in two isoforms: GS1, which is found mostly in the cytosol of leaves and non-photosynthetic tissues, and GS2, which is located in the chloroplasts, the plastids of roots, or etiolated plants (Cren and Hirel, 1999). In the majority of plants, GS2 is coded by a single nuclear gene while a small family of nuclear genes encodes GS1 (2–6 genes per haploid genome). The key factors important for the differential expression of GS1 genes are the phases of development and external stimuli (Coruzzi, 2003). A different set of genes regulates the subunits in homo- or hetero-octameric form (Inokuchi et al., 2002). Ammonium obtained from various physiological processes is assimilated by these isoforms. GS1 assimilates ammonium in roots obtained from the soil and reassimilates ammonium in cotyledons obtained by the degradation of nitrogenous resources during germination. The chloroplast form, GS2, helps primarily in the assimilation of the ammonia reduced from nitrate in the chloroplast and/or reassimilation of photorespiratory ammonium (Lam et al., 1996). The GS2 in maize is restricted within bundle sheath and mesophyll cells (Coruzzi, 2003). In rice cytosolic GS is found in the vascular bundle of leaf tissue where it helps in the export of nitrogen to sink tissues (Masclaux-Daubresse et al., 2010). In the root nodules, the main function of the GS is the fast acquirement of ammonium released into the plant cytosol by the N2-fixing bacteroides. The cytosolic and chloroplastic activities are affected by various environmental conditions and growing seasons.

5.6.2 GLUTAMATE SYNTHASE

The formation of two glutamate molecules from glutamine with ketoglutarate is catalyzed by glutamate synthase (glutamine: oxoglutarate aminotransferase; GOGAT). One glutamate molecule is used as a substrate by glutamine synthetase and the other is used as a carrier for shifting nitrogen to other tissues or donors for other metabolic processes:

$$\text{Glutamine} + 2\text{-Oxoglutarate} + 2 \text{ ferredoxin reduced} \rightarrow 2 \text{ glutamate} + 2 \text{ ferredoxin oxidised}$$

$$\text{Glutamine} + 2\text{- Oxoglutarate} + \text{NAD(P)H} \rightarrow 2 \text{ glutamate} + \text{NAD(P)}^+$$

Depending on the electron donor, glutamate synthase has two isoforms. Fd GOGAT receives electrons from reduced ferredoxin (Fd red) whereas NADH GOGAT accepts electrons from NADH. These enzymes are Fe-S flavoenzymes that differ from one another in terms of physicochemical, immunological, and regulatory properties and genes encoding them (Coruzzi, 2003; Suzuki and Knaff, 2005; Temple et al., 1998). Fd-GOGAT is predominantly found in chloroplasts and plays a role in ammonium assimilation, which is obtained from nitrate assimilation and photorespiration (Forde and Lea, 2007). Studies show the presence of this enzyme in the mesophyll cells of rice (Hayakawa et al., 1994). Fd-GOGAT is encoded by a single gene in the majority of the plants, except in Arabidopsis where it is coded by two genes (Coschigano et al., 1998). While this enzyme is generally a monomeric one (145–165 kDa), the enzyme found in rice comprises two subunits of 115 kDa and consists of one [3Fe-4S] cluster and one FMN per molecule (Hirasawa et al, 1996). In leaves, the transcription of the Fd-GOGAT gene is regulated by light resulting in the increment of enzyme product as well as translational product (Anjana and Iqbal, 2007). Nitrogen dependence of transcription and protein activity of the Fd-GOGAT varies from one species to another. The presence of nitrogen is essential for the enzyme activity in some species (Sakakibara et al., 1992), and in some other species, it is not essential (Migge et al., 1998). NADH-GOGAT is a monomeric enzyme having a molecular mass of 190–240 kDa in land plants. It is found in the vascular bundles of developing leaf blades. The NADH-GOGAT is less active in green leaves than the Fd-GOGAT (Hecht et al., 1988), but proteins and activity are documented at relatively higher levels in the non-green and developing leaf blades of rice (Yamaya et al., 1992).

In a few plant species such as *Pisum arvense*, *Lycopersicon esculentum*, and *Bryopsis maxima* other isoforms of glutamate synthase, NADPH-specific GOGAT has also been documented (Rus-Alvarez et al., 1994).

5.6.3 GLUTAMATE DEHYDROGENASE

An alternative pathway is possible for assimilation of nitrate, involving the direct reductive amination of α-ketoglutarate by the enzyme glutamate dehydrogenase (GDH). A reversible reaction that synthesizes or deaminates glutamate is catalyzed by glutamate dehydrogenase:

$$2\text{-Oxoglutarate} + \text{NH}_4^+ + \text{NAD(P)H} \leftrightarrow \text{glutamate} + \text{H}_2\text{O} + \text{NAD(P)}^+$$

GDH exists in two isoforms: in mitochondria an NADH-dependent form of GDH is found, and an NADPH-dependent form is confined in the chloroplasts of photosynthetic organs. Apart from being abundant, they can't replace GS–GOGAT pathway for assimilation of ammonium, and their main role is to deaminate glutamate. GDH has a much lower affinity for NH^+_4 than does GS and can hardly compete with GS for available NH^{+4}. The cost of the glutamate synthase cycle is one ATP for each NH^{+4} assimilated but provides the advantage of fast assimilation.

In Arabidopsis, GDH is coded by two nuclear genes, GDH1 and GDH2, having two subunits, α- and β-subunit (Melo-Oliveira et al., 1996). These subunits associate differently and form seven active isozymes. All the isoforms may have similar metabolic functions, as they showed similar activity ratios and kinetic properties in grapevine leaf crude extracts (Loulakakis and Roubelakis-Angelakis, 1996). GDH isoforms can act in both vegetative and reproductive stages during plant development (Marchi et al., 2013). The third GDH gene, GDH3, which is encoded by γ-subunit, was studied in Arabidopsis a long time ago and plays a role in the formation of heteromeric complexes along with the α- and β-subunits (Yamada et al, 200, Fontaine et al, 2013). This enzyme is primarily found in root companion cells (Fontaine et al., 2013). GDH is prominently found in the mitochondria (Dubois et al, 2003). In conditions of high ammonium accumulation, this enzyme can also be seen in the cytosol (Tercé-Laforgue et al., 2004). Mitochondrial expression of GDH is linked with low N availability, whereas its cytosolic counterpart is correlated with ammonium beyond threshold levels (Tercé-Laforgue et al., 2004).

5.7 FIXED NITROGEN IS EXPORTED AS ASPARAGINE AND UREIDES

Exporting the fixed nitrogen from the nodule to other regions of the host plant is the final step in nitrogen fixation. Export of the organic nitrogen products from nodules occurs mostly through the xylem. Consequently, the form in which the nitrogen is exported has been known by studying the xylem sap. There are certain shortcomings to such analyses, however, as in Allantoin there is no assurance that all of the nitrogen present in sap represents the current production of the nodule. Some of the better studies have been done, either directly on detached nodules or by monitoring the flow of organic nitrogen following fixation of 15 N2. These studies have demonstrated that although glutamine is the principal organic product of nitrogen fixation, it hardly accounts for a significant fraction of the nitrogen exported, at least in legumes. In some legumes like pea and clover (mainly of temperate origins), asparagine is the principal amino acid to be translocated. While in legumes of tropical origins, for example, soybean and cowpea, ureides (derivates of urea) are the main form to be translocated.

The bio-synthetic pathway for asparagine occurs in two transamination reactions in nodules. The transfer of an amino group from an amino acid to the carboxyl group of a keto acid is called a transamination reaction. A class of enzymes known as aminotransferases catalyzes such reactions and enables nitrogen initially fixed in glutamate to be induced into other amino acids and, finally, into protein. Aminotransferases can be seen to occur throughout the plant in the cytosol, in chloroplasts, and microbodies, wherever protein synthesis activity is high. The enzymes involved in asparagine biosynthesis in nodules appear to be similar to those found elsewhere in the plant. The initial step involves the transfer of an amino group from glutamate to oxaloacetate, catalyzed by the enzyme aspartate aminotransferase:

$$\text{glutamate} + \text{oxaloacetate} \rightarrow \alpha\text{-ketoglutarate} + \text{aspartate}$$

The GS–GOGAT reactions in the nodule provide the glutamate for this reaction. For the synthesis and export of asparagine, a continuous source of the 4-carbon acid oxaloacetate is required by the nodule. Oxaloacetate is formed as an intermediate in the respiratory oxidation of glucose and hence could be provided through the oxidation of carbon in the nodule. However, high activities of the enzyme phosphoenolpyruvate carboxylase (PEP carboxylase) are reported from the nodules of several species. A carboxylation reaction involving the addition of carbon dioxide to 3-carbon phosphoenolpyruvate (PEP) is catalyzed by PEP carboxylase forming oxaloacetate. PEP carboxylase is involved in many important metabolic pathways in plants and animals, including respiration and photosynthesis:

$$PEP + CO_2 \rightarrow OAA$$

The next step of asparagine bio-synthesis involves the transfer of amide nitrogen from glutamine to aspartate:

$$Glutamine + aspartate + ATP \rightarrow glutamate + asparagine + ADP + Pi$$

Asparagine synthetase catalyzes this reaction and for each asparagine synthesized, one molecule of ATP is utilized as an energy source. The synthesis of ureides, both biochemically and concerning the division of labor between the microsymbiont and tissues of the host plant, is more complex. Ureid synthesis is a complex process for both aspects including bio-chemical and division of labor between the microsymbiont and tissues of the host plant. Active symbiosis is required for the formation of allantoin and allantoic acid from the oxidation of purine nucleotides. Ureides are used mainly for the transport of nitrogen. Xylem translocates them to other regions of the plant, where they are then rapidly metabolized. NH^{+4} is released during this, and is then reassimilated via GS and GOGAT in the target tissue.

Although the synthesis of asparagine, and especially the ureides, both appear to be complex processes, there are some advantages relating to the energy costs and efficiencies of nitrogen export. It has been estimated that the carbon metabolism associated with nitrogen export may consume as much as 20% of the photosynthate diverted to nitrogen fixation. One way to judge efficiency is to consider the amount of carbon required for each nitrogen exported. The ureides, with a carbon-to-nitrogen ratio of 1 (C: N 1), are the most economic in the use of carbon. Both asparagine and citrulline (C: N 2) require more carbon in their transport and glutamine (C: N 2.5) would be the least economic. The energy costs of ureides, asparagine, and citrulline, in terms of ATP consumed, are about the same, so the principal advantage to be gained by the ureide-formers appears to be a favorable carbon economy.

5.8 CONCLUSION

Nitrogen is one of the most crucial elements essential for all forms of organisms because it is a chief component of proteins, nucleic acids, and other bio-molecules. In plants, the amount of nitrogen present varies from 1.5 to 2% of plant dry matter and 16% of total protein. The activities of microorganisms make the nitrogen accessible to plants and because of these soils may show a striking seasonal discrepancy in the supply and form of readily available nitrogen. The primary route for the introduction of nitrogen into the biosphere is the assimilation of ammonia by a wide variety of organisms (diazotrophs). Most plants acquire their nitrogen from the soil in the form of nitrates, which are derived from either fertilizers or the mineralization of indigenous organic matter. Nitrate is transferred into root cells where it is stored in the vacuole, reduced in the cytoplasm, or bought to the leaves, where it can be stored or reduced. The nitrate assimilation process involves nitrate uptake and its reduction to ammonium, which is carried out by two highly regulated enzymes, one cytosolic nitrate reductase (NR) and the other plastidic nitrite reductase (NiR). To avoid ammonium toxicity, plant cells rapidly convert the ammonium generated from nitrate assimilation or photorespiration into amino acids. The chief pathway for this conversion involves the sequential actions of two enzymes, namely glutamine synthetase and glutamate synthase (Lea et al.,1992). These enzymes together form the GS/GOGAT cycle, which is the major path of ammonia assimilation in plants. Exporting the fixed nitrogen from the nodule to other regions of the host plant is the final step in nitrogen fixation. Export of the organic nitrogen products from nodules occurs mostly through the xylem. The principal amino acid translocated is either asparagine or in the form of ureides.

REFERENCES

Ali, A. (2020). Nitrate assimilation pathway in higher plants: Critical role in nitrogen signaling and utilization. *Plant Science Today*, *7*(2), 182–192.

Ali, A., Sivakami, S., & Raghuram, N. (2007). Regulation of activity and transcript levels of NR in rice (*Oryza sativa*): Roles of protein kinase and G-proteins. *Plant Science*, *172*(2), 406–413.

Anjana, S. U., & Iqbal, M. (2007). Nitrate accumulation in plants, factors affecting the process, and human health implications. A review. *Agronomy for Sustainable Development*, *27*(1), 45–57.

Becker, T. W., Foyer, C., & Caboche, M. (1992). Light-regulated expression of the nitrate-reductase and nitrite-reductase genes in tomato and the phytochrome-deficient area mutant of tomato. *Planta*, *188*(1), 39–47.

Bowsher, C. G., Hucklesby, D. P., & Emes, M. J. (1993). Induction of ferredoxin-NADP$^+$ oxidoreductase and ferredoxin synthesis in pea root plastids during nitrate assimilation. *The Plant Journal*, *3*(3), 463–467.

Burger, G., Strauss, J., Scazzocchio, C., & Lang, B. F. (1991b). nirA, the pathway-specific regulatory gene of nitrate assimilation in *Aspergillus nidulans*, encodes a putative GAL4-type zinc finger protein and contains four introns in highly conserved regions. *Molecular and Cellular Biology*, *11*(11), 5746–5755.

Burger, G., Tilburn, J., & Scazzocchio, C. (1991a). Molecular cloning and functional characterization of the pathway-specific regulatory gene nirA, which controls nitrate assimilation in Aspergillus nidulans. *Molecular and Cellular Biology*, *11*(2), 795–802.

Campbell, W. H. (1999). Nitrate reductase structure, function and regulation: bridging the gap between biochemistry and physiology. *Annual Review of Plant Biology*, *50*(1), 277–303.

Campbell, W. H. (2002). Molecular control of nitrate reductase and other enzymes involved in nitrate assimilation. In *Photosynthetic nitrogen assimilation and associated carbon and respiratory metabolism* (pp. 35–48). Dordrecht: Springer.

Cheng, C. L., Acedo, G. N., Cristinsin, M., & Conkling, M. A. (1992). Sucrose mimics the light induction of Arabidopsis nitrate reductase gene transcription. *Proceedings of the National Academy of Sciences*, *89*(5), 1861–1864.

Coruzzi, G. M. (2003). Primary N-assimilation into amino acids in Arabidopsis. *The Arabidopsis Book/American Society of Plant Biologists*, *2*, e0010. https://doi.org/10.1199/tab.0010

Coschigano, K. T., Melo-Oliveira, R., Lim, J., & Coruzzi, G. M. (1998). Arabidopsis gls mutants and distinct Fd-GOGAT genes: implications for photorespiration and primary nitrogen assimilation. *The Plant Cell*, *10*(5), 741–752.

Crawford, N. M. (1995). Nitrate: nutrient and signal for plant growth. *The Plant Cell*, *7*(7), 859.

Crawford, N. M., & Arst Jr., H. N. (1993). The molecular genetics of nitrate assimilation in fungi and plants. *Annual Review of Genetics*, *27*(1), 115–146.

Cren, M., & Hirel, B. (1999). Glutamine synthetase in higher plants regulation of gene and protein expression from the organ to the cell. *Plant and Cell Physiology*, *40*(12), 1187–1193.

Deng, M. D., Moureaux, T., Cherel, I., Boutin, J. P., & Caboche, M. (1991). Effects of nitrogen metabolites on the regulation and circadian expression of tobacco nitrate reductase. *Plant Physiology and Biochemistry (Paris)*, *29*(3), 239–247.

Dubois, F., Tercé-Laforgue, T., Gonzalez-Moro, M. B., Estavillo, J. M., Sangwan, R., Gallais, A., & Hirel, B. (2003). Glutamate dehydrogenase in plants: Is there a new story for an old enzyme?. *Plant Physiology and Biochemistry*, *41*(6–7), 565–576.

Duncanson, E., Gilkes, A. F., Kirk, D. W., Sherman, A., & Wray, J. L. (1993). nir1, a conditional-lethal mutation in barley causing a defect in nitrite reduction. *Molecular and General Genetics MGG*, *236*(2–3), 275–282.

Fariduddin, Q., Varshney, P., & Ali, A. (2018). Perspective of nitrate assimilation and bioremediation in (a non-nitrogen fixing cyanobacterium): An overview. *Journal of Environmental Biology*, *39*(5), 547–557.

Fedorova, E., Greenwood, J. S., & Oaks, A. (1994). In-situ localization of nitrate reductase in maize roots. *Planta*, *194*(2), 279–286.

Fontaine, J. X., Tercé-Laforgue, T., Bouton, S., Pageau, K., Lea, P. J., Dubois, F., & Hirel, B. (2013). Further insights into the isoenzyme composition and activity of glutamate dehydrogenase in Arabidopsis thaliana. *Plant Signaling & Behavior*, *8*(3), e23329.

Forde, B. G., & Lea, P. J. (2007). Glutamate in plants: metabolism, regulation, and signalling. *Journal of Experimental Botany*, *58*(9), 2339–2358.

Friemann, A., Brinkmann, K., & Hachtel, W. (1991). Sequence of a cDNA encoding the bi-specific NAD (P) H-nitrate reductase from the treeBetula pendula and identification of conserved protein regions. *Molecular and General Genetics MGG*, *227*(1), 97–105.

Glass, A. D., Shaff, J. E., & Kochian, L. V. (1992). Studies of the uptake of nitrate in barley: IV. Electrophysiology. *Plant Physiology*, *99*(2), 456–463.

Hase, T., Schürmann, P., & Knaff, D. B. (2006). The interaction of ferredoxin with ferredoxin-dependent enzymes. In *Photosystem I* (pp. 477–498). Springer, Dordrecht.

Hayakawa, T., Nakamura, T., Hattori, F., Mae, T., Ojima, K., & Yamaya, T. (1994). Cellular localization of NADH-dependent glutamate-synthase protein in vascular bundles of unexpanded leaf blades and young grains of rice plants. *Planta*, *193*(3), 455–460.

Hecht, U., Oelmüller, R., Schmidt, S., & Mohr, H. (1988). Action of light, nitrate and ammonium on the levels of NADH-and ferredoxin-dependent glutamate synthases in the cotyledons of mustard seedlings. *Planta*, *175*(1), 130–138.

Heldt, H. W. (1997). Nitrate assimilation. In: *Plant Biochemistry and molecular Biology*. New York: Oxford University Press.

Hirasawa, M., Hurley, J. K., Salamon, Z., Tollin, G., & Knaff, D. B. (1996). Oxidation–reduction and transient kinetic studies of spinach ferredoxin-dependent glutamate synthase. *Archives of Biochemistry and Biophysics*, *330*(1), 209–215.

Hoff, T., Truong, H. N., & Caboche, M. (1994). The use of mutants and transgenic plants to study nitrate assimilation. *Plant, Cell & Environment*, *17*(5), 489–506.

Huber, J. L., Huber, S. C., Campbell, W. H., & Redinbaugh, M. G. (1992). Reversible light/dark modulation of spinach leaf nitrate reductase activity involves protein phosphorylation. *Archives of Biochemistry and Biophysics*, *296*(1), 58–65.

Huber, J. L., Redinbaugh, M. G., Huber, S. C., & Campbell, W. H. (1994). Regulation of maize leaf nitrate reductase activity involves both gene expression and protein phosphorylation. *Plant Physiology*, *106*(4), 1667–1674.

Inokuchi, R., Kuma, K. I., Miyata, T., & Okada, M. (2002). Nitrogen-assimilating enzymes in land plants and algae: phylogenic and physiological perspectives. *Physiologia Plantarum*, *116*(1), 1–11.

Jackson, W.A., Pan, W.A., Moll, R.H., & Kamprath, E.J. (1986). Uptake, translocation and reduction of nitrate. In C. Neyra, ed., *Biochemical Basis of Plant Breeding*, Boca Raton, FL: CRC Press, pp. 73–108.

Kaiser, W. M., & Huber, S. (1994). Modulation of nitrate reductase in vivo and in vitro: Effects of phosphoprotein phosphatase inhibitors, free Mg 2+ and 5′-AMP. *Planta*, *193*(3), 358–364.

Kishorekumar, R., Bulle, M., Wany, A., & Gupta, K. J. (2020). An overview of important enzymes involved in nitrogen assimilation of plants. In *Nitrogen Metabolism in Plants* (pp. 1–13). New York, NY: Humana.

Kudla, B., Caddick, M. X., Langdon, T., Martinez-Rossi, N. M., Bennett, C. F., Sibley, S.,... & Arst Jr., H. N. (1990). The regulatory gene areA mediating nitrogen metabolite repression in *Aspergillus nidulans*. Mutations affecting specificity of gene activation alter a loop residue of a putative zinc finger. *The EMBO Journal*, *9*(5), 1355–1364.

LaBrie, S. T., & Crawford, N. M. (1994). A glycine to aspartic acid change in the MoCo domain of nitrate reductase reduces both activity and phosphorylation levels in Arabidopsis. *Journal of Biological Chemistry*, *269*(20), 14497–14501.

Lam, H. M., Coschigano, K. T., Oliveira, I. C., & Melo-Oliveira, R. (1996). MeloOliveira R, Coruzzi GM. The molecular-genetics of nitrogen assimilation into amino acids in higher plants. *Annual Review of Plant Physiology and Plant Molecular Biology*, *47*, 569–93.

Larsson, C. M., & Ingemarsson, B. (1989). Molecular aspects of nitrate uptake in higher plants. In *Molecular and Genetic Aspects of Nitrate Assimilation*, edited by John L. Wray and James R. Kinghorn.

Lea, P. J., Blackwell, R. D., & Joy, K. W. (1992). Ammonia assimilation in higher plants. In *Nitrogen Metabolism of Plants* (Proceedings of the Phytochemical Society of Europe 33) (pp. 153–186). Oxford: Clarendon.

Lillo, C. (1994). Light regulation of nitrate reductase in green leaves of higher plants. *Physiologia Plantarum*, *90*(3), 616–620.

Lin, Y., Hwang, C. F., Brown, J. B., & Cheng, C. L. (1994). 5 [prime] Proximal regions of Arabidopsis nitrate reductase genes direct nitrate-induced transcription in transgenic tobacco. *Plant Physiology*, *106*(2), 477–484.

Loulakakis, K. A., & Roubelakis-Angelakis, K. A. (1996). The seven NAD (H)-glutamate dehydrogenase isoenzymes exhibit similar anabolic and catabolic activities. *Physiologia Plantarum*, *96*(1), 29–35.

MacKintosh, C. (1992). Regulation of spinach-leaf nitrate reductase by reversible phosphorylation. *Biochimica et Biophysica Acta (BBA)-Molecular Cell Research*, *1137*(1), 121–126.

Maia, L. B., & Moura, J. J. (2014). How biology handles nitrite. *Chemical Reviews*, *114*(10), 97.

Marchi, L., Degola, F., Polverini, E., Tercé-Laforgue, T., Dubois, F., Hirel, B., & Restivo, F. M. (2013). Glutamate dehydrogenase isoenzyme 3 (GDH3) of Arabidopsis thaliana is regulated by a combined effect of nitrogen and cytokinin. *Plant Physiology and Biochemistry*, *73*, 368–374.

Masclaux-Daubresse, C., Daniel-Vedele, F., Dechorgnat, J., Chardon, F., Gaufichon, L., & Suzuki, A. (2010). Nitrogen uptake, assimilation and remobilization in plants: challenges for sustainable and productive agriculture. *Annals of Botany*, *105*(7), 1141–1157.

McClure, P. R., Kochian, L. V., Spanswick, R. M., & Shaff, J. E. (1990a). Evidence for cotransport of nitrate and protons in maize roots: I. Effects of nitrate on the membrane potential. *Plant Physiology*, *93*(1), 281–289.

McClure, P. R., Kochian, L. V., Spanswick, R. M., & Shaff, J. E. (1990b). Evidence for cotransport of nitrate and protons in maize roots: II. Measurement of NO_3^- and H^+ fluxes with ion-selective microelectrodes. *Plant Physiology*, *93*(1), 290–294.

Melo-Oliveira, R., Oliveira, I. C., & Coruzzi, G. M. (1996). Arabidopsis mutant analysis and gene regulation define a nonredundant role for glutamate dehydrogenase in nitrogen assimilation. *Proceedings of the National Academy of Sciences*, *93*(10), 4718–4723.

Migge, A., Carrayol, E., Hirel, B., Lohmann, M., Meya, G., & Becker, T. W. (1998). Influence of UV-A or UV-B light and of the nitrogen source on the induction of ferredoxin-dependent glutamate synthase in etiolated tomato cotyledons. *Plant Physiology and Biochemistry*, *36*(11), 789–797.

Mohr, H., Neininger, A., & Seith, B. (1992). Control of nitrate reductase and nitrite reductase gene expression by light, nitrate and a plastidic factor. *Botanica Acta*, *105*(2), 81–89.

Näsholm, T., Kielland, K., & Ganeteg, U. (2009). Uptake of organic nitrogen by plants. *New Phytologist*, *182*(1), 31–48. https://doi.org/10.1111/j.1469-8137.2008.02751.x

Pilgrim, M. L., Caspar, T., Quail, P. H., & McClung, C. R. (1993). Circadian and light-regulated expression of nitrate reductase in Arabidopsis. *Plant Molecular Biology*, *23*(2), 349–364.

Pouteau, S., Cherel, I., Vaucheret, H., & Caboche, M. (1989). Nitrate reductase mRNA regulation in *Nicotiana plumbaginifolia* nitrate reductase-deficient mutants. *The Plant Cell*, *1*(11), 1111–1120.

Qu, C. P., Xu, Z. R., Hu, Y. B., Lu, Y., Yang, C. J., Sun, G. Y., & Liu, G. J. (2016). RNA-SEQ reveals transcriptional level changes of poplar roots in different forms of nitrogen treatments. *Frontiers in Plant Science*, *7*, 51.

Riens, B., & Heldt, H. W. (1992). Decrease of nitrate reductase activity in spinach leaves during a light-dark transition. *Plant Physiology*, *98*(2), 573–577.

Ruiz-Cristin, J., & Briskin, D. P. (1991). Characterization of a H^+ NO_3^- symport associated with plasma membrane vesicles of maize roots using $36ClO_3^-$ as a radiotracer analog. *Archives of Biochemistry and Biophysics*, *285*(1), 74–82.

Rus-Alvarez, A., & Guerrier, G. (1994). Proline metabolic pathways in calli from *Lycopersicon esculentum* and *L. pennellii* under salt stress. *Biologia Plantarum*, *36*(2), 277–284.

Sakakibara, H., Kawabata, S., Hase, T., & Sugiyama, T. (1992). Differential effects of nitrate and light on the expression of glutamine synthetases and ferredoxin-dependent glutamate synthase in maize. *Plant and Cell Physiology*, *33*(8), 1193–1198.

Sakano, K. (1990). Proton/phosphate stoichiometry in uptake of inorganic phosphate by cultured cells of *Catharanthus roseus* (L.) G. Don. *Plant Physiology*, *93*(2), 479–483.

Sakano, K., Yazake, Y., & Mimura, T. (1992). Cytoplasmic acidification induced by inorganic phosphate uptake in suspension cultured *Catharanthus roseus* cells. *Plant Physiology*, *99*, 672–680.

Schnell, R. A., & Lefebvre, P. A. (1993). Isolation of the Chlamydomonas regulatory gene NIT2 by transposon tagging. *Genetics*, *134*(3), 737–747.

Schuster, C., & Mohr, H. (1990a). Appearance of nitrite-reductase mRNA in mustard seedling cotyledons is regulated by phytochrome. *Planta*, *181*(3), 327–334.

Schuster, C., & Mohr, H. (1990b). Photooxidative damage to plastids affects the abundance of nitrate-reductase mRNA in mustard cotyledons. *Planta*, *181*(1), 125–128.

Schwarz, G., & Mendel, R. R. (2006). Molybdenum cofactor biosynthesis and molybdenum enzymes. *Annual Review of Plant Biology*, *57*, 623–647.

Sechley, K. A., Yamaya, T., & Oaks, A. (1992). Compartmentation of nitrogen assimilation in higher plants. *International Review of Cytology*, *134*, 85–163.

Sivasankar, S., Rothstein, S., & Oaks, A. (1997). Regulation of the accumulation and reduction of nitrate by nitrogen and carbon metabolites in maize seedlings. *Plant Physiology*, *114*(2), 583–589.

Suzuki, A., & Knaff, D. B. (2005). Glutamate synthase: structural, mechanistic and regulatory properties, and role in the amino acid metabolism. *Photosynthesis Research*, *83*(2), 191–217.

Temple, S. J., Vance, C. P., & Gantt, J. S. (1998). Glutamate synthase and nitrogen assimilation. *Trends in Plant Science*, *3*(2), 51–56.

Tercé-Laforgue, T., Dubois, F., Ferrario-Méry, S., de Crecenzo, M. A. P., Sangwan, R., & Hirel, B. (2004). Glutamate dehydrogenase of tobacco is mainly induced in the cytosol of phloem companion cells when ammonia is provided either externally or released during photorespiration. *Plant Physiology*, *136*(4), 4308–4317.

Ullrlch, C.I., & Novacky, A.J. (1990). Extra- and intracellular pH and membrane potential changes induced by K+, CI-, H2P04-, NO3- uptake and fusicoccin in root hairs of *Limnobium storoniferum*. *Plant Physiology*, *94*, 1561–1567.

Vaughn, K. C., & Campbell, W. H. (1988). Immunogold localization of nitrate reductase in maize leaves. *Plant Physiology*, *88*(4), 1354–1357.

Vincentz, M., Moureaux, T., Leydecker, M. T., Vaucheret, H., & Caboche, M. (1993). Regulation of nitrate and nitrite reductase expression in *Nicotiana plumbaginifolia* leaves by nitrogen and carbon metabolites. *The Plant Journal*, *3*(2), 315–324.

Wang, L., & Macko, S. A. (2011). Constrained preferences in nitrogen uptake across plant species and environments. *Plant, Cell & Environment*, *34*(3), 525–534.

Wang, Y. Y., Cheng, Y. H., Chen, K. E., & Tsay, Y. F. (2018). Nitrate transport, signaling, and use efficiency. *Annual Review of Plant Biology*, *69*, 85–122.

Yamaya, T., Hayakawa, T., Tanasawa, K., Kamachi, K., Mae, T., & Ojima, K. (1992). Tissue distribution of glutamate synthase and glutamine synthetase in rice leaves: occurrence of NADH-dependent glutamate synthase protein and activity in the unexpanded, nongreen leaf blades. *Plant Physiology*, *100*(3), 1427–1432.

Yoneyama, T., & Suzuki, A. (2019). Exploration of nitrate-to-glutamate assimilation in non-photosynthetic roots of higher plants by studies of 15N-tracing, enzymes involved, reductant supply, and nitrate signaling: A review and synthesis. *Plant Physiology and Biochemistry*, *136*, 245–254.

6 Factors Affecting Nitrogen Uptake, Transport, and Assimilation

Junaid Ahmad Magray, Shabir Ahmad Zargar,*
Tajamul Islam, and Hanan Javid
Department of Botany, University of Kashmir, Srinagar, Jammu and Kashmir, India
*Corresponding author: Shabir Ahmad Zargar
E-mail Id: shabirbotany786@gmail.com

CONTENTS

- 6.1 Introduction 69
- 6.2 Nitrogen Uptake and Transport 70
  - 6.2.1 Forms of Nitrogen 70
  - 6.2.2 Nitrogen Uptake Sites 70
  - 6.2.3 Nitrogen Uptake Rates 71
  - 6.2.4 Transport 71
- 6.3 Factors Affecting Uptake and Transport Of Nitrogen 73
  - 6.3.1 Nitrate Uptake Repression by Ammonium 73
  - 6.3.2 Rhizosphere pH 74
  - 6.3.3 Interactions among Ions 74
  - 6.3.4 Supply of Photosynthates 75
  - 6.3.5 Temperature 75
  - 6.3.6 Mycorrhizal Associations 76
- 6.4 Nitrogen Assimilation 76
- 6.5 Factors Affecting Assimilation 77
  - 6.5.1 Nitrogen Levels 77
  - 6.5.2 Ultraviolet B Radiation 77
  - 6.5.3 Gaseous Factors 78
  - 6.5.4 Soil Salinity 78
  - 6.5.5 Metals 79
- 6.6 Conclusion 79
- References 80

6.1 INTRODUCTION

One of the most important mineral nutrients for plants is nitrogen (N). It is normally taken up by plant roots as nitrate and ammonium or amino acids from the soil solution. Leaf surfaces can absorb minor quantities of nitrogen directly from the atmosphere after deposition on leaves, but the contribution of this is very little. Through symbiotic interactions with microorganisms, gaseous N is directly acquired from the air by some plants. Nitrogen fixation occurs in root nodules in legumes, but this mechanism can also happen in some other structures, and the N captured by bacteria residing either within the plant or on the surface of roots can be subsequently used by plants. Under conditions of low availability, these N sources become quite significant as the ability of plants to utilize these encounters with fungi and bacteria provide significant competitive advantages. In soil, the N concentrations available differ significantly from micromolar to millimolar and may depend on a broad spectrum of physical and chemical characteristics of the soil and conditions of the environment such as temperature and water supply (Miller et al., 2007). The type and the amount of N available are determined by bacterial activity. In agricultural soils, N fertilizers are added directly as ammonium or nitrate, or more commonly as urea. Nitrate is the most important source of N for crops growing in aerobic temperate soils because microbes present in these soils readily transform all N forms into nitrate. The supply of ammonium and amino acids increases in extreme conditions when temperatures of soil are either colder or hotter than in temperate environments. In bulk soil solution, the free amino acid concentration varies greatly, with the highest concentrations in the organic matter-rich surface layer of soils (Jones et al., 2002). In agricultural soils most of the organic N is obtained from the breakdown of plant matter by microbes, as the amino acid concentration in plant tissues is high, making them essential sources of organic nitrogen for the soil. In plants that lack the uptake mechanism of amino acids at the root level, the use of this amino source of N was demonstrated nicely (Svennerstam et al., 2007). Once nitrate passes the cell membranes it will either be metabolized in the root cytosol or passively exported through the xylem gradient to the shoot along the gradient (Forde, 2002; Orsel et al., 2002).

6.2 NITROGEN UPTAKE AND TRANSPORT

N being a main constituent of proteins, amino acids, chlorophyll, and nucleic acids, plays a crucial role in plant productivity. Despite variation with age, species, and plant organ, nearly 5% of the dry weight of plants is contributed by organic N. In stems and leaves nearly 60% of the N is found as membrane protein or an enzyme and most is present in free amino acid nitrogen form (Parsons and Tinsley, 1975). In plants, over 90% of the N is present as a storage protein.

6.2.1 Forms of Nitrogen

Ammonium and nitrate are the only main ionic forms of N that are actively taken up by plants both in unfertilized and fertilized soils (Haynes and Goh, 1978). The nutrient elements uptake in the form of ions by roots is a physiologically active process. Thus, all environmental and soil factors that have an impact on the uptake process, as well as nutrient availability, affect nutrient absorption from the soil solution.

6.2.2 Nitrogen Uptake Sites

Due to the availability of various forms of N, plants use distinct pathways for the uptake of N into cells. In roots, initial absorption of organic and mineral N takes place through plasmalemma of cortical and epidermal cells. Root hairs present on the apical part of roots maximize the surface area for nutrient uptake (Marschner, 1995), and before death, these root hairs stay functional for a few days only. Therefore, in root hairs, nutrient uptake requires, on one side, a transport system with high turnover but on the other side, assures the efficient nutrient extraction from soil, which is very essential for nutrients less mobile in soil. Thus, N-transporters present on the root hair cell membrane require special attention. Generally, the systems for mineral N uptakes could also be found at the leaf cell plasmalemma, as N-deficient plants can be recovered from chlorosis by application of sprays containing NOs or NOs on foliage (Bowman and Paul, 1992).

6.2.3 Nitrogen Uptake Rates

Vegetation uptake rates of N are unique to a given ecosystem and ecosystem condition. Generally, plants with lower rates of production commonly have a lower demand for N-uptake, although it is not always clear whether a low production rate limits N uptake or whether a low uptake limits production (Cole, 1981). However, plants in N-deficient environments tend to make more effective use of the N they have already taken up, either by holding leaves for extended periods or by transporting back to tissues that are alive at the time of tissue senescence (Cole, 1981).

Under climates that are not favorable to vigorous plant growth, such as tundra and desert habitats, the rate of N uptake by plants per unit land area is significantly lower than that of temperate forests and grasslands. In general, the absorption rate is much higher in intensively controlled agricultural ecosystems than in most natural forest and grassland ecosystems.

The available soil N varies considerably both in time and space, because of various factors like temperature, precipitation, soil type, wind, and pH. Thus, the form of N preferably taken up by the plants is dependent on adaptations of plants to soil conditions. In general, plants found in reducing and low pH soils as present in well-grown forests or tundra mostly absorb amino acids or ammonium, whereas plants found in aerobic and high pH soils take up nitrate (Maathuis, 2009).

6.2.4 Transport

In roots, nitrate is absorbed and they have been shown to co-exist in plants with two nitrate transport systems to function coordinately to absorb and allocate nitrate from soil solution into the whole plant body (Figure 6.1) (Daniel-Vedele et al., 1998; Tsay et al., 2007).

The NRT1 gene family is commonly believed to mediate the low-affinity transport systems (LATS) in the root, except that of AtNRT1.1, both a nitrate sensor (Ho et al., 2009) and dual affinity transporter (Wang et al., 1998; Liu et al., 1999). The family of NRT1 in *Arabidopsis* has 53 genes of which 51 genes are expressed in the entire plant and exhibit various tissue expression patterns (Tsay et al., 2007), indicating that at least some of them have a specific and specialized role. The AtNRT1.1 (formerly Chl1) gene was the first to be isolated and the most widely studied gene (Tsay et al.,

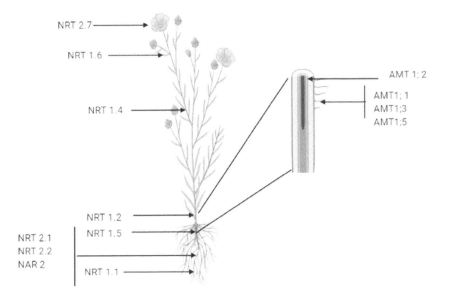

FIGURE 6.1 Localization of NRT1, NRT2, and AMT genes in Arabidopsis.

1993). The protein is present on the cell membrane and the gene is expressed in the epidermal cells of the root tips and more developed portion of the root cortex and endodermis (Huang et al., 1999). In the epidermis of the root, only AtNRT1.2 is constitutively expressed and participates actively in the low-affinity constitutive system (Huang et al., 1999). The nitrate is transported through various cell membranes and distributed in various tissues, after absorption by root cells. AtNRT1.5 is present on the cell membrane of root pericycle cells near the xylem and is involved in the long-distance transport of nitrate from the root to shoot (Lin et al., 2008). In leaf petiole the only gene expressed is AtNRT1.4, and the nitrate amount in the petiole of mutant-type is half that found in wild variety (Chiu et al., 2004). The nitrate is delivered to the developing embryo from maternal tissue through the action of the AtNRT1.6 gene expressed in the funiculus and silique vascular tissue (Almagro et al., 2008).

One prominent feature of the NRT1 transporter family is the ability of some Group II members (Tsay et al., 2007) to transport di- or tripeptides in the heterologous systems, in addition to nitrate transport whereas tetra/pentapeptides are transported by OPT (oligo-peptide transporter) proteins.

The high-affinity transport system (HATS) depends on the activities of the so-called NRT2 transporter family, operating at low external nitrate concentrations (Williams and Miller, 2001). Even though almost all NRT2 transporters of higher plants are not yet functionally characterized AtNRT2.1 has now been reported to be a main constituent of the HATS in *Arabidopsis* in association with the protein called NAR2 (Orsel et al., 2006), as shown by lower nitrate content of leaf and a loss of nearly 75% in NO_3^- high-affinity uptake activity in AtNRT2.1 gene-disrupted mutant (Filleur et al., 2001). As a result, at low NO_3^- concentration growth of such mutants is sternly restricted (Orsel et al., 2004, 2006). In clce mutant, a protein that belongs to the AtCLC (ChLoride Channel) family of *Arabidopsis* is affected leading to lesser content of nitrate in them. De Angeli et al. (2006) showed that the AtCLCa protein of vacuoles is also an exchanger of NO_3^-/H^+, enabling nitrate to accumulate inside the vacuole. The AtCLCa gene insertion mutant shows decreased nitrate storage potential but normal development. After altering AtNRT2.7, such phenotype was also recently discovered. In aerial organs, AtNRT2 is expressed and also strongly induced in dried seeds. In seeds of two allelic mutants for atnrt2.7, there is an accumulation of little nitrate. On the other hand, the seeds of plants overexpressing AtNRT2.7 acquire a higher amount of nitrate and show less dormancy than the seeds of wild-type (Chopin et al., 2007).

Both electrophysiological and pH-dependent equilibrium studies involving the charged NH_4^+ and uncharged NH_3 groups indicate that in all physiological circumstances the ion predominates and is also the dominant species for regulated transport across the membrane (Ludewig et al., 2007). Methylammonium uptake-defective mutant of yeast with functional complementation and genome sequencing of model plant species revealed that at least six genes of the similar ammonium transporters family were found in *Arabidopsis* species as well (Gazzarini et al., 1999), 10 in rice species (Sonoda et al., 2003) feeding on ammonium, and 14 in poplars (Couturier et al., 2007). Physiological and ammonium influx experiments were performed on single, double, triple, and quadruple mutants, obtained by T-DNA insertion or RNAi method, or by supplementing the quadruple mutant with single genes, to analyze the function of each of the AMT (ammonium transporter) genes separately in plants (Yuan et al., 2007). Each protein was shown to contribute synergistically to ammonium transport, with AMT1.1 and AMT1.3 giving a comparable capacity of 30–35%, while AMT1.2 gave a lower capacity of 18–25%. The AMT1.5 gene is believed to code a second transport system having a very lower capacity and lower Km value of 4.5 mM. From these studies, an intricate picture is now emerging. AMT1 proteins are spatially organized (Figure 6.1), with transporters located in root hairs or outer root cells uptaking ammonium from the soil with maximum ammonium affinities (AMT1.1, AMT1.3, AMT1.5). The lower affinity of AMT1.1, and its location in the endodermis along the root hair zone, suggests a function in the retrieval of ammonium that is released from the cortex or that enters the root via the apoplastic route. The low affinity and location of AMT1.1 in the root hair zone endodermis indicate a role in the recovery of ammonium entering the root via the apoplastic route or released from the cortex. NH4+ export out and NH_3 import into the vacuole are driven by the electrochemical gradient existing between the cytosol and vacuole. Indeed, in the transport of NH_3 into the vacuole, tonoplast intrinsic proteins belonging to the TIP family were shown to play a vital role (Loque et al., 2005). An unidentified electrogenic ammonium transporter is required for vacuolar loading with NH_4^+.

Plant amino acid transporters identified so far belong to at least five gene families, comprising nearly 67 genes in *Arabidopsis* (Rentsch et al., 2007). Gene expression patterns as well as protein subcellular localization or substrate specificities can provide a clear indication of the role of each protein. Transporters involved in amino acid uptake in the root are identified through forward and reverse genetic approaches (Hirner et al., 2006; Svennerstam et al., 2007). All these studies have shown that for acidic and neutral amino acid uptake by roots, LHT1 (lysine/histidine transporter), a member of ATF (amino acid transport family), is very important. When uncharged amino acids are present at higher concentrations in the outside medium they are also transported by protein, namely AAP1 (amino acid permease 1) (Lee et al., 2007). At concentration range relevant to field condition positively charged amino acids like L-arginine or L-lysine are taken up by AAP5 (Svennerstam et al., 2008). Näsholm et al. (2009) proposed a hypothetical model for the uptake of amino acid in roots of non-mycorrhizal plants keeping in view the precise localization of all mRNA transporters inside different types of cells in the root.

After nitrate has been taken up by the plant, it may be subject to immediate reduction and assimilation into organic form in root tissue or transported in the xylem to the leaves for reduction and further metabolism there. Nitrate may also be stored in the roots, in vacuoles for varying periods, before reduction or transport. Amino acids can be synthesized in roots and transported in the xylem to developing shoots, or formed in leaf tissue and transported in the phloem to sinks, such as developing seeds or apices. A picture of this can be obtained by analysis of the xylem and phloem sap compositions (pate and Layzell, 1990). Xylem sap analysis reveals a range of molecules used for nitrogen transport-including as nitrate (common in the grasses), and amino acids and their derivatives (pate, 1989). Some plants, such as *Xanthium* spp., *Cucumis* spp., and others, reduce all of their nitrates in their shoots, with nitrate making up over 95% of their xylem nitrogen, and nitrate accumulation often being seen in leaf tissue. In contrast with this group are the legumes, and some gymnosperms, which usually have very little nitrate in their xylem sap. In these plants, all the nitrate is reduced in the root tissue, and nitrogen transport is in organic form (Pate and Layzell, 1990). Between these two extremes is a wide variety of plants that show intermediate behavior, reducing some nitrate in the roots and transporting some for shoot reduction (Andrews, 1986). The location of nitrate reduction in these species depends on factors such as root temperature, which can affect enzyme activity, and nitrate supply, which can saturate the root reduction system, forcing nitrate transport. Any organic nitrogen compounds made in the roots rely on photosynthates transported down from the leaves to provide carbon skeletons for their synthesis. Thus, the relative amounts of nitrate and organic nitrogen transported depend on the species concerned, soil temperature, developmental stage, soil nitrate concentration, and other factors (Andrews, 1986).

6.3 FACTORS AFFECTING UPTAKE AND TRANSPORT OF NITROGEN

6.3.1 Nitrate Uptake Repression by Ammonium

Normally the rates of NH_4^+ uptake are not affected by the absence or presence of NO_3^- in nutrient solutions (Munns and Jackson, 1978; Youngdahl et al., 1982), but in many plant species high NH_4^+ limits net NO_3^- uptake (Buczek, 1979; MacKown et al., 1982; Youngdahl et al., 1982). However, in some plants, NH_4^+ had little or no effect on NO_3^- uptake (Schrader et al., 1972; Neyra and Hageman, 1975; Edwards and Barber, 1976a).

In most of the cases, NH_4^+ inhibits NO_3^- uptake without affecting the activity of enzyme NO_3^- reductase (e.g., Rao and Rains, 1976; MacKown et al., 1982). Also, Rao and Rains (1976) had shown that NO_3^- uptake in short-term experiments is reduced at ambient NH_4^+ level without any visible effect on nitrate reductase activity. Some researchers are of the view that mechanism of NO_3^- uptake is inhibited by the endogenous cytoplasmic NH_4^+ level in root tissue (Jackson et al., 1976b; Jackson, 1978) while others believe that during NH_4^+ nutrition-accumulated amino acids in the roots exert end-product regulation of NO_3^- uptake (Heimer and Filner, 1971; Doddema and Otten, 1979). Evidence suggests that both modes of action are present (Doddema and Otten, 1979) and to know the exact mechanism(s) involved further investigation needs to be done.

6.3.2 Rhizosphere pH

At the cell membrane of roots, active uptake of cations involves the excretion of HCO_3^- or OH^- ions actively (Nye, 1981). As no net charge can cross the soil root boundary these processes are important to maintain electroneutrality. In NH_4^+-rich soils, plants absorb more cations than anions (N being the element often absorbed in the largest amounts) and plant growth leads to a net H^+ ion efflux into the rhizosphere. With NH_4^+ nutrition the pH values can drop up to 2.8 in sand culture or nutrient solutions (Maynard and Barker, 1969). Similarly, a decrease in soil pH near the root surface is shown to be caused by the growth of plants grown in a container or field, having soils supplied with NH_4^+- instead of NO_3^--N (Miller et al., 1970; Riley and Barber, 1971; Smiley, 1974).

In comparison, the rhizospheric pH increases due to absorption of more anions and net efflux of HCO_3^- or OH^- ions when the main N form supplied is NO_3^- (Pierre et al., 1970; Bagshaw et al., 1972; Pierre and Banwart, 1973; Smiley, 1974). In Table 6.1 the distinct effects of the NH_4^+ or NO_3^- uptake by plants on pH of leachates from sand cultures are shown clearly. The effects of NO_3^- and NH_4^+ nutrition on rhizospheric pH have adverse effects on the respective ion absorption. In nutrient solution, for example, maximum uptake of NO_3^- by plants occurs at a pH of around 4–5 while that for NH_4^+ occurs at around pH 7–8 (Sheat et al., 1959; Rao and Rains, 1976). Therefore, the pH at which the rhizosphere tends to and the pH for optimum uptake of the respective ions, following their absorption, are at two ends of the pH scale. It is interesting to note that the rhizosphere acidification associated with NH_4^+ uptake is also at least partially responsible for the toxic effect of NH_4^+ nutrition.

6.3.3 Interactions among Ions

During the accumulation of ions by plants competition among ions during the uptake process and enhancement of one ion by another are common physiological events. While the effects of NH_4^+ and NO_3^- nutrition on the uptake of other ions by plants are well known, little is known about the effects of the other ions on NO_3^- and NH_4^+ uptake.

a. Ammonium. NH_4^+, Mg^{2+}, Ca^{2+}, and K^+ typically tend to compete with each other during plant ion accumulation (Haynes and Goh, 1978; Reisenauer, 1978). NH_4^+ uptake typically results in a significant decrease in K^+ uptake and vice versa to a lesser degree (Ajayi et al., 1970; Moraghan and Porter, 1975). Rufty et al. (1982), however, observed that while increasing ambient NH_4^+ concentrations steadily inhibited K^+ uptake, rising K^+ concentrations had little or no effect on NH_4^+ uptake. There are also a few exceptions where low ambient NH_4^+ concentrations had no effect or may increase cation uptake (Blair et al., 1970; Rayar and van Hai, 1977). The increase in ambient NH_4^+ levels that accompany the reduced cation uptake is generally discussed in terms of the competition at the cell membrane either with NH_4^+ ions per se or with H^+ ions excreted during active uptake of NH_4^+ (Cox and Reisenauer, 1973; Haynes and Goh, 1978). Within the free space of roots, high H^+ ion concentrations also inhibit the movement of other cations (Haynes, 1980).

NH_4^+ nutrition typically results in improved phosphate and sulfate absorption as compared to NO_3^- nutrition (Blair et al., 1970). NH_4^+ nutrition leads to a decrease in rhizosphere pH due to which the ratio of $H_2PO_4^-$ to HPO_4^{2-} ions decreases in soil conditions (Soon and Miller, 1977). The $H_2PO_4^-$ ion is taken up many times quicker than $H_2PO_4^-$ and HPO_4^{2-} salts also appear to precipitate on the root–soil surface (Miller et al., 1970).

b. Nitrate. Generally, the effect of nitrate nutrition is opposite to ammonium nutrition as indicated by inhibition of anion uptake and stimulation of cation uptake by nitrate (Haynes and Goh, 1978). Even though NO_3^- uptake is a specific process and not significantly influenced by the presence of different anions (e.g.,., Cl^-, Br^-, and SO_4^{2-}) in the surrounding medium (Rao and Rains, 1976), NO_3^- uptake does by and large bring about a reduced absorption of different anions (Blair et al., 1970). Several workers reported that in plants an inverse relationship exists between NO_3^- and Cl^- accumulation (Weigel et al., 1973; Hiatt and Leggett, 1974; Kafkafi et al., 1982). A significant increase of the NO_3^- level can partly reduce the harmful impact of salinity on the growth of the plant by diminishing the uptake of Cl^- by the plant (Kafkafi et al., 1982). It is also well documented that with increasing concentration of ambient NO_3^- cation uptake also increases (Jackson and Williams, 1968; Cox and Reisenauer, 1973; Jackson et al., 1974). The uptake of both monovalent and divalent cations is stimulated, and it may be that the increase in rhizosphere pH, as a consequence of rapid NO_3^- uptake, that produces favorable conditions for cation uptake (Maas, 1969). Jackson et al. (1974) observed that an association exists between the onset of the accelerated phase of NO_3^- uptake and the enhancement of K^+ uptake. Conversely, the rate of NO_3^- uptake accelerates when the supply of Ca^{2+} and K^+ increases in nutrient solution (Minotti et al., 1968, 1969).

6.3.4 Supply of Photosynthates

As both NH_4^+ and NO_3^- uptake seem to be active processes, both processes are likely to be affected by carbohydrate (energy) supply. Indeed, Sasakawa and Yamamoto (1978) observed that both NH_4^+ and NO_3^- uptake was suppressed by removal of the endosperm from *Oryza* seedlings (Figure 6.1) while sucrose supplied exogenously restored uptake. The uptake of both forms of N is significantly reduced by low light (Ta and Ohira, 1981). Many workers showed that the root strongly needs a continuous energy supply to support the NO_3^- uptake system (Pearson and Steer, 1977). For example, the uptake of N is inhibited more than Cl^- when endosperm from Zea seedlings is removed (Jackson et al., 1973). The daily pattern of plant NO_3^- uptake, which is highest at noon (Pearson and Steer, 1977; Clement et al., 1978), coincides closely with photosynthate translocation away from leaves (Pearson and Steer, 1977). NH_4^+ uptake by plants often shows diurnal rhythms (Van Egmond, 1978) that can be interrupted during darkness by adding glucose to the nutrient medium or by providing continuous light. The drop in uptake of NH_4^+ in darkness is attributed to the loss of root carbohydrate (sugar) reserves used for NH_4^+ assimilation (Reisenauer, 1978). NH_4^+ uptake seems to be limited by the detoxification of ingested NH_4^+ and supply of C skeletons for the immediate assimilation, but during NO_3^- absorption, such an effect is less significant, because NO_3^- is less toxic than NH_4^+ and may be translocated to the shoots or accumulated in root vacuoles. Michael et al., 1970 thus showed that a decrease in the carbohydrate supply to the roots by removing phloem rings in *Phaseolus vulgaris* resulted in a reduction in N uptake; the uptake of NH_4^+ was substantially lower than that of NO_3^-.

6.3.5 Temperature

As compared to NH_4^+, the uptake of NO_3^- is very much impeded by lower temperatures (Sasakawa and Yamamoto, 1978; Clarkson and Warner, 1979). For example, Frota and Tucker (1972) found that NH_4^+ uptake by *Lactuca sativa* was greater than that of NO_3^- at 8°C and reached an optimum at 25°C in the rooting medium. At about 23°C, nitrate absorption normally becomes greater than NH_4^+ and rises to 35°C (Frota and Tucker, 1972). Grasmanis and Nicholas (1971) observed that apple trees growing in nutrient solutions containing both NH_4^+- and NO_3^- in 1:7 ratio taken up both ions in almost equal quantities, except for the wintertime when NH_4^+ absorption was more that of NO_3^-. This disparity was due to comparatively lower NO_3^- uptake rates at the lower winter temperatures.

6.3.6 Mycorrhizal Associations

The vascular arbuscular (VA), ericaceous, and ectotrophic mycorrhizae are three main types of mycorrhizal associations between higher plants and fungi, and these associations are symbiotic in nature. Only some families of trees have ectotrophic mycorrhizas and the main feature of this type of association is a fungal sheath that completely covers the lateral root tips and penetrates between the cortex cells. The sheath of fungi has a pronounced effect on the quantities or forms of compounds that enter the roots and also improve rootlet longevity (Smith, 1980). Mycelial strands produced by ectotrophic mycorrhiza can penetrate up to 12 cm of soil and litter (Skinner and Bowen, 1974).

The growth of fungal hyphae between and inside the cells of the root cortex is rigorous in both VA and ericaceous mycorrhiza, but no fungal sheath forms. A wide network of loosely arranged hyphae penetrates the substrate up to 8 cm (Read and Stribley, 1975; Rhodes and Gerdemann, 1975).

The mycorrhizae growing on the external side of the root in the soil helps plants in exploring higher volumes of soil, thus enabling the plants to use nutrients present in the soil more efficiently. This is critically essential in case of plants feeding on non-mobile nutrients like phosphorus (Nye and Tinker, 1977). Mycorrhizas also influence plant N nutrition, especially when the main source of N available to plants is less mobile NH_4^+ instead of more mobile NO_3^- (Bowen and Smith, 1981).

The culture-grown ericoid and ectomycorrhizas generally prefer NH_4^+-N over NO_3^--N (Stribley and Read, 1974). Moreover, ericoid mycorrhizal fungi have been reported to enhance N uptake by ericoid plants like *Vaccinium spp.* (Stribley and Read, 1974, 1976). These ericaceous plants thrive in soils having low rates of nitrification and NH_4^+ as the main form of available nitrogen and low rates of nitrification. In *Pinus* spp. having ectotrophic infection this is also true (Smith, 1980). The use of NO_3^- may be most significant in terms of the VA mycorrhizae because the plants associated with these mycorrhizae are widespread in soils with high nitrification rates. However, the NO_3^- uptake is least influenced by mycorrhizal associations because NO_3^- is highly mobile in the soil. In plants with VA mycorrhizas, N content does not increase more often (Possingham and Groot Obbink, 1971). However, the addition of NO_3^--N to the soil generally decreases VA mycorrhizal contamination (Kruckelmann, 1975; Azcon et al., 1982) by reducing mycorrhiza propagule infectivity (spores or infectious parts) (Chambers et al., 1980). The growth of ectotrophic mycorrhizas is also suppressed by the application of nitrogen fertilizers. Mycorrhizae play a role in the absorption and utilization of amino acids by plants. *Vaccinium macrocarpon* having an association with ericaceous fungi can consume alanine and glutamic acid more efficiently than NH_4^+ (Pearson and Read, 1975). The amino acid uptake and transport in ericaceous (Stribley and Read, 1975, 1980) and ectotrophic (Melin and Nilsson, 1952) mycorrhizae were also shown. The mycorrhizal fungi can also compete effectively with other soil microorganisms for basic organic nitrogen compounds via the penetration of organic matter, thus consequently benefiting the host plant (Bowen and Smith, 1981).

6.4 NITROGEN ASSIMILATION

The reduction of nitrate to ammonium and its incorporation into amino acids comprises N assimilation. Basic nitrate assimilation occurs chiefly in the plant roots and depends strongly on the age of root and available space for root growth. Nitrate taken up by plants is reduced by nitrate reductase (NR) to nitrite (Kuoadio et al., 2007; Rosales et al., 2011). In higher plants, the reduction of nitrate to nitrite with pyridine nucleotide in N assimilation is catalyzed by NR (Ahmad and Abdin, 1999). The nitrite is transported immediately from the cytosol into leaf chloroplasts and root plastids because it is highly reactive. Nitrite reductase further reduces nitrite to NH_4^+ in these organelles (Rosales et al., 2011).

Processes such as photorespiration, reduction of nitrate, N_2 fixation, direct absorption, or deamination of nitrogen compounds like asparagine produce ammonium in plants like *Arabidopsis* (Wickert et al., 2007). Ammonium is the only reduced form of N available to plants for assimilation into amino acids, thus at first all inorganic nitrogen is reduced to ammonium (Ruiz et al., 2007). Glutamine and glutamate are then produced due to the assimilation of ammonia in legumes and non-legumes and these are organic N from which are translocated from source to sink. Glutamate synthase, or glutamine2-oxoglutarate aminotransferase (GOGAT), and glutamate dehydrogenase (GDH) are the major enzymes involved (Frechilla et al., 2002; Esposito et al., 2005; Wickert et al., 2007). Also, GS plays a critical role in NH_4^+ assimilation and its activity is considered crucial and probably a rate-limiting step assimilation of NH_4^+. Temple et al. (1998) further reported that GS and GOGAT act together in primary NH_4^+ assimilation into amino acids.

For the growth of many organisms, including fungi, bacteria, and plants, urea is a readily accessible N source and in nature occurs ubiquitously. Urea plays a role as a primary N source because of its extensive use as N fertilizer in agriculture. Plants take up urea actively from the soil, and in arginine catabolism of plants, it is also an intermediate (Witte, 2011). In the plant cells, urea assimilation after uptake is catalyzed by two different non-degrading enzymes, namely urease (or urea amidohydrolase) and urea amidolase and both these enzymes lead to hydrolysis of urea to NH_3 and CO_2 in the cytosol (Wang et al., 2008).

6.5 FACTORS AFFECTING ASSIMILATION

6.5.1 Nitrogen Levels

Plants exhibiting low root NO_3^- levels transfer the majority of uptaken NO_3^- to shoots with its subsequent reduction and incorporation into amino acids (Aslam et al., 2001). Certain NRA variations occurred regardless of overall N concentrations. According to a study of Maighany and Ebrahimzadeh (2004) carried under salt stress on two wheat varieties exhibiting variations in N level and assimilation of nitrates, NRA was reported to be very closely linked to root N levels than leaf N levels. Mostly an inverse relationship between NRA and N content was observed but during tillering only a direct relation between them existed in the leaves. The effect of salt in the leaves, however, was much more visible than in roots in changing the NR activity and N level. This indicates that sodic stress changes the N content and NR activity and this effect can vary with organs, cultivar, stage of development, and salt stress level. When ammonium was used instead of nitrate as an N source, the activity of glutamine synthetase declined. A slight decrease in GSA at higher ammonium concentrations and the stimulation of GDH in pea plant roots at lower ammonium concentrations were also observed (Frechilla et al., 2002).

Aslam et al. (2001) stated that while NO_3^- uptake system activity was inhibited by amino acids, irrespective of NO_3^- availability, the restriction of NRA induction was due to a reduction in uptake of NO_3^- and as a result NO_3^- availability was reduced. The induction of NO_3^- reduction and uptake systems are normally downregulated by amino acids, whether accumulated internally or given externally to the plant. Since NO_3^- concentration was reduced by amino acids, this suggests that the inadequate NO_3^- causes the inhibition. Aslam et al. (2001) also stated that in roots provided with 10 mM NO_3^- the increase in RNA was not hindered by amino acids. This implies that NO_3^- should always be sufficiently present.

6.5.2 Ultraviolet B Radiation

Increased UV-B solar radiations are believed to harm agricultural plants. UV-B radiation can restrict all main photosynthesis processes, including thylakoid membrane photochemical reactions, enzyme processes in the Calvin cycle, CO2-diffusion limits in the stomata, amino acid residue destruction as well as oxygen-mediated damages in the plant cell membrane to unsaturated fatty acids. Present terrestrial UV-B rays vary from 2 and 12 kJm^{-2} d^{-1} every day with higher doses in the near equator and mid-latitudes, with a rise of 6–14% from the 1980s (Surabhi et al., 2009). Although plants need light for producing food production, higher radiation exposures, in general, pose a threat to plant health.

The UV-B radiation, particularly at high levels, significantly inhibited the uptake and assimilation of essential minerals for plant nitrate, and consequently, the seedling growth was limited. GDH activity was improved to avert ammonia toxicity under lower UV-B radiation levels, when the GS and GOGAT activity declined significantly, resulting in ammonia accumulation in plants, which is toxic at higher levels. Also, the GDH structure or other associated genes may be disrupted under higher levels of UV-B radiation, reducing GDH activity.

Compared with control seedlings, both roots and leaves displayed reduced NRA values after UV-B exposure (Quaggiotti et al., 2004), in line with data recorded for barley (Ghisi et al., 2002) and *Vigna unguiculata* L. (Balakumar et al., 1999), respectively.

6.5.3 Gaseous Factors

The lower NO2 levels do not affect organic N content in plants or organic N concentration in leaves and roots, but the organic N concentration in plant leaves grown at low levels of nitrate was marginally increased. The plant-absorbed nitrogen dioxide can be turned into nitrate and nitrite. As a result, assimilation is carried out. There were no major impacts of exposure to low NO2 levels on per plant nitrate uptake by the plant and the rates of uptake. Exposure to high NO2, however, induces a higher nitrate level (Qiao and Murray, 1998). High rates of CO_2 assimilation in cucumber leaves increased nitrate reduction by inducing activity and synthesis of NR, and the carbohydrates obtained from CO_2 assimilation are believed to serve as positive regulatory metabolites. This is similar in sunflower leaves, in which the expression and GSA have been modulated by CO_2 assimilation rates after short exposures to high atmospheric carbon, and photosynthesized sugars possibly act as regulatory metabolites (Aguera et al., 2006). In addition, CO_2 assimilation can be simultaneously controlled by N assimilation as assimilation of N into glutamate is a very essential sink for redox equivalents of photosynthetic electron flow. In the roots when N is quantitatively assimilated, plants grown in ammonium save up to 11.9% of total growth cost as opposed to plants cultivated in nitrate; in the shoot, it is decreased to 6.1% as a result of the use of photons directly. Assimilation of leaf NO_3 into amino acids is due to the use of carbon skeletons and ATP and (NADPH) reductants produced photosynthetically (Guo et al., 2005).

Also, the higher concentrations of CO_2 (or low O_2) in the atmosphere decrease the photorespiration rates and increase photosynthesis initially and growth rates in most plants by as much as 35% (C_3 plants) (Guo et al., 2005). That is in line with the observation that enrichment of atmospheric CO_2 is not responsible for stimulating wheat plant growth that receives NO_3 as the only source of N to the equal degree as NO_4^+ recipients (Bloom et al., 2010); thus, the conditions that minimize photorespiration (i.e., elevated CO_2 (or low O_2) levels) prevents assimilation of NO_3 in C_3 plant shoots.

6.5.4 Soil Salinity

A significant factor that alters the soil and makes it unfit for agriculture is soil salinity. Cordovilla et al. (1999), studying the impacts of NaCl on N fixation, growth, and assimilation of *Vicia faba* L and *Pisum Sativum* L. plants fertilized and inoculated with KNO_3, observed that Fababean response to salt was normally in line with the findings recorded for *Medicago trunculata* and *Cicer arietinum* and other *V. faba* genotypes in that plants fixing N were more salt-sensitive than plants fertilized with N.

The salt present in the irrigation water can intervene with the use of N by plants, as salt content has recently been shown to be the main factor responsible for low N accessibility (Debouba et al., 2006), also because the activity of nitrate reductase in leaves is heavily dependent on root nitrate flux and is strictly affected by NaCl-induced osmotic shock (Silveira et al., 2001). In both the leaves and roots of tomato (*Lycopersicon esculentum*) seedlings under high salinity, NO3 levels were greatly reduced. A significant nitrate decrease, from 60% and 80%, occurred in the leaves at 100 mM and 50 mM NaCl, respectively. A similar decrease was found in the root for NO_3^- (Debouba et al., 2006).

In legumes, the salt concentration of 50 to 200 mM of NaCl affects the production by interfering with the growth of the plants (Cordovilla et al., 1999). During salt stress there occurs the accumulation of ammonium nitrate and free amino acids and reduces their role in the formation of nitrogen compounds.

ClO_3^- is one of the alarming contaminants present in the waste discharged from industrial waste and disinfectants of drinking water. When it is released in the environment inhibition of nitrate reductase and growth occurs, and the toxic effect is well-built if the ambient nitrate concentration is low (Tischner, 2000).

Plants have developed a well-organized system for absorption and utilization of N. The core processes or molecules involved in these methods, however, remain unclear. Yet more research is needed to determine the roles and dynamics of different environmental factors that work on the ability of the plant to absorb N.

6.5.5 Metals

Because of potential negative ecological impacts, heavy metal contamination of agricultural soils has become a critical environmental issue. Due to their broad occurrence and their chronic and acute negative impact on plants grown in such soils, such toxic elements are called soil contaminants (Yadav, 2010). In addition to being harmful, several studies, depending on the amount of these metals in soil, have reported their chronic and acute negative impacts on the metabolic processes regulating growth and development.

La could facilitate nitrate absorption and transport in plants, which could be attributed to the fact that La has a beneficial impact on seedling transpiration, encouraging the uptake and transportation of water and minerals. Lower amounts of La encouraged GDH activity, but Cd significantly decreased the content of nitrate absorbed from the nutrient solution. The net uptake of nitrate in 50 µM Cd treated plants was nearly 20% of the absorbed nitrate in control plants; the decrease in total uptake of nitrate was greater in 100 µM Cd-treated plants (12% of the nitrate uptake in control plants) (Gouia et al., 2000). In most plant species studied to date, high Al levels explicitly limit NO_3 uptake and thus induce a deficit in N metabolism. Around the same time, Al was found to be able to directly act on the assimilation of NO_3 within the plant, impairing the action of NR. The same is found true for excessive Mo use, which leads to a decrease in the normal growth of plants (Ruiz et al., 2007).

The nutrients are absorbed by plants through the soil, where H^+ ions are released into the soil by root hairs. Cations linked with negatively charged soil particles are displaced by these hydrogen ions so that these cations are accessible for absorption by the roots. Some metals can stall nitrate absorption, thereby affecting NR activity.

6.6 CONCLUSION

In this chapter, we discussed various factors (soil pH, temperature, salinity, nitrogen concentration, etc.) affecting the uptake, transport, and assimilation of nitrogen (N). Plant growth and development is affected by N, one of the major macronutrients. In plants both inorganic and organic forms are metabolized, however, ammonium and nitrate are the only main ionic forms of N that are actively taken up by all the plants from the soil. Some other forms of nitrogen such as urea, nitrite, and amino acids are also metabolized in plants. Plants use distinct pathways for the uptake of N into cells. In roots, initial absorption of organic and mineral N takes place through plasmalemma of cortical, epidermal, and leaf cells. The uptaken N is carried by various transporters via the xylem to the whole body, where the nitrogen is finally assimilated into amino acids.

REFERENCES

Agüera, E., Ruano, D., Cabello, P., & de la Haba, P. (2006). Impact of atmospheric CO_2 on growth, photosynthesis and nitrogen metabolism in cucumber (*Cucumis sativus* L.) plants. *Journal of Plant Physiology*, *163*(8), 809–817.

Ahmad, A., & Abdin, M. Z. (1999). NADH: nitrate reductase and NAD (P) H: nitrate reductase activities in mustard seedlings. *Plant Science*, *143*(1), 1–8.

Ajayi, O., Maynard, D. N., & Barker, A. V. (1970). The effects of potassium on ammonium nutrition of tomato (*Lycopersicon esculentum* Mill.) 1. *Agronomy Journal*, *62*(6), 818–821.

Almagro, A., Lin, S. H., & Tsay, Y. F. (2008). Characterization of the Arabidopsis nitrate transporter NRT1. 6 reveals a role of nitrate in early embryo development. *The Plant Cell*, *20*(12), 3289–3299.

Andrews, M. (1986). The partitioning of nitrate assimilation between root and shoot of higher plants. *Plant, Cell & Environment*, *9*(7), 511–519.

Aslam, M., Travis, R. L., & Rains, D. W. (2001). Differential effect of amino acids on nitrate uptake and reduction systems in barley roots. *Plant Science*, *160*(2), 219–228.

Azcón, R., Gomez-Ortega, M., & Barea, J. M. (1982). Comparative effects of foliar-or soil-applied nitrate on vesicular-arbuscular mycorrhizal infection in maize. *New Phytologist*, *92*(4), 553–559.

Bagshaw, R., Vaidyanathan, L. V., & Nye, P. H. (1972). The supply of nutrient ions by diffusion to plant roots in soil. *Plant and Soil*, *37*(3), 627–639.

Balakumar, T., Sathiameena, K., Selvakumar, V., Ilanchezhian, C. M., & Paliwal, K. (1999). UV-B radiation mediated alterations in the nitrate assimilation pathway of crop plants 2. Kinetic characteristics of nitrite reductase. *Photosynthetica*, *37*(3), 469–475.

Blair, G. J., Miller, M. H., & Mitchell, W. A. (1970). Nitrate and ammonium as sources of nitrogen for corn and their influence on the uptake of other ions 1. *Agronomy Journal*, *62*(4), 530–532.

Bloom, A. J., Burger, M., Asensio, J. S. R., & Cousins, A. B. (2010). Carbon dioxide enrichment inhibits nitrate assimilation in wheat and Arabidopsis. *Science*, *328*(5980), 899–903.

Bowen, G. D. (1974). The uptake and translocation of phosphate by mycelial strands of pine mycorrhizas. *Soil Biology and Biochemistry*, *6*(1), 53–56.

Bowen, G. D., & Smith, S. E. (1981). The effects of mycorrhizas on nitrogen uptake by plants. *Ecological Bulletins*, 237–247.

Bowman, D. C., & Paul, J. L. (1992). Foliar absorption of urea, ammonium, and nitrate by perennial ryegrass turf. *Journal of the American Society for Horticultural Science*, *117*(1), 75–79.

Buczek, J. (1979). Ammonium and potassium effect on nitrate assimilation in cucumber seedlings. *Acta Societatis Botanicorum Poloniae*, *48*(2), 157–169.

Chambers, C. A., Smith, S. E., & Smith, F. A. (1980). Effects of ammonium and nitrate ions on mycorrhizal infection, nodulation and growth of *Trifolium subterraneum*. *New Phytologist*, *85*(1), 47–62.

Chiu, C. C., Lin, C. S., Hsia, A. P., Su, R. C., Lin, H. L., & Tsay, Y. F. (2004). Mutation of a nitrate transporter, AtNRT1: 4, results in a reduced petiole nitrate content and altered leaf development. *Plant and Cell Physiology*, *45*(9), 1139–1148.

Chopin, F., Orsel, M., Dorbe, M. F., Chardon, F., Truong, H. N., Miller, A. J., Krapp, A., & Daniel-Vedele, F. (2007). The Arabidopsis ATNRT2. 7 nitrate transporter controls nitrate content in seeds. *The Plant Cell*, *19*(5), 1590–1602.

Clarkson, D. T., & Warner, A. J. (1979). Relationships between root temperature and the transport of ammonium and nitrate ions by Italian and perennial ryegrass (*Lolium multiflorum* and *Lolium perenne*). *Plant Physiology*, *64*(4), 557–561.

Clement, C. R., Hopper, M. J., Jones, L. H. P., & Leafe, E. L. (1978). The uptake of nitrate by *Lolium perenne* from flowing nutrient solution: II. Effect of light, defoliation, and relationship to CO_2 flux. *Journal of Experimental Botany*, *29*(5), 1173–1183.

Cole, D. W. (1981). Nitrogen uptake and translocation by forest ecosystems. *Ecological Bulletins*, 219–232.

Cordovilla, M. D. P., Ligero, F., & Lluch, C. (1999). Effects of NaCl on growth and nitrogen fixation and assimilation of inoculated and KNO3 fertilized *Vicia faba* L. and *Pisum sativum* L. plants. *Plant science*, *140*(2), 127–136.

Couturier, J., Montanini, B., Martin, F., Brun, A., Blaudez, D., & Chalot, M. (2007). The expanded family of ammonium transporters in the perennial poplar plant. *New Phytologist*, *174*(1), 137–150.

Cox, W. J., & Reisenauer, H. M. (1973). Growth and ion uptake by wheat supplied nitrogen as nitrate, or ammonium, or both. *Plant and Soil*, *38*(2), 363–380.

Daniel-Vedele, F., Filleur, S., & Caboche, M. (1998). Nitrate transport: A key step in nitrate assimilation. *Current Opinion in Plant Biology*, *1*(3), 235–239.

De Angeli, A., Monachello, D., Ephritikhine, G., Frachisse, J. M., Thomine, S., Gambale, F., & Barbier-Brygoo, H. (2006). The nitrate/proton antiporter AtCLCa mediates nitrate accumulation in plant vacuoles. *Nature*, *442*(7105), 939–942.

Debouba, M., Gouia, H., Suzuki, A., & Ghorbel, M. H. (2006). NaCl stress effects on enzymes involved in nitrogen assimilation pathway in tomato "Lycopersicon esculentum" seedlings. *Journal of Plant Physiology*, *163*(12), 1247–1258.

Doddema, H. E. N. K., & Otten, H. (1979). Uptake of nitrate by mutants of Arabidopsis thaliana, disturbed in uptake or reduction of nitrate: III. Regulation. *Physiologia Plantarum*, *45*(3), 339–346.

Edwards, J. H., & Barber, S. A. (1976). Nitrogen uptake characteristics of corn roots at low n concentration as influenced by plant age 1. *Agronomy Journal*, *68*(1), 17–19.

Esposito, S., Guerriero, G., Vona, V., Di Martino Rigano, V., Carfagna, S., & Rigano, C. (2005). Glutamate synthase activities and protein changes in relation to nitrogen nutrition in barley: the dependence on different plastidic glucose-6P dehydrogenase isoforms. *Journal of Experimental Botany*, *56*(409), 55–64.

Filleur, S., Dorbe, M. F., Cerezo, M., Orsel, M., Granier, F., Gojon, A., & Daniel-Vedele, F. (2001). An Arabidopsis T-DNA mutant affected in Nrt2 genes is impaired in nitrate uptake. *FEBS Letters*, *489*(2–3), 220–224.

Forde, B. G. (2002). Local and long-range signaling pathways regulating plant responses to nitrate. *Annual Review of Plant Biology*, *53*(1), 203–224.

Frechilla, S., Lasa, B., Aleu, M., Juanarena, N., Lamsfus, C., & Aparicio-Tejo, P. M. (2002). Short-term ammonium supply stimulates glutamate dehydrogenase activity and alternative pathway respiration in roots of pea plants. *Journal of Plant Physiology*, *159*(8), 811–818.

Frota, J. N. E., & Tucker, T. C. (1972). Temperature influence on ammonium and nitrate absorption by lettuce. *Soil Science Society of America Journal*, *36*(1), 97–100.

Gazzarrini, S., Lejay, L., Gojon, A., Ninnemann, O., Frommer, W. B., & von Wirén, N. (1999). Three functional transporters for constitutive, diurnally regulated, and starvation-induced uptake of ammonium into Arabidopsis roots. *The Plant Cell*, *11*(5), 937–947.

Ghisi, R., Trentin, A. R., Masi, A., & Ferretti, M. (2002). Carbon and nitrogen metabolism in barley plants exposed to UV-B radiation. *Physiologia Plantarum*, *116*(2), 200–205.

Gouia, H., Ghorbal, M. H., & Meyer, C. (2000). Effects of cadmium on activity of nitrate reductase and on other enzymes of the nitrate assimilation pathway in bean. *Plant Physiology and Biochemistry*, *38*(7–8), 629–638.

Grasmanis, V. O., & Nicholas, D. J. D. (1971). Annual uptake and distribution of N 15-labelled ammonia and nitrate in young Jonathan/MM104 apple trees grown in solution cultures. *Plant and Soil*, *35*(1–3), 95–112.

Guo, S., Schinner, K., Sattelmacher, B., & Hansen U. (2005). Different apparent CO_2 compensation points in nitrate- and ammonium-grown Phaseolus vulgaris and the relationship to non-photorespiratory CO_2 evolution. *Physiologia Plantarum*, *123*, 288–301.

Haynes, R. J. (1980). Ion exchange properties of roots and ionic interactions within the root apoplasm: Their role in ion accumulation by plants. *The Botanical Review*, *46*(1), 75–99.

Haynes, R. J., & Goh, K. M. (1978). Ammonium and nitrate nutrition of plants. *Biological Reviews*, *53*(4), 465–510.

Haynes, R. J., & Goh, K. M. (1978). Ammonium and nitrate nutrition of plants. *Biological Reviews*, *53*(4), 465–510.

Heimer, Y. M., & Filner, P. (1971). Regulation of the nitrate assimilation pathway in cultured tobacco cells: III. The nitrate uptake system. *Biochimica et Biophysica Acta (BBA)-General Subjects*, *230*(2), 362–372.

Hiatt, A. J., & Leggett, J. E. (1974). Ionic interactions and antagonisms in plants. In: E.W. Carson (Ed.), *The Plant Root and its Environment*. University Press of Virginia, Charlottesville, pp. 101–134.

Hirner, A., Ladwig, F., Stransky, H., Okumoto, S., Keinath, M., Harms, A., Frommer, W.B. & Koch, W. (2006). Arabidopsis LHT1 is a high-affinity transporter for cellular amino acid uptake in both root epidermis and leaf mesophyll. *The Plant Cell*, *18*(8), 1931–1946.

Ho, C. H., Lin, S. H., Hu, H. C., & Tsay, Y. F. (2009). CHL1 functions as a nitrate sensor in plants. *Cell*, *138*(6), 1184–1194.

Huang, N. C., Liu, K. H., Lo, H. J., & Tsay, Y. F. (1999). Cloning and functional characterization of an Arabidopsis nitrate transporter gene that encodes a constitutive component of low-affinity uptake. *The Plant Cell*, *11*(8), 1381–1392.

Jackson, W. A. (1978). CRITIQUE-OF "Factors influencing nitrate acquisition by plants: assimilation and fate of reduced nitrogen". In *Soil–Plant–Nitrogen Relationships* (pp. 45–88). Academic Press.

Jackson, W. A., & Williams, D. C. (1968). Nitrate-stimulated uptake and transport of strontium and other cations. *Soil Science Society of America Journal*, *32*(5), 698–704.

Jackson, W. A., Flesher, D., & Hageman, R. H. (1973). Nitrate uptake by dark-grown corn seedlings: some characteristics of apparent induction. *Plant Physiology*, *51*(1), 120–127.

Jackson, W. A., Johnson, R. E., & Volk, R. J. (1974). Nitrite uptake patterns in wheat seedlings as influenced by nitrate and ammonium. *Physiologia Plantarum*, *32*(2), 108–114.

Jackson, W. A., Kwik, K. D., & Volk, R. J. (1976). Nitrate uptake during recovery from nitrogen deficiency. *Physiologia Plantarum*, *36*(2), 174–181.

Jones, D. L., Owen, A. G., & Farrar, J. F. (2002). Simple method to enable the high resolution determination of total free amino acids in soil solutions and soil extracts. *Soil Biology and Biochemistry*, *34*(12), 1893–1902.

Kafkafi, U., Valoras, N., & Letey, J. (1982). Chloride interaction with nitrate and phosphate nutrition in tomato (*Lycopersicon esculentum* L.). *Journal of Plant Nutrition*, *5*(12), 1369–1385.

Krapp, A. (2015). Plant nitrogen assimilation and its regulation: a complex puzzle with missing pieces. *Current Opinion in Plant Biology*, *25*, 115–122.

Kruckelmann, H. W. (1975). Effects of fertilizers, soils, soil tillage, and plant species on the frequency of Endogone chlamydospores and mycorrhizal infection in arable soils. In: F.E. Sanders, B. Mosse and P.B. Tinker (eds.), *Endomycorrhizas. Proceedings of a Symposium*. Academic Press: London, pp. 511–525.

Lee, Y. H., Foster, J., Chen, J., Voll, L. M., Weber, A. P., & Tegeder, M. (2007). AAP1 transports uncharged amino acids into roots of Arabidopsis. *The Plant Journal*, *50*(2), 305–319.

Lin, S. H., Kuo, H. F., Canivenc, G., Lin, C. S., Lepetit, M., Hsu, P. K., Tillard, P., Lin, H.L., Wang, Y.Y., Tsai, C.B. and Gojon, A., & Tsay, Y. F. (2008). Mutation of the Arabidopsis NRT1. 5 nitrate transporter causes defective root-to-shoot nitrate transport. *The Plant Cell*, *20*(9), 2514–2528.

Liu, K. H., Huang, C. Y., & Tsay, Y. F. (1999). CHL1 is a dual-affinity nitrate transporter of Arabidopsis involved in multiple phases of nitrate uptake. *The Plant Cell*, *11*(5), 865–874.

Loqué, D., Ludewig, U., Yuan, L., & von Wirén, N. (2005). Tonoplast intrinsic proteins AtTIP2; 1 and AtTIP2; 3 facilitate NH3 transport into the vacuole. *Plant physiology*, *137*(2), 671–680.

Ludewig, U., Neuhäuser, B., & Dynowski, M. (2007). Molecular mechanisms of ammonium transport and accumulation in plants. *FEBS Letters*, *581*(12), 2301–2308.

Maas, E. V. (1969). Calcium uptake by excised maize roots and interactions with alkali cations. *Plant Physiology*, *44*(7), 985–989.

Maathuis, F. J. (2009). Physiological functions of mineral macronutrients. *Current Opinion in Plant Biology*, *12*(3), 250–258.

MacKown, C. T., Jackson, W. A., & Volk, R. J. (1982). Restricted nitrate influx and reduction in corn seedlings exposed to ammonium. *Plant Physiology*, *69*(2), 353–359.

Maighany, F., & Ebrahimzadeh, H. (2004). Intervarietal differences in nitrogen content and nitrate assimilation in wheat (*Triticum aestivum* L.) under salt stress. *Pakistan Journal of Botany*, *36*(1), 31–40.

Márquez, A. J., Betti, M., García-Calderón, M, Credali, A., Díaz, P., Borsani, O., & Monza, J. (2007). Nitrogen metabolism in *Lotus japonicus* and the relationship with drought stress. *Lotus Newsletter*, 99.

Marschner, H. (1995). *Mineral Nutrition of Higher Plants* (No. BOOK). Academic Press.

Maynard, D. N., & Barker, A. V. (1969). Studies on the tolerance of plants to ammonium nutrition. *Journal of the American Society for Horticultural Science*, *94*, 235–239.

Melin, E., & Nilsson, H. (1952). Transport of labelled nitrogen from an ammonium source to pine seedlings through mycorrhizal mycelium. *Svensk Botanisk Tidskrift*, *46*, 281–285.

Michael, G., Martin, P., & Owassia, I. (1970). The uptake of ammonium and nitrate from labelled ammonium nitrate in relation to the carbohydrate supply of the roots. *Zeitsch rift für Planzenernahrung, Dungen und Bodenkunde*, *110*, 225–238.

Miller, A. J., & Cramer, M. D. (2005). Root nitrogen acquisition and assimilation. In *Root Physiology: From Gene to Function* (pp. 1–36). Springer.

Miller, A. J., Fan, X., Orsel, M., Smith, S. J., & Wells, D. M. (2007). Nitrate transport and signalling. *Journal of Experimental Botany*, *58*(9), 2297–2306.

Miller, M. H., Mamaril, C. P., & Blair, G. J. (1970). Ammonium effects on phosphorus absorption through pH changes and phosphorus precipitation at the soil-root interface 1. *Agronomy Journal*, *62*(4), 524–527.

Minotti, P. L., Williams, D. C., & Jackson, W. A. (1968). Nitrate uptake and reduction as affected by calcium and potassium. *Soil Science Society of America Journal*, *32*(5), 692–698.

Minotti, P. L., Williams, D. C., & Jackson, W. A. (1969). Nitrate uptake by wheat as influenced by ammonium and other cations 1. *Crop Science*, *9*(1), 9–14.

Moraghan, J. T., & Porter, O. A. (1975). Maize growth as affected by root temperature and form of nitrogen. *Plant and Soil*, *43*(1–3), 479–487.

Munn, D. A., & Jackson, W. A. (1978). Nitrate and ammonium uptake by rooted cuttings of sweet potato 1. *Agronomy Journal*, *70*(2), 312–316.

Näsholm, T., Kielland, K., & Ganeteg, U. (2009). Uptake of organic nitrogen by plants. *New Phytologist*, *182*(1), 31–48.

Neyra, C. A., & Hageman, R. H. (1975). Nitrate uptake and induction of nitrate reductase in excised corn roots. *Plant Physiology*, *56*(5), 692–695.

Nye, P. H. (1981). Changes of pH across the rhizosphere induced by roots. *Plant and Soil*, *61*(1–2), 7–26.

Nye, P. H., & Tinker, P. B. (1977). *Solute Movement in the Soil–Root System* (Vol. 4). California: University of California Press.

Orsel, M., Chopin, F., Leleu, O., Smith, S. J., Krapp, A., Daniel-Vedele, F., & Miller, A. J. (2006). Characterization of a two-component high-affinity nitrate uptake system in Arabidopsis. Physiology and protein-protein interaction. *Plant Physiology*, *142*(3), 1304–1317.

Orsel, M., Eulenburg, K., Krapp, A., & Daniel-Vedele, F. (2004). Disruption of the nitrate transporter genes AtNRT2. 1 and AtNRT2. 2 restricts growth at low external nitrate concentration. *Planta*, *219*(4), 714–721.

Orsel, M., Filleur, S., Fraisier, V., et al. (2002) Nitrate transport in plants: Which gene and which control? *Journal of Experimental Botany*, *53*, 825–833.

Parsons, J. W., & Tinsley, J. (1975). Nitrogenous substances. In *Soil Components* (pp. 263–304). Springer.

Pate, J. S. (1989). Synthesis, transport, and utilization of products of symbiotic nitrogen fixation. In *Plant Nitrogen Metabolism* (pp. 65–115). Springer.

Pate, J. S., & Layzell, D. B. (1990). Energetics and biological costs of nitrogen assimilation. *The Biochemistry of Plants*, *16*, 1–42.

Pearson, C. J., & Steer, B. T. (1977). Daily changes in nitrate uptake and metabolism in Capsicum annuum. *Planta*, *137*(2), 107–112.

Pearson, V., & Read, D. J. (1975). The physiology of the mycorrhizal endophyte of Calluna vulgaris. *Transactions of the British Mycological Society*, *64*(1), 1–7.

Pierre, W. H., & Banwart, W. L. (1973). Excess-base and excess-base/nitrogen ratio of various crop species and parts of plants 1. *Agronomy Journal*, *65*(1), 91–96.

Pierre, W. H., Meisinger, J., & Birchett, J. R. (1970). Cation-anion balance in crops as a factor in determining the effect of nitrogen fertilizers on soil acidity 1. *Agronomy Journal*, *62*(1), 106–112.

Possingham, J. V., & Groot Obbink, J. (1971). Endotrophic mycorrhiza and the nutrition of grape vines. *Vitis 10*, 120–130.

Qiao, Z., & Murray, F. (1998). The effects of NO_2 on the uptake and assimilation of nitrate by soybean plants. *Environmental and Experimental Botany*, *39*(1), 33–40.

Quaggiotti, S., Trentin, A. R., Dalla Vecchia, F., & Ghisi, R. (2004). Response of maize (*Zea mays* L.) nitrate reductase to UV-B radiation. *Plant Sience*, *167*(1), 107–116.

Rao, K. P., & Rains, D. W. (1976). Nitrate absorption by barley: I. Kinetics and energetics. *Plant Physiology*, *57*(1), 55–58.

Rayar, A. J., & Van Hai, T. (1977). Effect of ammonium on uptake of phosphorus, potassium, calcium and magnesium by intact soybean plants. *Plant and Soil*, *48*(1), 81–87.

Read, D. J., & Stribley, D. P. (1975). Some mycological aspects of the biology of mycorrhiza in the Ericaceae. In *Endomycorrhizas; Proceedings of a Symposium*.

Reisenauer, H. M. (1978). Absorption and utilization of ammonium nitrogen by plants. In *Soil–Plant–Nitrogen Relationships* (pp. 157–170). Academic Press.

Rentsch, D., Schmidt, S., & Tegeder, M. (2007). Transporters for uptake and allocation of organic nitrogen compounds in plants. *FEBS Letters*, *581*(12), 2281–2289.

Rhodes, L. H., & Gerdemann, J. W. (1975). Phosphate uptake zones of mycorrhizal and non-mycorrhizal onions. *New Phytologist*, *75*(3), 555–561.

Riley, D., & Barber, S. A. (1971). Effect of ammonium and nitrate fertilization on phosphorus uptake as related to root-induced pH changes at the root-soil interface. *Soil Science Society of America Journal*, *35*(2), 301–306.

Rosales, E. P., Iannone, M. F., Groppa, M. D., & Benavides, M. P. (2011). Nitric oxide inhibits nitrate reductase activity in wheat leaves. *Plant Physiology and Biochemistry*, *49*(2), 124–130.

Rufty Jr, T. W., Jackson, W. A., & Raper Jr, C. D. (1982). Inhibition of nitrate assimilation in roots in the presence of ammonium: the moderating influence of potassium. *Journal of Experimental Botany*, *33*(6), 1122–1137.

Rui, C., Huang, X. H., Qing, Z. H. O. U., & Cheng, X. Y. (2007). Effects of lanthanum (III) on nitrogen metabolism of soybean seedlings under elevated UV-B radiation. *Journal of Environmental Sciences*, *19*(11), 1361–1366.

Ruiz, J. M., Rivero, R. M., & Romero, L. (2007). Comparative effect of Al, Se, and Mo toxicity on NO_3^- assimilation in sunflower (*Helianthus annuus* L.) plants. *Journal of Environmental Management*, *83*(2), 207–212.

Sasakawa, H., & Yamamoto, Y. (1978). Comparison of the uptake of nitrate and ammonium by rice seedlings: influences of light, temperature, oxygen concentration, exogenous sucrose, and metabolic inhibitors. *Plant Physiology*, *62*(4), 665–669.

Schrader, L. E., Domska, D., Jung Jr, P. E., & Peterson, L. A. (1972). Uptake and Assimilation of Ammonium-N and Nitrate-N and their Influence on the growth of corn (*Zea mays* L.) 1. *Agronomy Journal*, *64*(5), 690–695.

Sheat, D. E. G., Fletcher, B. H., & Street, H. E. (1959). Studies on the growth of excised roots: VIII. The growth of excised tomato roots supplied with various inorganic sources of nitrogen. *New Phytologist*, *58*(2), 128–141.

Shi-Wei, G., Yi, Z., Ying-Xu, G., Yong, L., & Qi-Rong, S. (2007). New insights into the nitrogen form effect on photosynthesis and photorespiration. *Pedosphere*, *17*(5), 601–610.

Silveira, J. A. G., Melo, A. R. B., Viégas, R. A., & Oliveira, J. T. A. (2001). Salinity-induced effects on nitrogen assimilation related to growth in cowpea plants. *Environmental and Experimental Botany*, *46*(2), 171–179.

Smiley, R. W. (1974). Rhizosphere pH as influenced by plants, soils, and nitrogen fertilizers. *Soil Science Society of America Journal*, *38*(5), 795–799.

Smith, S. S. (1980). Mycorrhizas of autotrophic higher plants. *Biological Reviews*, *55*(4), 475–510.

Sonoda, Y., Ikeda, A., Saiki, S., Yamaya, T., & Yamaguchi, J. (2003). Feedback regulation of the ammonium transporter gene family AMT1 by glutamine in rice. *Plant and Cell Physiology*, *44*(12), 1396–1402.

Soon, Y. K., & Miller, M. H. (1977). Changes in the rhizosphere due to NH_4^+ and NO_3^- fertilization and phosphorus uptake by corn seedlings (*Zea mays* L.). *Soil Science Society of America Journal*, *41*(1), 77–80.

Stribley, D. P., & Read, D. J. (1974). The biology of mycorrhiza in the ericaceae iv. the effect of mycorrhizal infection on uptake of 15n from labelled soil by vaccinium macrocarpon ait. *New Phytologist*, *73*(6), 1149–1155.

Stribley, D. P., & Read, D. J. (1975). Some nutritional aspects of the biology of ericaceous mycorrhizas. In *Endomycorrhizas; Proceedings of a Symposium*.

Stribley, D. P., & Read, D. J. (1976). The biology of mycorrhiza in the Ericaceae: VI. The effects of mycorrhizal infection and concentration of ammonium nitrogen on growth of cranberry (*Vaccinium macrocarpon* Ait.) in sand culture. *New Phytologist*, *77*(1), 63–72.

Stribley, D. P., & Read, D. J. (1980). The biology of mycorrhiza in the Ericaceae: VII. The relationship between mycorrhizal infection and the capacity to utilize simple and complex organic nitrogen sources. *New Phytologist*, *86*(4), 365–371.

Surabhi, G. K., Reddy, K. R., & Singh, S. K. (2009). Photosynthesis, fluorescence, shoot biomass and seed weight responses of three cowpea (*Vigna unguiculata* (L.) Walp.) cultivars with contrasting sensitivity to UV-B radiation. *Environmental and Experimental Botany*, *66*(2), 160–171.

Svennerstam, H., Ganeteg, U., & Näsholm, T. (2008). Root uptake of cationic amino acids by Arabidopsis depends on functional expression of amino acid permease 5. *New Phytologist*, *180*(3), 620–630.

Svennerstam, H., Ganeteg, U., Bellini, C., & Näsholm, T. (2007). Comprehensive screening of Arabidopsis mutants suggests the lysine histidine transporter 1 to be involved in plant uptake of amino acids. *Plant Physiology*, *143*(4), 1853–1860.

Ta, T. C., & Ohira, K. (1981). Effects of various environmental and medium conditions on the response of Indica and Japonica rice plants to ammonium and nitrate nitrogen. *Soil Science and Plant Nutrition*, *27*(3), 347–355.

Temple, S. J., Vance, C. P., & Gantt, J. S. (1998). Glutamate synthase and nitrogen assimilation. *Trends in Plant Science*, *3*(2), 51–56.

Tischner, R. (2000). Nitrate uptake and reduction in higher and lower plants. *Plant, Cell & Environment*, *23*(10), 1005–1024.

Tsay, Y. F., Chiu, C. C., Tsai, C. B., Ho, C. H., & Hsu, P. K. (2007). Nitrate transporters and peptide transporters. *FEBS Letters*, *581*(12), 2290–2300.

Tsay, Y. F., Schroeder, J. I., Feldmann, K. A., & Crawford, N. M. (1993). The herbicide sensitivity gene CHL1 of Arabidopsis encodes a nitrate-inducible nitrate transporter. *Cell*, *72*(5), 705–713.

Van Egmond, F. (1978). Nitrogen nutritional aspects of the ionic balance of plants. *Nitrogen in the Environment*, *2*, 171–189.

Wang, R., Liu, D., & Crawford, N. M. (1998). The Arabidopsis CHL1 protein plays a major role in high-affinity nitrate uptake. *Proceedings of the National Academy of Sciences*, *95*(25), 15134–15139.

Wang, W. H., Köhler, B., Cao, F. Q., & Liu, L. H. (2008). Molecular and physiological aspects of urea transport in higher plants. *Plant Science*, *175*(4), 467–477.

Weigel Jr, R. C., Schillinger, J. A., McCaw, B. A., Gauch, H. G., & Hsiao, E. (1973). Nutrient-nitrate levels and the accumulation of chloride in leaves of snap beans and roots of soybeans 1. *Crop Science*, *13*(4), 411–412.

Wickert, E., Marcondes, J., Lemos, M. V., & Lemos, E. G. (2007). Nitrogen assimilation in Citrus based on CitEST data mining. *Genetics and Molecular Biology*, *30*(3), 810–818.

Williams, L. E., & Miller, A. J. (2001). Transporters responsible for the uptake and partitioning of nitrogenous solutes. *Annual Review of Plant Biology*, *52*(1), 659–688.

Witte, C. P. (2011). Urea metabolism in plants. *Plant Science*, *180*(3), 431–438.

Yadav, S. K. (2010). Heavy metals toxicity in plants: an overview on the role of glutathione and phytochelatins in heavy metal stress tolerance of plants. *South African Journal of Botany*, *76*(2), 167–179.

Youngdahl, L. J., Pacheco, R., Street, J. J., & Vlek, P. L. G. (1982). The kinetics of ammonium and nitrate uptake by young rice plants. *Plant and Soil*, *69*(2), 225–232.

Yuan, L., Loqué, D., Ye, F., Frommer, W. B., & von Wirén, N. (2007). Nitrogen-dependent posttranscriptional regulation of the ammonium transporter AtAMT1; 1. *Plant Physiology*, *143*(2), 732–744.

Yun, C. A. O., Xiao-Rong, F. A. N., Shu-Bin, S. U. N., Guo-Hua, X. U., Jiang, H. U., & Qi-Rong, S. H. E. N. (2008). Effect of nitrate on activities and transcript levels of nitrate reductase and glutamine synthetase in rice. *Pedosphere*, *18*(5), 664–673.

7 Role of Nitrogen in Photosynthesis

Afrozah Hassan[1], Shabana Gulzar[2], and Irshad Ahmad Nawchoo[1]

[1]Plant Reproductive Biology, Genetic Diversity, and Phytochemistry Research Laboratory, Department of Botany, University of Kashmir, Srinagar, India
[2]Govt Degree College for Women Cluster University, Srinagar, India
Corresponding author: malikaafreen6@gmail.com

CONTENTS

7.1 Introduction 87
7.2 Role of Nitrogen to Scale up Photosynthesis 87
7.3 Molecular Approach 89
7.4 Conclusion 91
Acknowledgments 91
References 91

7.1 INTRODUCTION

All of the vital processes in plants are correlated with proteins, of which nitrogen is an important constituent; hence nitrogen occupies a prominent place in plant metabolism. Generally, nitrogen is a limiting factor to plant growth and therefore considerable attention has been paid to the costs and benefits of nitrogen acquisition and uses at the leaf, plant, and stand level (Chapin et al.,, 1987). Nitrogen plays a significant role in various physiological processes. It imparts dark-green color in plants and stimulates vegetative growth and development. It also promotes root growth. Nitrogen enhances the growth of leafy vegetables, produces rapid early growth, improves fruit quality, and increases the protein content of fodder crops. It helps in the uptake and utilization of nutrients like potassium and phosphorous and controls the overall growth of the plant (Hemerly, 2016).

Nitrogen is widely utilized from roots and leaves, and hence increases the leaf size and color of spinach (Wang et al., 2008). It is a vital constituent of nucleic acids, proteins, chlorophyll, and growth hormones (Baker et al., 1974). Nitrogen increases the greenness of plants, CO_2 assimilation rate, and crop yield and quality and improves resistance to various environmental stresses such as limited water availability and saline soil conditions (Yousuf et al., 2016; Bondada et al., 1996; Chen et al., 2010). For successful crop production application of nitrogen is more important than the other major essential nutrients and has revealed promising results at a rate of 120 kg ha-1. It helps in increasing plant height of rice, number of tillers, dry weight, length of panicle, number of filled grains, straw yield, biological yield harvest index, benefit-cost ratio, and grain yield (Malik et al., 2014). To increase crop production application of nitrogen is necessary (Massignam et al., 2009).

7.2 ROLE OF NITROGEN TO SCALE UP PHOTOSYNTHESIS

Photosynthesis is a process in which light energy is captured and stored by a series of reactions that convert light energy into the free energy needed to power life (Blankenship, 2014). Photosynthesis provides raw materials for all plant products and hence is essential for the production of food and fiber (Evans, 1993). The distribution of nitrogen varies from one plant organelle to another. Total nitrogen distribution in the chloroplast is 75%, mitochondria 5%, peroxisomes 2.5%, cytosol 7.5%, and cell walls 10% (Makino and Osmon, 1991; Li et al., 2017; Onoda et al., 2017). In C3 plants, nitrogen distribution within leaves has been reported in spinach, rice, and Arabidopsis (Li et al., 2017; Zhong et al., 2019).

Nitrogen has a close relationship with stomatal conductance and movement. Stomatal conductance is affected by a series of environmental factors including nutrient status, light, water, CO2 levels, and temperature. Nitrate, as the main source of nitrogen for plants, could regulate stomatal movements such as CLCa as a NO3 –/H+ exchanger is localized to the vacuole membrane and can specifically accumulate nitrate (De Angeli et al., 2006). However, stomatal opening is reduced in NRT1.1 mutant lacking the plasma membrane nitrate/H+ co-transporter (Guo et al., 2003). The presence of nitrate promotes the opening of major anion channel SLAC1 and thus drives stomatal closure (Müller et al., 2017).

Many crop scientists have believed that enhancing photosynthesis at the level of the single leaf would increase yield because greater than 90% of crop biomass is derived from photosynthetic products. Therefore, by enhancing nitrogen production leaf photosynthesis can be increased, and hence photosynthesis and nitrogen have been regarded as revenue and cost in leaf economics (Mooney and Gulmon, 1979). Equivalent to 75% of leaf nitrogen is found in the chloroplasts (Hak et al., 1993) and most of it is invested in ribulose bisphosphate carboxylase alone. Therefore, lower rates of photosynthesis in conditions of nitrogen limitation are often ascribed to the reduction in chlorophyll content and Rubisco activity (Toth et al., 2002; Verhoeven et al., 1997).

The efficacy with which plants use nitrogen and water in photosynthesis varies significantly among species, over time, and among leaves for a given species. These differences are important in defining the distribution, dispersal, and survival of species, particularly in a rapidly changing climate. In agriculture, the sensitivity of plant growth to nitrogen fertilization is of great importance. For example, nitrogen deficiency reduces leaf production, individual leaf area, and total leaf area (Vos et al., 1992).

For many C3 and C4 species an increase in nitrogen availability results in higher leaf nitrogen content, which results in a strong positive correlation between photosynthesis and leaf nitrogen content (Connor et al., 1993). Moreover, the specific leaf nitrogen (SLN) content positively affects photosynthesis, which is partially related to nitrogen (N) partitioning in photosynthetic enzymes, the size, number, composition of chloroplasts, and pigment content. Leaf senescence is also influenced by nitrogen and is correlated to a decline in photosynthetic capacity (Allison et al., 1997). Moreover, the photosynthetic proteins of plastids are extensively degraded in an early phase of senescence compared to other proteins (Daubresse et al., 2010; Hortensteiner, 2002). Therefore, for both annual and perennial plants, this process contributes to nitrogen remobilization from senescent leaves to growing organs and seeds (Cliquet et al., 1990; Maillard et al., 2015).

The most significant features of high-light-grown leaves in contrast with lowlight ones are (i) a higher chlorophyll a: b ratio; (ii) less chlorophyll per unit nitrogen; (iii) an increased electron transport capacity per unit chlorophyll; and (iv) a slightly greater ratio of electron transport capacity to Rubisco activity. Hence, nitrogen partitioning within leaves changes with growth irradiance in such a way that it almost maximizes photosynthesis (Evans, 1989). In both natural and agricultural environments, nitrogen nutrition plays a crucial role in determining plant photosynthetic capacity (Abrol, 1993; Abrol et al., 1999).

The association between the constituents of the photosynthetic system may change over the range of nitrogen content, reflecting the adaptation of the photosynthetic system. Leaf nitrogen affects the morphology (Figure 7.1) and size of chloroplasts. The abundance of nitrogen increases the number of chloroplasts per mesophyll cell and their cross-sectional area and length compared to nitrogen-deficient chloroplasts, which have to some extent more thylakoid membrane but lower stromal volume (Sivasankar et al., 1998).

FIGURE 7.1 Increase in nitrogen content increases plant size and structure.

Similarly, the density of protein (predominantly Rubisco) in the stroma is greater with a high nitrogen supply (Kutik et al., 1995). Nitrogen deficiency brings equivalent losses in the LHC, reaction centers, the plastoquinone pool, and cytochrome 'f' (Leong and Anderson, 1984). The capacity for nitrogen fixation exclusively depends upon the nitrogenase enzyme system, which hydrolyzes 16 ATPs per N2 fixed, and hence carries out one of the most metabolically expensive processes in biology (Simpson and Burris, 1984).

The quantitative increase in yield due to nitrogen acquisition by plants revealed that nitrogen is essential for new cell formation. The central process of cell division needs a sustained supply of nitrogen to meristematic centers for the formation of proteins and nucleic acids. Therefore, nitrogen drives vegetative development and growth (Wann et al., 1979, Thornley et al., 1976). Nitrogen is also essential for reproductive development, because with a limited nitrogen supply there is diminished seed set, increased flower abortion, and decreased seed growth. Bloom et al. (1989) proposed that nitrate assimilation up to 25% influenced either photosynthetic or mitochondrial electron transport capacity.

Efficient use of nitrogen is supposed to contribute to the fitness of the plant (Field and Mooney, 1986; Aerts and Chapin, 1999). Photosynthetic nitrogen-use efficiency (PNUE), which is termed as the photosynthetic capacity per unit nitrogen, decreases with decreasing nitrogen content (Field and Mooney, 1986; Hikosaka et al., 1998). To mitigate the problem of food security in China, chemical nitrogen input is the major element for the continuous increase of food production (Zhu et al., 2002). Therefore, the low NUE all over the world especially in the agriculture sector is not only a wastage of resources but also problematic for environmental pollution and inconsistent with sustainable agricultural productivity (Kwong et al., 2002; Wang et al., 2017; Xing et al., 2002). Hence, leaf photosynthesis rate depends on both nitrogen nutrition status and environmental conditions (Sinclair and Horie, 1989).

7.3 MOLECULAR APPROACH

The seedling growth in Arabidopsis is regulated by carbon: Nitrogen availability, storage lipid mobilization, ribulose bisphosphate carboxylase (small subunit gene transcript levels), and photosynthetic gene expression (Martin et al., 2002). Therefore, the carbon-to-nitrogen ratio more than carbohydrate status alone appears to play the predominant role in regulating various aspects of seedling growth, including storage reserve mobilization and photosynthetic gene expression. Sekhar et al. (2020) reported that under fast-changing global climate conditions and multiple gene-editing approaches, improved stomatal and mesophyll conductance and NUE for enhanced crop productivity are key events in photosynthetic processes. This would assist in developing the best strategy to generate resilient crop plants with improved productivity under a fast-changing climate. Thus, the availability of nitrogen is an essential environmental parameter influencing seedling growth and development.

Paul and Stitt (1993) reported that the growth of *Nicotiana tabacum* seedlings under nitrogen-limiting conditions results in a dramatic redirection of biomass allocation to roots vs. shoots. The GATA transcription factor GNC plays an important role in photosynthesis and growth in poplar trees. In light of phytohormone development, and stress responses, the predicted cis-elements suggested potential roles of poplar GATA genes. A poplar GATA gene, PdGATA19/PdGNC (GATA nitrate-inducible carbon-metabolism-involved), was identified from a fast-growing poplar clone. PdGNC expression was significantly upregulated in leaves under both high (50 mM) and low (0.2 mM) nitrate concentrations. Transcriptomic analysis revealed that PdGNC was involved in photosynthetic electron transfer and carbon assimilation in the leaf, cell division and carbohydrate utilization in the stem, and nitrogen uptake in the root. This showed that PdGNC plays a crucial role in plant growth and is potentially useful in tree molecular breeding (An et al., 2020).

In crops, grain number is linearly correlated with total plant nitrogen content. This is because nitrogen is an important resource, both limiting yield and contributing to the determination of grain number (Sinclair and Jamieson, 2006). Yoshida et al. (2006) reported that the grain number per unit of plant nitrogen content is higher in semi-dwarf indica rice genotypes than in japonica rice genotypes; because the semi-dwarf indica rice genotypes preferentially tend to differentiate grains on secondary and tertiary rachis branches. These traits may be governed by the indica Gn1 allele that increases grain number (Ashikari et al., 2005). Tree species with different requirements for PFD show that shade-tolerant species may have denser canopies than sun-demanding species because of smaller amounts of non-photosynthetic structural nitrogen or supporting tissue in their leaves (Kull and Jarvis, 1995).

Makino (2011) studied nitrogen utilization, photosynthesis, and grain yield in Rice and xWheat1 and reported that the semi-dwarf cultivars can use large inputs of nitrogen fertilizer without lodging. The introduction of dwarfing genes permitted the production of varieties with high leaf nitrogen content and improved sink capacity. Therefore, large inputs of nitrogen fertilizer have drawn much attention to the environmental impact of nitrogen fertilization practices (Cassman et al., 1998). Cechin and Fumis (2004) studied the effects of nitrogen availability on growth and photosynthesis in plants of sunflower (*Helianthus annuus* L., var. CATISSOL-01) grown in the greenhouse under natural photoperiod. The plants were grown in vermiculite under contrasting nitrogen supply, with nitrogen supplied as ammonium nitrate and reported that higher nitrogen concentration resulted in higher shoot dry matter production per plant and the effect was apparent from 29 days after sowing.

Da Matta et al. (2002) studied the limitations to photosynthesis in *Coffea canephora* as a result of nitrogen and water availability and reported that photosynthetic NUE was considerably low. Limited nitrogen but not water slightly decreased the maximum photochemical efficiency of photosystem II (PSII). Under continuous irrigation, limited nitrogen plants had a smaller quantum yield of electron transport (ϕ_{PSII}) through slight decreases of photochemical quenching (QP). Various plant molecules such as amino acids, chlorophyll, nucleic acids, ATP, and phytohormones, which contain nitrogen as a structural part, are necessary to complete the biological processes, involving carbon and nitrogen metabolisms, photosynthesis, and protein production (Frink et al., 1999; Crawford and Forde, 2002).

Many critical candidate genes also have been overexpressed and knocked out to test for biomass and plant nitrogen status. Nitrate influx increased due to over-expression of HATS-like NRT2.1 but at the same time, NUE and its utilization phenotypically remain unchanged (Olson et al., 1979). Based on the transcriptomic profiling Asparagine has crucial importance for nitrogen uptake in roots and is considered an ideal nitrogen-transporting molecule (Todd et al., 2008; Curci et al., 2017)

According to Curci et al. (2018) genes encoding asparagine were downregulated in leaves and roots of durum wheat under limited nitrogen (Wan et al., 2017). It has been observed that genes were downregulated in roots and leaves that were involved in carbon, nitrogen, amino acid metabolisms, and photosynthetic activity for plants grown under nitrogen-free conditions (Gelli et al., 2014). Genome and transcript sequences are comprised of long strings of nucleotide monomers (A, C, G, and T/U) that require different quantities of nitrogen atoms for biosynthesis. The strength of selection acting on transcript nitrogen content is influenced by the number of nitrogen plants required to conduct photosynthesis.

Specifically, plants that require more nitrogen to conduct photosynthesis experience stronger selection on transcript sequences to use synonymous codons that cost less nitrogen to biosynthesize. Plants that use the C4 photosynthetic pathway exhibit higher NUE when compared with plants that use C3 photosynthesis. The cohort of changes that facilitated C4 evolution enabled plants to reduce resource allocation to photosynthetic machinery without causing a corresponding reduction in photosynthetic rate. Thus, C4 plants can achieve 50% higher rates of photosynthesis than C3 plants given the same amount of nitrogen (Evans and von Caemmerer, 2000).

Meinzer and Zhu (1998) observed that photosynthesis increases linearly with the increase in leaf nitrogen content in sugarcane. The remobilization of nitrogen from leaves to stalks during the reproductive stage causes a reduction in photosynthesis (Park et al., 2005). More recently, biochemical induction of photosynthesis in response to changing irradiance was faster under high nitrogen conditions in rice (Sun et al.,, 2016).

Nitrogen stimulated leaf growth through the synthesis of proteins involved in cell growth, cell division, cell wall, and cytoskeleton synthesis (Lawlor et al., 2002). The photosynthetic capacity can be predicted by combining nitrogen content and the leaf dry mass, both expressed per unit leaf area. This has proved a useful way of parameterizing photosynthesis over the large areas of a natural ecosystem that is necessary for global models (Rogers et al., 2017). In the relation between photosynthesis and leaf nitrogen content, the species vary. An increase in carbon gain per unit of photosynthetic nitrogen could free up nitrogen for investment in new tissues elsewhere and increase growth. This is observed when plants are grown under elevated atmospheric CO_2 (Ainsworth and Long, 2005). Hence, nitrogen and photosynthesis are strongly correlated.

7.4 CONCLUSION

Nitrogen plays an important role in various physiological processes. The photosynthetic capacity mainly depends on leaf nitrogen. Nitrogen is a vital constituent of chlorophyll, which is necessary for photosynthesis. Photosynthesis is an important biochemical process and is essential to mitigate the problem of food security. Recent research has revealed that nitrogen application is more important than the other major essential fertilizers/nutrients for successful crop production. Various plant molecules such as amino acids, chlorophyll, nucleic acids, ATP, and phytohormones, which contain nitrogen as a structural part, are necessary to complete the biological processes, involving photosynthesis, carbon, nitrogen metabolism, and protein production. Many critical candidate genes also have been overexpressed and knocked out to test for biomass and plant nitrogen status. Nitrogen is thus the most limiting nutrient for crop production, and its efficient use is important for meeting the agricultural crisis, which is one of the utmost scientific challenges of the coming decade.

ACKNOWLEDGMENTS

We are highly thankful to the Department of the Botany University of Kashmir for providing the necessary facilities during the present study.

REFERENCES

Abrol, Y. P. (1993). *Nitrogen: Soils, Physiology, Biochemistry, Microbiology, Genetics*. Indian National Science Academy, New Delhi.

Abrol, Y. P. Chatterjee, S. R. Kumar, P. A. and Jain, V. (1999). Improvement in nitrogen use efficiency: physiological and molecular approaches. Current Science 76, 1357–1364.

Aerts, R. and Chapin III, F. S. (1999). The mineral nutrition of wild plants revisited: A re-evaluation of processes and patterns. Advances in Ecological Research 30, 1–67.

Ainsworth, E. A. and Long, S. P. (2005). What have we learned from 15 years of free air CO2 enrichment (FACE)? A meta-analytic review of the responses of photosynthesis, canopy properties and plant production to rising CO2. New Phytologist 165, 351–372.

Allison, J. C. S. Williams, H. T. and Pammenter, N. W. (1997). Effect of specific leaf nitrogen content on photosynthesis of sugarcane. Annals of Applied Biology 13, 339–350.

An, Y. Zhou, Y. Han, X. Shen, C. Wang, S. Liu, C. and Xia, X. (2020). The GATA transcription factor GNC plays an important role in photosynthesis and growth in poplar. Journal of Experimental Botany 71, 1969–1984.

Ashikari, M. Sakakibara, H. Lin, S. Yamamoto, T. Takashi, T. Nishimura, A. and Matsuoka, M. (2005). Cytokinin oxidase regulates rice grain production. Science 309, 741–745.

Barker, A. V. Maynard, D. N. and Mills, H. A. (1974). Variations in nitrate accumulation among spinach cultivars. *Journal of the American Society for Horticultural Science* 99, 132–134.

Blankenship, R. E. (2014). *Molecular Mechanisms of Photosynthesis*. John Wiley & Sons, Chichester, UK.

Bloom, A. J. Caldwell, R. M. Finazzo, J. Warner, R. L. and Weissbart, J. (1989). Oxygen and carbon dioxide fluxes from barley shoots depend on nitrate assimilation. Plant Physiology 91, 352–356.

Bondada, B. R. Oosterhuis, D. M. Norman, R. J. and Baker, W. H. (1996). Canopy photosynthesis, growth, yield, and boll 15N accumulation under nitrogen stress in cotton. Crop Science 36, 127–133.

Cassman, K. G. Peng, S. Olk, D. C. Ladha, J. K. Reichardt, W. Dobermann, A. and Singh, U. (1998). Opportunities for increased nitrogen-use efficiency from improved resource management in irrigated rice systems. Field Crops Research 56, 7–39.

Cechin, I. and de Fátima Fumis, T. (2004). Effect of nitrogen supply on growth and photosynthesis of sunflower plants grown in the greenhouse. Plant Science 166, 1379–1385.

Chapin, F. S. Bloom, A. J. Field, C. B. and Waring, R. H. (1987). Plant responses to multiple environmental factors. Bioscience 37, 49–57.

Chen, W., Hou, Z. Wu, L. Liang, Y. and Wei, C. (2010). Effects of salinity and nitrogen on cotton growth in arid environment. Plant and Soil 326, 61–73.

Cliquet, J. B. Deleens, E. and Mariotti, A. (1990). C and N mobilization from stalk and leaves during kernel filling by 13C and 15N tracing in *Zea mays* L. Plant Physiology 94, 1547–1553.

Connor, D. J. Hall, A. J. and Sadras, V. O. (1993). Effect of nitrogen content on the photosynthetic characteristics of sunflower leaves. Functional Plant Biology 20, 251–263.

Crawford, N.M. and Forde, B.G. (2002). Molecular and developmental biology of inorganic nitrogen nutrition. *Arabidopsis Book. Crops Research* 105, 22–26.

Curci, P. L. Berges, H. Marande, W. Maccaferri, M. Tuberosa, R. and Sonnante, G. (2018). Asparagine synthetase genes (AsnS1 and AsnS2) in durum wheat: structural analysis and expression under nitrogen stress. Euphytica 214, 1–13.

Curci, P. L. Cigliano, R. A. Zuluaga, D. L. Janni, M. Sanseverino, W. and Sonnante, G. (2017). Transcriptomic response of durum wheat to nitrogen starvation. Scientific Reports 7, 1–14.

DaMatta, F. M. Loos, R. A. Silva, E. A. and Loureiro, M. E. (2002). Limitations to photosynthesis in *Coffea canephoraas* a result of nitrogen and water availability. Journal of Plant Physiology 159, 975–981.

De Angeli, A. Monachello, D. Ephritikhine, G. Frachisse, J. M. Thomine, S. Gambale, F. and Barbier-Brygoo, H. (2006). The nitrate/proton antiporter AtCLCa mediates nitrate accumulation in plant vacuoles. Nature 442, 939–942.

Evans, J. R. (1989). Photosynthesis and nitrogen relationships in leaves of C 3 plants. Oecologia 78, 9–19.

Evans, J. R. (1993). Photosynthetic acclimation and nitrogen partitioning within a lucerne canopy. I. Canopy characteristics. Functional Plant Biology 20, 55–67.

Evans, J. R. and von Caemmerer, S. (2000). Would C4 rice produce more biomass than C3 rice? In *Studies in Plant Science* (Vol. 7, pp. 53–71). Elsevier.

Field, C.B. and Mooney, H.A. (1986).The photosynthesis–nitrogen relationship in wild plants. In: T.J. Givnish (Eds.), *On the Economy of Plant Form and Function*. Cambridge University Press, Cambridge (pp. 25–55). Cultivars. Journal American Society for Horticultural Science 99, 32–134.

Frink, C. R. Waggoner, P. E. and Ausubel, J. H. (1999). Nitrogen fertilizer: retrospect and prospect. Proceedings of the National Academy of Sciences 96, 1175–1180.

Gelli, M. Duo, Y. Konda, A. R. Zhang, C. Holding, D. and Dweikat, I. (2014). Identification of differentially expressed genes between sorghum genotypes with contrasting nitrogen stress tolerance by genome-wide transcriptional profiling. BMC Genomics 15, 1–16.

Guo, F. Q. Young, J. and Crawford, N. M. (2003). The nitrate transporter AtNRT1. 1 (CHL1) functions in stomatal opening and contributes to drought susceptibility in Arabidopsis. The Plant Cell 15, 107–117.

Hak, R. Rinderle-Zimmer, U. Lichtenthaler, H. K. Natr, L. (1993). Chlorophyll a fluorescence signatures of nitrogen deficient barley leaves. Photosynthetica 28, 151–159.

Hemerly, A. (2016). Genetic controls of biomass increase in sugarcane by association with beneficial nitrogen-fixing bacteria. In *Plant and Animal Genome XXIV Conference. Plant and Animal Genome*, during month of January.

Hikosaka, K. Hanba, Y. T. Hirose, T. and Terashima, I. (1998). Photosynthetic nitrogen use efficiency in leaves of woody and herbaceous species. Functional Ecology 12, 896–905.

Hörtensteiner, S. and Feller, U. (2002). Nitrogen metabolism and remobilization during senescence. Journal of Experimental Botany 53, 927–937.

Kull, O. and Jarvis, P. G. (1995). The role of nitrogen in a simple scheme to scale up photosynthesis from leaf to canopy. Plant, Cell and Environment 18, 1174–1182.

Kutík, J. Nátr, L. Demmers-Derks, H. H. and Lawlor, D. W. (1995). Chloroplast ultrastructure of sugar beet (*Beta vulgaris* L.) cultivated in normal and elevated CO_2 concentrations with two contrasted nitrogen supplies. Journal of Experimental Botany 46, 1797–1802.

Kwong, K.N.K. Bholah, A. Volcy, L. and Pynee, K. (2002). Nitrogen and phosphorus transport by surface runoff from a silty clay loam soil under sugarcane in the humid tropical environment of Mauritius. Agriculture, Ecosystems & Environment 91, 147–57.

Lawlor, D. W. (2002). Carbon and nitrogen assimilation in relation to yield: mechanisms are the key to understanding production systems. Journal of Experimental Botany 53, 773–787.

Leong, T. Y. Anderson, J. M. (1984). Adaptation of the thylakoid membranes of pea chloroplasts to light intensities. I. Study on the distribution of chlorophyll-protein complexes. Photosynthesis Research 5, 105–115.

Li, L. Nelson, C. J. Trösch, J. Castleden, I. Huang, S. and Millar, A. H. (2017). Protein degradation rate in *Arabidopsis thaliana* leaf growth and development. The Plant Cell 29, 207–228.

Maillard, A. Diquélou, S. Billard, V. Laîné, P. Garnica, M. Prudent, M. and Ourry, A. (2015). Leaf mineral nutrient remobilization during leaf senescence and modulation by nutrient deficiency. Frontiers in Plant Science 6, 317.

Makino, A. (2011). Photosynthesis, grain yield, and nitrogen utilization in rice and wheat. Plant Physiology 155, 125–129.

Makino, A., Osmond, B. (1991). Effects of nitrogen nutrition on nitrogen partitioning betwee chloroplasts and mitochondria in pea and wheat. Plant Physiology 96, 355–362.

Malik, T. H. Lal, S. B. Wani, N. R. Amin, D. and Wani, R. A. (2014). Effect of different levels of nitrogen on growth and yield attributes of different varieties of basmati rice (*Oryza sativa* L.). International Journal of Scientific and Technology Research 3, 444–448.

Martin, T. Oswald, O. and Graham, I. A. (2002). Arabidopsis seedling growth, storage lipid mobilization, and photosynthetic gene expression are regulated by carbon: nitrogen availability. Plant Physiology 128, 472–481.

Masclaux-Daubresse, C. Daniel-Vedele, F. Dechorgnat, J. Chardon, F. Gaufichon, L. and Suzuki, A. (2010). Nitrogen uptake, assimilation and remobilization in plants: challenges for sustainable and productive agriculture. Annals of Botany 105, 1141–1157.

Massignam, A. M. Chapman, S. C. Hammer, G. L. and Fukai, S. (2009). Physiological determinants of maize and sunflower grain yield as affected by nitrogen supply. Field Crops Research 113, 256–267.

Meinzer, F. C.and Zhu, J. (1998). Nitrogen stress reduces the efficiency of the $C_4 CO_2$ concentrating system, and therefore quantum yield, in Saccharum (sugarcane) species. Journal of Experimental Botany 49, 1227–1234.

Mooney, H. A. and Gulmon, S. L. (1979). Environmental and evolutionary constraints on the photosynthetic characteristics of higher plants. In: Solbrig, O.T., Jain, S., Johnson, G.B., and Raven, P.H. (Eds), *Topics in Plant Population Biology* (pp. 316–337). Palgrave, London..

Müller, H. M. Schäfer, N. Bauer, H. Geiger, D. Lautner, S. Fromm, J. and Hedrich, R. (2017). The desert plant *Phoenix dactylifera* closes stomata via nitrate-regulated SLAC 1 anion channel. New Phytologist 216, 150–162.

Olson, R. V. Murphy, L. S. Moser, H. C. and Swallow, C. W. (1979). Fate of tagged fertilizer nitrogen applied to winter wheat. Soil Science Society of America Journal 43, 973–975.

Onoda, Y. Wright, I.J. Evans, J.R. Hikosaka, K. Kitajima, K. Niinemets, Ü. Poorter, H. Tosens, T. and Westoby, M. (2017). Physiological and structural tradeoffs underlying the leaf economics spectrum. New Phytologist 214, 1447–1463.

Park, S. E. Robertson, M. and Inman-Bamber, N. G. (2005). Decline in the growth of a sugarcane crop with age under high input conditions. Field Crops Research 92, 305–320.

Paul, M.J. and Stitt, M. (1993) Effects of nitrogen and phosphorus deficiencies on levels of carbohydrates, respiratory enzymes and metabolites in seedlings of tobacco and their response to exogenous sucrose. Plant Cell Environment 16, 1047–1057

Rogers, A. Medlyn, B. E. Dukes, J. S. Bonan, G. Von Caemmerer, S. Dietze, M. C. and Zaehle, S. (2017). A roadmap for improving the representation of photosynthesis in Earth system models. New Phytologist 213, 22–42.

Sekhar, K. M. Kota, V. R. Reddy, T. P. Rao, K. V. and Reddy, A. R. (2020). Amelioration of plant responses to drought under elevated CO_2 by rejuvenating photosynthesis and nitrogen use efficiency: implications for future climate-resilient crops. Photosynthesis Research 1–20.

Simpson, F. B. and Burris, R. H. (1984). A nitrogen pressure of 50 atmospheres does not prevent evolution of hydrogen by nitrogenase Science 224, 1095–1097.

Sinclair, T. R. and Jamieson, P. D. (2006). Grain number, wheat yield, and bottling beer: An analysis. Field Crops Research 98, 60–67.

Sinclair, T.R. and Horie, T. (1989) Leaf nitrogen, photosynthesis, and crop radiation use efficiency: a review. Crop Science 29: 90–98

Sivasankar, A. Lakkineni, K. C. Jain, V. Kumar, P. A. and Abrol, Y. P. (1998). Differential response of two wheat genotypes to nitrogen supply. I. Ontogenic changes in laminae growth and photosynthesis. Journal of Agronomy and Crop Science 181, 21–27.

Sun, J. Ye, M. Peng, S. and Li, Y. (2016). Nitrogen can improve the rapid response of photosynthesis to changing irradiance in rice (*Oryza sativa* L.) plants. Scientific Reports 6, 1–10.

Thornley, J.H.M. (1976). *Mathematical Models in Plant Physiology*. Academic Press.

Todd, J., Screen, S. Crowley, J. Peng, J. Andersen, S. Brown, T. and Duff, S. M. (2008). Identification and characterization of four distinct asparagine synthetase (AsnS) genes in maize (*Zea mays* L.). Plant Science 175, 799–808.

Tóth, V. R. Mészáros, I. Veres, S. and Nagy, J. (2002). Effects of the available nitrogen on the photosynthetic activity and xanthophyll cycle pool of maize in field. *Journal of Plant Physiology* 159, 627–634.

Verhoeven, J. T. A. and Liefveld, W. M. (1997). The ecological significance of organo chemical compounds in Sphagnum. *Acta Botanica Neerlandica* 46, 117–130.

Vos, J. and Biemond, H. (1992). Effects of nitrogen on the development and growth of the potato plant. 1. Leaf appearance, expansion growth, life spans of leaves and stem branching. *Annals of Botany* 70, 27–35.

Wan, Y. King, R. Mitchell, R.A.C. Hassani, K. and Hawkesford, M.J (2017). Spatiotemporal expression patterns of wheat amino acid transporters reveal their putative roles in nitrogen transport and responses to abiotic stress. Scientific Reports. 27:5461–5474.

Wang, X. Deng, X. Pu, T. Song, C. Yong, T. Yang, F. and Yang, W. (2017). Contribution of interspecific interactions and phosphorus application to increasing soil phosphorus availability in relay intercropping systems. Field Crops Research 204, 12–22.

Wang, Z. H. Li, S. X. and Malhi, S. (2008). Effects of fertilization and other agronomic measures on nutritional quality of crops. Journal of the Science of Food and Agriculture 88, 7–23.

Wann, M. and Raper, Jr., C. D. (1979). A dynamic model for plant growth: Adaptation for vegetative growth of soybeans 1. Crop Science 19, 461–467.

Xing, G.and Zhu, Z. (2002). Regional nitrogen budgets for China and its major watersheds. Biogeochemistry 57:405–427.

Yoshida, H. Horie, T. and Shiraiwa, T. (2006). A model explaining genotypic and environmental variation of rice spikelet number per unit area measured by cross-locational experiments in Asia. Field Crops Research 97, 33.

Yousuf, P. Y, Ganie, A. H., Khan, I., Qureshi, M. I., Ibrahim, M. M., Sarwat, M., Iqbal, M., Ahmad, A. (2016) Nitrogen-efficient and nitrogen-inefficient indian mustard showed differential expression pattern of proteins in response to elevated CO^- and low nitrogen. Frontiers in Plant Science 7, 1074.

Zhong, C. Jian, S. F. Huang, J. Jin, Q. Y. and Cao, X. C. (2019). Trade-off of within-leaf nitrogen allocation between photosynthetic nitrogen-use efficiency and water deficit stress acclimation in rice (*Oryza sativa* L.). Plant Physiology and Biochemistry 135, 41–50.

Zhu, Z. L. Chen, D. L. (2002). Nitrogen fertilizer use in China–Contributions to food production, impacts on the environment and best management strategies. Nutrient Cycling in Agroecosystems 63, 117–127.

# 8 Recycling and Remobilization of Nitrogen during Senescence

Aadil Farooq War, Ishfaq Ahmad Sheergojri,
Subzar Ahmad Nanda, Mohd Asgar Khan,
Ishfaq Ul Rehmaan, Zafar Ahmad Reshi, and Irfan Rashid*
Department of Botany, University of Kashmir, Srinagar, India
*Correspondance: waraady20@gmail.com

CONTENTS

8.1	Introduction	97
8.2	Internal Redistribution of Nitrogen During the Plant's Life Cycle	99
8.3	Protein Degradation is Universal to N Remobilization	99
8.4	The Chloroplast Degradation Pathway	100
	8.4.1 Degradation of the Stromal Proteins	101
	8.4.2 Degradation of Thylakoid-Bound Proteins	102
8.5	The Ubiquitin/Proteosome Pathway	103
8.6	Autophagic and Vacuolar Pathway of Protein Degradation	104
8.7	Remobilization of Nitrogenous Compounds	105
	8.7.1 The GS/GOGAT Pathway	106
	8.7.2 The PPDK-GS/GOGAT Pathway	106
8.8	Translocation of Nitrogen Compounds to the Sink Tissues	107
8.9	Targets to Improve Nitrogen Fertilization	108
8.10	Conclusion	110
References		110

8.1 INTRODUCTION

Senescence is a final and crucial developmental stage of the plant that results in the death of either cells, tissues, or the entire plant. Plants perform two modes of senescence: mitotic or replicative senescence and post-mitotic senescence (Gan, 2003). Mitotic/replicative senescence occurs in meristematic tissues after the arrest of a cell division or replication event. In contrast, post-mitotic senescence typically takes place in plant organs, such as leaves and flower petals after attaining cellular differentiation and maturation, adopting an active and programmed degenerative regimen (Ben-Porath and Weinberg, 2005; Lim et al., 2010). The hypothesis of developmental senescence emerged from the fact that all plant species and genotypes don't senesce at the same time and in identical ways, and that it is primarily caused by endogenous factors unique to each plant. Among the senescence of different plant organs, leaf senescence is more developmentally controlled. Leaf senescence is fundamentally a mechanism of nutrient recycling required for the management of plant resources, during which nutrients stored in tissues undergoing senescence are diverted to other tissues of the plant, where they can be used for the development of new vegetative, reproductive, or storage organs (Yousuf et al., 2021; Gregersen et al., 2008). The leaf senescence is likely to respond to plant source/sink requirements and its regulation integrates stress response, growth hormone, and nutrient-sensing systems (Guiboileau et al., 2010). When a leaf approaches the stage of senescence, its cells experience systemic cellular component disorganization and coordinated metabolic and gene expression changes (Schippers et al., 2015). During senescence, the anabolic reactions of carbon and nitrogen assimilation are metabolically overtaken by the catabolism of proteins, chlorophyll, nucleic acids, and membrane lipids (Schippers et al., 2015; Lim et al., 2007). This elevated catabolic behavior is possibly responsible for transforming the cellular materials acquired during the leaf growth into exportable metabolites (Woo et al., 2019). Ultrastructural studies demonstrate that during senescence organelle like chloroplasts are the first to get dismantled, while the nucleus and the mitochondria remain in place until the late stages of leaf senescence, creating a pool of nutrients for remobilization. The cellular disintegration and remobilization operations, which occur concurrently, are therefore likely to be systemic and highly coordinated (Buchanan-Wollaston et al., 2003). Thus, the interconnectedness of extremely complex regulatory networks and numerous cross-talking channels must be functioning to carry out leaf senescence, and these networks must be flexible enough to control and epitomize the nutrient remobilization during leaf senescence (Lim et al., 2007). Thus, given the fact that leaf senescence is a detrimental mechanism for the leaf, it is an altruistic process, leading to plant tolerance to prevailing environmental conditions, ensuring optimum spring production, and enhanced plant survival in its temporal regions (Fischer, 2012; Gregersen et al., 2013).

The young leaves are typically major net importers ("sinks") of all the nutrients required to establish the cellular and molecular components (Good et al., 2004). However, the advent of leaf senescence is correlated with the transition from net importers to net exporters of "mobile" compounds, as a result of which the total nutrient content of leaves starts to decline (Marschner, 1995). The empirical evidence often refers to this situation as "redistribution," "retranslocation," "resorption" or "remobilization" (Marschner, 1995; Killingbeck, 2004). Various methods, including phloem sap sampling and examination using radioactive tracer tests, have shown that macronutrients except for calcium (i.e. N, P, K, S, and Mg) are typically highly mobilized, while micronutrients except for manganese (i.e. Fe, Cu, Zn, B, Cl, Mo, and Ni) display at least moderate mobility (Himelblau E, Amasino RM. 2001). Nitrogen being central to the majority of biomolecules constitutes a major portion of the cargo that is remobilized from senescing leaves to nascent tissues or storage organs (Himelblau and Amasino, 2001). In most plant cells, the greatest reservoir of nitrogen potentially accessible for remobilization during senescence is present as proteins. In photoautotrophic tissues, over 50% of this protein is contained in soluble (stromal) and insoluble (thylakoid) chloroplast proteins (Ho'rtensteiner and Feller., 2002). Protein degradation is, therefore, the main objective during N-resorption, as the bulk of N-resorption is due to the hydrolysis of proteins to amino acids, which are likely to be inter-converted, hydrolyzed, catabolized, or exported without any transformation (Brouquisse et al., 2001; Hortensteiner and Feller, 2002). This was supported by temporal shifts in the pattern of protease expression observed to take place in mitochondria, chloroplasts, and nucleus to ensure collaborative degradation of proteins into amino acids, amides, and ammonium (Roberts et al., 2012; Diaz-Mendoza et al., 2014). As a consequence, complex traffic of amino acids, peptides, and proteins appears to occur among cellular compartments involving cytosol, chloroplasts, lytic vacuoles, and special vesicles, thereby delivering the cargo to the sink tissues (Roberts et al., 2012, Avila-Ospina et al., 2014, Diaz-Mendoza et al., 2014). This remobilization and recycling of nitrogenous compounds during senescence thereby redirect nutrients to the plant and thus makes a major contribution to its longevity and the developmental program. Emphasizing the quantitative significance and intricacies of the chemical processes associated with its remobilization, nitrogen metabolism particularly during senescence has received the most scientific attention and will be the center of attention in this chapter. Therefore, in this chapter, we will briefly describe the internal remobilization of nitrogen, pathways involved in nitrogen remobilization, and targets to improve nitrogen remobilization to increase NUE of plants.

8.2 INTERNAL REDISTRIBUTION OF NITROGEN DURING THE PLANT'S LIFE CYCLE

The nitrogen metabolism in plants includes several steps, including internalization, assimilation, mobilization, and when plant ages, remobilization, and recycling (Jin et al., 2015; Xu et al., 2012; Masclaux-Daubresse et al., 2010; Kichey et al., 2007). The internal redistribution of nitrogen can be grouped into four phases: (1) the primary distribution of nitrogen from the N-assimilating tissues to the sink tissue during the growing phase (2) the reallocation of nitrogenous compounds from the metabolic remobilization of N during the growing phase (3) the resorption of nitrogen from the senescing leaves and its carriage to the perennial tissue; and (4) remobilization of N from perennating tissues to actively growing tissues. The last two steps constitute seasonal N cycling. Plants use N derived from a variety of potential sources typically grouped as external or internal resources. Externally plants get the nitrogen through mineralization of soil organic matter (or fertilizers), diazotrophic nitrogen fixation, organic N via mycorrhizal symbionts and in some ecosystems direct deposition of atmospheric N. Internally, plant derives the nitrogen from storage tissues by physiological processes of recycling and remobilization. The availability of soil nitrogen, which is typically minimal, may fluctuate significantly in space and time due to variables such as soil type and pH, precipitation, temperature, and wind (Maathuis, 2009). Therefore, the preferred form of N taken up by the plant depends on its adaptation to the soil conditions. Plants adapted to reduced soils and low pH as seen in mature forest ecosystems or arctic tundra usually pick up ammonium or amino acids, whereas plants suited to more acidic soils with higher pH prefer nitrates (Maathuis, 2009). Nitrate uptake occurs at the root level and two nitrate transport proteins of the NRT1 and NTR2 gene families have been shown to coexist in plants that function in a concerted manner to extract nitrate from the soil solution and distribute it across the plant (Tsay et al., 2007). Nitrogen assimilation then involves the action of nitrate reductase to reduce this nitrate to ammonium, followed by ammonium assimilation. The NH_4^+ and NO_3^- in roots and leaves are incorporated into amino acids, predominantly by the conglomeration of glutamine synthetase (GS) and glutamate synthase more specifically called GOGAT. Compared to carbon autotrophy, which is primarily operated by photosynthetic carbon dioxide fixation, this is called nitrogen autotrophy, operated by nitrogen fixation (Yoneyama et al., 2003). In addition to nitrogen autotrophy, nitrogen to developing parts—such as growing leaves, grain filling, and growing roots—is supplied through N reallocation, transported through the phloem from mature reserve organs (Good et al., 2004). Such an N-nutrition structure involving the supply of nitrogen from assimilating tissues and mature reserve organs through xylem and phloem, respectively, is very critical for the systemic growth of different plant organs (Yoneyama et al., 2003). Radioactive tracing using 15N in rapeseed, legumes, and cereals has shown that the most important phase of plant growth is the onset of grain filling, as nitrogen fixation and uptake decreases during final stages of plant development such as seed filling (Salon et al., 2001). This supply of nitrogen during grain filling is usually inadequate to meet the high demand for seeds, which means that progressive and numerous remobilization paths, which occur successively in the various plant organs, are needed to deliver nitrogen to the seeds (Masclaux et al., 2001). However, in the case of perennial plants, the N remobilization at the time of senescence is not only directed to seeds but also the perennating tissues for storage. Moreover, during the spring flush/regrowth of shoots, the N (stored proteins and amino acids) reserved in stem bases and roots is optimized for transfer to growing tissues via vascular bundles (Yoneyama et al., 2003). Thus, the absorption and allocation of nitrogen in plants are well controlled to ensure balanced growth.

8.3 PROTEIN DEGRADATION IS UNIVERSAL TO N REMOBILIZATION

Proteins, especially those localized in the chloroplast, are the primary source of nitrogen for remobilization (Peoples and Dalling, 1988). The protein degradation during nitrogen remobilization predominantly recognizes chloroplastic enzymes and structural proteins, particularly RuBisCO as

TABLE 8.1
Potential Roles of Leaf Senescence-enhanced Genes Found by Differential Screening and Other Methods

The function of encoded protein/enzyme	Encoded enzyme	references
Protein degradation	Cysteine protease	Frank et al., 2019
	Aspartic protease	Diaz-Mendoza et al., 2016
	Clp proteases	Deschênes-Simard et al., 2014
	Ubiquitin	
	F-Box protein	
Protein processing	Vacuolar processing enzymes	Hara-Nishimura et al., 2005
Autophagy	phosphatidylinositol 3-kinase	Yoshimoto et al., 2004
	ATG9	Thompson and Doelling, 2005
N mobilization	Glutamine synthetase	Pageau et al., 2006
	Aminotransferase	Diab et al., 2016
	Vegetative storage protein	Noquet et al., 2004
	Branched-chain α ketoacid dehydrogenases	Buchanan-Wollaston, 1994
Cell wall degradation	Endoxyloglucan transferase	Park et al., 1998
	β glucosidase	Yoshida et al., 2001
Transcriptional regulation	WRKY factors	Robotzek et al., 2001
	Leucine zipper proteins	Chang et al., 2014
Signaling pathways	Receptor kinase SARK	Xu et al., 2011
	Receptor kinase SIRK	Buchanan-Wollaston, 2003
	Calmodulin binding protein	Nie et al., 2012

remobilization substrates. This is evident from the fact that RuBisCo contributes 50% and 20% of the total soluble protein content in leaves of C3 and C4 plants, respectively (Mae et al., 1983; Ho¨rtensteiner and Feller., 2002). All other stocks of cellular nitrogen such as cytosolic and other proteins, chlorophyll, nucleic acids, and free amino acids constitute comparatively minor organic nitrogen stocks. This emphasis was consistent with the fact that chloroplasts are always the first subcellular organelles to get dismantled, leading to the most apparent visually quantified phenotype, which is gradual leaf yellowing, triggered by chlorophyll and protein degradation (Tamary et al., 2019). It is also apparent that degradation of proteins is the primary function of N-resorption, where the degradation products are likely to be inter-converted, catabolized, hydrolyzed, or exported without modification (Brouquisse et al., 2001; Hortensteiner and Feller, 2002; Krupinska, 2007). It has been reported that most of the potential roles of leaf senescence-enhanced genes are associated with protein degradation and transport (Table 8.1). Therefore, the biochemistry of protein degradation has been the focus of efforts to understand nitrogen remobilization during leaf senescence. Three major pathways for protein degradation during leaf senescence have been revealed by functional and homologous analyses of SAGs—the chloroplast degradation pathway, the ubiquitin/proteasome system, and the vacuolar and autophagic (APG) pathway.

8.4 THE CHLOROPLAST DEGRADATION PATHWAY

The chloroplast is a well-known organelle present in all photosynthetic cells that provides food and energy to plants in the form of sugar or starch (Jarvis and López-Juez, 2013). Chloroplastic (CIP) proteins, contributing more than 70% of all leaf proteins, are known to be a significant source of nitrogen for remobilization during leaf senescence, hence their breakdown is the first step in

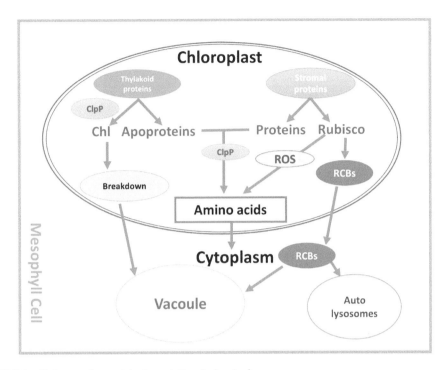

FIGURE 8.1 Pathways for protein degradation during leaf senescence.

N-resorption (Hörtensteirer and Feller, 2002). The degradation of chloroplast proteins specifically starts early in the progression of senescence and the subsequent peptides and amino acids are transported out of the senescent tissue (Masclaux et al., 2000). In addition, the degradation of RuBisCo and other stromal proteins can happen within discrete chloroplasts, indicating the existence of plastidial proteases. The chloroplast protein degradation has been divided into sections based on proteins present in their sub-cellular compartments: stromal protein degradation and thylakoid bound protein degradation (Figure 8.1) (Adam, 2001). Several studies are supporting the assumption that at least the initial steps of chlorophyll and chloroplast protein degradation take place in intact organelles. While some studies suggest that a few steps might take place outside the plastids it must be assumed that various catabolic pathways may occur that might be organized in a way that has not yet been explored.

8.4.1 Degradation of the Stromal Proteins

The enzymes present in the stroma are those engaged in carbon and nitrogen assimilation and are depleted early during senescence, leading to a decrease in photosynthetic ability. The proteolysis of the major stromal enzymes in C3-plants (i.e., RuBisCo) has been the focus of many assays (Mitsuhashi et al., 1992; Ishida et al., 2002). The massively acquired RuBisCo represents the largest intracellular nitrogen reservoir in vascular plants that can be mobilized during senescence or nitrogen deficiency through proteolysis (Feller et al., 2008; Hortensteiner and Feller, 2002). While chloroplast harbors a large protease repository responsible for the proteolysis of stromal proteins, no single protease has been reported to break down RuBisCo (Lin and Wu, 2004; Gregersen and Holm, 2007). These findings demonstrate that chloroplast proteases (Clp) are not actively engaged in Rubisco proteolysis, which renders the function of stromal Clp and its potential substrates uncertain in higher plants (Adam and Clarke, 2002). In addition, these findings indicate the involvement of other pathways that work concurrently to degrade RuBisCo. The proteolysis of RuBisCo has been proposed to be orchestrated by reactive oxygen species that lead to the direct (non-enzyme) breakdown of the large RuBisCo subunit into two 37 kDa and 16 kDa fragments (Ishida et al., 1997). However, ultrastructural examination of wheat leaves has demonstrated that cytoplasmic vesicles include Rubisco and/or Rubisco-containing bodies (RCBs) and other stromal proteins assumed to be degraded through autophagy (Wad et al., 2009).

Glutamine synthetase, the rate-limiting enzyme concerned with the transformation of ammonia, is yet another enzyme hydrolyzed early in the senescence. Glutamine synthetase is more susceptible to hydrolysis in isolated chloroplasts and is degraded much faster than Rubisco and other stromal enzymes (Mitsuhashi and Feller, 1992; Thoenen and Feller, 1998). However, the cytosolic isoform of this enzyme seems to be stable and remains active for longer as compared to its plastidial form, which is degraded early in the senescence (Diaz et al., 2008; Orsel et al., 2014). These results suggest that proteolysis doesn't only depend upon protease activity but also on substrate susceptibility, thereby indicating that chloroplast contains more fragile and susceptible enzymes than cytosol. The protein catabolism in chloroplast stroma relies on the activity of endo- and exopeptidases (Brouquisse et al., 2001). There are various types of peptidases that have been found to exist in plastids. Metalloendopeptidase such as FtsH6 and M58 are crucial enzymes involved in the degradation of stromal proteins (Zelisko et al., 2005; Díaz-Mendoza et al., 2014). Metalloprotease FtsH6 is plastidial, found to degrade Lhcb3 in vivo as well as in vitro, while M58 is localized in plastoglobules (i.e., lipid bodies that develop in plastids during senescence), however, the role of the latter remains uncertain (Roberts et al., 2012; Lundquist et al., 2012). Aminopeptidases are another class of protease found in plastids that pertain to the complete breakdown of stromal proteins by degrading peptides generated by the action of endopeptidases (Waditee-Sirisattha et al., 2011). Together, this data suggests that metalloendopeptidase and aminopeptidase are crucial enzymes required for the breakdown of stromal proteins.

8.4.2 Degradation of Thylakoid-Bound Proteins

Thylakoids are protein-rich, the most common being chlorophyll apoproteins. Apoprotein, LHCP II, along with stromal Rubisco, is the main source of nitrogen in senescing leaves (Zienkiewicz et al., 2012). The morphological study of senescing chloroplasts reveals dramatic shifts in the thylakoid membrane framework. The connection among grana stacks is stabilized and the membranes gradually vanish in combination with a rise in the number and size of plastoglobules (Hörtensteiner, and Feller, 2002; Lucinski et al., 2011) The dissolution of membrane structure is concomitant with the breakdown of membrane components such as chlorophyll (Chl), proteins, and lipids (Matile, 1992). The chlorophyll degradation is the prerequisite for the degradation of other thylakoid bound proteins like LHCP II (Komenda et al., 2012). Chlorophyll is degraded by a signaling cascade explicitly caused by the start of senescence. The degradation of chlorophyll is driven by the pathway called phyllobilin/pheophorbide oxygenase (PAO) pathway, involving a collection of chlorophyll catabolic genes (CCGs) (Hortensteiner and Krautler, 2011). The process commences with the reduction of chlorophyll 'b' to chlorophyll 'a' by chlorophyll 'b' reductase enzyme (encoded by NONYELLOW COLORING1 (NYC1) and NONYELLOW COLORING1–LIKE (NOL)) and 7-hydroxymethyl chlorophyll and reductase (HCAR). The chlorophyll 'a' is then transformed to pheophytin-a by the removal of magnesium by magnesium-dechetalase encoded by NONYELLOWINGS/STAYGREENs (NYES/SGRs), which is subsequently hydrolyzed into pheophorbide-a and phytol by the action of PHEOPHYTINASE (PPH). The pheophorbide-a porphyrin ring is broken by PAO, and the green color of chlorophyll catabolites is lost, resulting in oxidized red chlorophyll catabolite, which is consequently converted to produce primary fluorescent chlorophyll catabolite (pFCC) by the chlorophyll catabolite reductase. The pFCC is further modified and transported to the vacuole where it is isomerized by acid pH into non-fluorescent components (Hortensteiner and Krautler, 2011). Studies have found that disruption in any of the chlorophyll catalyzing enzymes induces mobilization dysfunction of LHCP II during foliar senescence (Ougham et al., 2005). One of the most exquisite pieces of shreds of evidence of this codependence comes from the stay-green mutant studies of phaseolus vulgaris (Bachmann et al., 1994). Usually, LHCP II and Rubisco in wild-type plants are degraded at similar rates during foliar senescence, but *P. vulgaris* with abrasion in the chlorophyll degradation pathway not only mobilizes chlorophyll at a much slower rate but also retains LHCP II well past the stage where it is degraded in respective wild-type plants. By comparison, mutants that retained their Chl until late senescence also retained LHCP II, while there was no effect on the catabolism of Rubisco subunits (Bachmann et al., 1994). There is conclusive proof that LHCP II is stabilized by Chl inclusion, which prevents proteolysis as long as these protein-pigment complexes remain in place (Komenda et al., 2012). The disposal of protein pigment complexes stimulates the release of possibly hazardous Chl, which may lead to photo-oxidative damage. The breakdown of this photosynthetic apparatus is therefore not only crucial to mitigate photo-oxidative damage to cells, but also to remobilize the nitrogen from the Chl-apoproteins (Thomas and Donnison, 2000). The breakdown of apoproteins during senescence happens only after the destruction of Chl. For instance, when Chl catabolism in the stay-green mutants is inhibited, LHCP II retains stability and proteolytic cleavage is confined to a narrow N-terminal region protruding into the stroma (Thomas and Donnison, 2000). The pathway for the breakdown of Chl has been explored almost exclusively during the last decade, with the identification of the first chlorophyll catabolite from barley (Kra¨utler et al., 1991). The non-fluorescent Chl catabolites (NCCs) identified from different species exemplify the final breakdown products of Chl that deposit in the central vacuole (Mu¨hlecker and Kra¨utler, 1996; Curty and Engel, 1996). These NCCs are chemically tetrapyrroles extracted from Chl by oxygenolytic breakdown of porphyrin and are not accessible for remobilization during senescence. As a result, the four moles of nitrogen present in each Chl molecule are not remobilized but are wasted when leaves are shed (Eckhardt et al., 2004). Since the initial steps in the breakdown of proteins and chlorophylls take place within chloroplasts, catabolites formed within chloroplasts may be further metabolized after release from chloroplasts suggesting the role of the ubiquitin/proteasome pathway and the vacuolar and autophagic pathway.

8.5 THE UBIQUITIN/PROTEOSOME PATHWAY

The ubiquitin-26S proteasome pathway is essential for selective proteolysis both as a normal routine and in response to specific stimuli (Sullivan et al., 2003). This pathway is crucial to remove dysfunctional cytoplasmic proteins and to rapidly degrade short-lived proteins by identifying the covalent connexion of 76-amino acid ubiquitins, which serves as a guide for selectively targeting particular proteins for degradation (Smalle and Vierstra, 2004). The ubiquitin-26S multi-subunit complex involves three enzymes with distinct functions: 1) ubiquitin-activating enzyme (El), 2) ubiquitin-conjugating enzyme (E2), and 3) ubiquitin-protein ligase (E3). The genes responsible for coding the proteins of this multi-protein complex and several others involved in this process are induced by senescence, suggesting that ubiquitin-dependent protein breakdown is indeed an essential part of non-chloroplastic proteolysis during senescence (Woo et al., 2001; Park et al., 1998). For example, genes encoding E2 and E3 are induced in senescing cotton cotyledons and the expression of certain genes encoding proteosome-constituent proteins, such as 26S ATP/ubiquitin-dependent proteinase chain and 26S ATPase subunit, have also been found to be upregulated in these tissues during senescence (Shen et al., 2006). In addition, the F-box protein ORE9, part of the ubiquitin pathway SCF-complex (Skp1/Cul1/F-box protein), appears to have a positive effect on senescence in Arabidopsis, presumably by promoting degradation of the transcriptional repressor of senescence-associated-genes (Woo et al., 2001). Studies investigating ubiquitin-26S proteasome pathway expression profiles in senescing leaves of Arabidopsis have demonstrated that the predominant polyubiquitin genes upregulated during senescence are U BQ4, U BQ3, pSEN3, At1g14400 (ubiquitin carrier protein), At2g21950 (SKP1 interacting partner 6), and At1g53750 (26S proteasome ATPase subunit) (Gepstein et al., 2003; Lin and Wu, 2004). In addition, transcriptome analysis of wheat flag senescing leaves has identified genes such as kelch repeat protein F-box, 20S proteasome a5 subunit, ubiquitin-like protein 5, and ubiquitin-conjugating enzyme E2, which are found to be associated with ubiquitin pathways, further strengthening the role of ubiquitin/proteasome pathway in protein degradation during senescence (Gregersen and Holm, 2007). Although there is a great deal of evidence to date that senescing membrane proteins are tagged by ubiquitin before degradation, there are other signs that conformational changes can also activate their proteolysis. For instance, the D1 protein of thylakoids is degraded after undergoing conformation change in a light-dependent manner. The proteolysis relies on the presence of an α-helix 14-amino acid considered as destabilization sequence. This stabilization sequence is attacked by light-activated oxygen, especially radical hydroxyl, and directs the protein to undergo conformation change, which is eventually hydrolyzed by FtsH and add up to target for N remobilization (Yoshioka et al., 2006).

8.6 AUTOPHAGIC AND VACUOLAR PATHWAY OF PROTEIN DEGRADATION

For non-specific protein and organelle turnover, autophagy, the process of eating oneself is important, and can be triggered by several stress factors including leaf senescence, nutrient deficiency, cellular injury, or pathogen attack (Levine and Klionsky, 2004). The proteins to be destroyed are not individually targeted in proteolysis by autophagy, as in the case of degradation by the ubiquitin/26S proteasome pathway. Instead, autophagy involves encapsulation of parts of the cytoplasm inside double-membrane vesicles called autophagosomes, which are believed to originate from the endoplasmic reticulum (Marshal et al., 2016; Masclaux-Daubresse et al., 2017; Wang et al., 2018). The outer membrane of the autophagosome then fuses with the tonoplast and releases the inner vesicle of the autophagosome into the vacuole, where the cargo is degraded (Li and Vierstra, 2012; Masclaux-Daubresse et al., 2017; Wang et al., 2018). Thus, autophagic degradation of proteins is indiscriminate and is therefore expected to be triggered when nutrients need to be rapidly released and resorbed, as in stress responses and senescence (Doelling et al., 2002). Yeast studies reported the identification of two conjugating pathways mediated by APG8 and APG12 that culminate in autophagy. These two gene products are bound to other cell factors like ubiquitination (Mizushima et al., 1998; Ichimura et al., 2000). The interesting fact is those possible orthologs for all yeast conjugation pathway components of the APG8 and APG12 are present in Arabidopsis (Doelling et al., 2002). The argument that autophagy plays a key role in senescence is reinforced by the observation that one of these orthologists, APG7, is necessary for normal leaf senescence (Doelling et al., 2002). Interruption of APG7 gene expression by insertional mutation has been found to have little effect on the growth and development of Arabidopsis but induces hypersensitivity to nutrient-limiting conditions together with premature leaf senescence (Doelling et al., 2002). Moreover, APG7 gene transcripts are preferably found to accumulate in senescent leaves of Arabidopsis, further emphasizing the role of autophagy in senescence (Doelling et al., 2002). The AtATG18a upregulation finding in senescent leaves and also its sequence homology with the yeast autophagy gene ATG18 provides additional evidence to support the role of autophagy in senescence (Xiong et al., 2005). Furthermore, the AtTG18a gene when disrupted by RNA interference disrupts the normal development of autophagosomes and leads to premature leaf senescence. (Xiong et al., 2005). In senescence, autophagy has special significance as a process to destroy cytoplasmic proteins (Kim and Klionsky, 2000). However, it is not clear whether autophagy has any contribution to the digestion of membrane proteins, like LHCP II of thylakoids or Rubisco, which is present in the chloroplast stroma. However, the presence of vesicles in the cytoplasm that include Rubisco and/or Rubisco degradation products (Rubisco-containing bodies; RCB) and other stromal proteins have been reported in the ultrastructural analyses of senescing leaves of wheat (Chiba et al., 2003). In comparison to the colored globules seen during the senescence of soybean leaf and broccoli flower, these wheat vesicles emerge at an early stage of the process, probably at the onset of Rubisco degradation (Ishida et al., 2008; Wada et al., 2009). RCBs are therefore believed to be involved in the degeneration of stromal proteins at the beginning of senescence. Further ultrastructural studies have shown that autophagocytosis identical to autophagy in yeast can destroy these RCBs (Ishida et al., 2008). In senescent wheat leaves, recent immunocytochemical electron microscopy has shown that Rubisco and degradation products of other stromal enzymes like glutamate synthase are clustered in small spherical particles inside the cytosol and the vacuole (Chiba et al., 2003). Cytoplasmic lipid-protein particles formed during the removal of lipoprotein membranes by senescence are also plausibly depleted by autophagy (Thompson et al., 1998).

The proteins involved in autophagy can be divided into many major complexes or processes, such as protein kinases involved in the regulation of autophagosome formation, two ubiquitin-like conjugation systems, phosphatidylinositol 3-kinase complex, and the ATG9 complex, which can aid in membrane recruitment at the site for the formation of autophagosome (Yoshimoto et al., 2004 Suzuki et al., 2005; Thompson and Doelling, 2005). The confirmation of the presence and necessity of all these complexes in Arabidopsis leaf senescence comes from the result of knockout mutants of several corresponding genes that exhibit the sensitivity to starvation and early-senescence phenotypes, which is similar to the phenotype expected for autophagy disruption (Doelling et al., 2002; Yoshimoto et al., 2004 Suzuki et al., 2005 Thompson and Doelling, 2005). In addition to these, several other autophagic proteins were identified in senescing leaves of other organisms, including Hsp70-related proteins, which help in transporting individual proteins directly to the vacuole without using vesiculated intermediate and AUT1-like autophagocytosis proteins (Cuervo et al., 2000; Shen et al., 2006, Gregersen and Holm, 2007). These findings indicate that during leaf senescence, autophagic proteins are needed to sustain cellular activity, likely by eliminating and degrading abnormal proteins. After the transport of cytoplasmic proteins into the vacuoles through autophagy, they are degraded by vacuolar exo- and endoproteases, such as cysteine protease (SAG12), serine proteinase, aspartic proteinase, cathepsin B-like Cys proteinase, peptidases, papain-like Cys proteinase, aminopeptidase, and endopeptidase. The activation of these vacuolar proteases in senescent plants has been well documented in Arabidopsis, several populus species, cotton, and wheat (Gregersen and Holm, *2007*; Buchanan-Wollaston et al., 2003, 2005; Lin and Wu, 2004; Andersson et al., 2004; Bhalerao et al., 2003; Shen et al., 2006). Although several factors can affect their activity during leaf senescence, these factors will then overlap in a single degrading pathway (Martfnez et al., 2007).

8.7 REMOBILIZATION OF NITROGENOUS COMPOUNDS

A large variety of amino acids is generated by protein degradation, but not all of them appear to be successfully mobilized and loaded into the blaze (Tegeder, 2014). The main amino acids translocated by phloem during nitrogen removal are glutamine or asparagine (Taylor et al., 2012). Watanabe et al. (2013) reported a spike in the ratios Gln/Glu and Asn/Asp when carrying out Arabidopsis metabolite profiling during senescence. This indicates that the Asp into Asn and Glu was intentionally transformed into Gln during leaf senescence. This is evident from the analysis of phloem that indicated the presence of high levels of glutamine and asparagine along with serine and proline (Masclaux-Daubresse et al., 2008). These amino acids are known to be present in the senescing leaf blade phloem and are all found to be obtained from glutamate and aspartate. Based on the recent detection and characterization of SAGs, two similar biochemical cycles for interconverting amino acids and N-export have been proposed: the GS/GOGAT and the PPDK-GS/GOGAT pathway (Lin and Wu, 2004; Shen et al., 2006; Gregersen and Holm, 2007).

8.7.1 THE GS/GOGAT PATHWAY

Amino acids generated by protein degradation can be further modified by the action of aminotransferase to generate free ammonia. Nucleotide degradation is another source of free ammonia, which has been confirmed by examining the leaf senescence-induced gene expression of enzymes, such as cytidine deaminase and urease. The ammonium possessing phytotoxic activity is immediately transformed into glutamate by the amination of 2-oxoglutarate by glutamate dehydrogenase via GS-GOGAT cycle (NADH-GDH; EC 1.4.1.2) and eventually into glutamine by cytosolic glutamine synthetase (GS1; EC 6.3.1.2). GDH and cytosolic GS1 play a prominent role in amino acid synthesis and redistribution in the senescing leaves (Bernard and Habash, 2009). The experiments on the source-sink relationship have shown that GDH is induced in old leaves when N-remobilization is maximum, indicating that GDH's physiological function is to produce glutamate for translocation (Masclaux et al., 2000; Bernard and Habash, 2009). However, studies have shown that the rate of formation of glutamate is as low as 0.2% of the overall ammonium release or 1.2% of the oxidative deamination rate of glutamate, suggesting that GDH must catalyze the oxidative deamination of glutamate reversibly to 2-oxoglutarate and ammonium (Aubert et al., 2001). It is thus plausible that GDH provides 2-oxoglutarate through oxidation of glutamate for nitrogen remobilization in old leaves. GS, a key enzyme in the metabolism of nitrogen, is however classified according to its cellular distribution into cytosolic (GS1) and plastidial (GS2) forms. GS1 plays a prominent role in the glutamine synthesis for the remobilization of organic leaf N, while GS2 is active in the re-assimilation of ammonia from photorespiration of photosynthetic tissues (Terco-Laforgue et al., 2004; Martin et al., 2005; Kamachi et al., 1992). Experiments with a rice knock-out mutant have specifically established that GS1 is accountable for the N-remobilization of senescent tissues (Tabuchi et al., 2007). In addition to GS1, two types of GOGATs are also present in plants: ferredoxin-independent (Fd-GOGAT; EC 1.4.7.1) and NADH-dependent GOGAT (NADH-GOGAT; EC 1.4.1.14). The Fd-independent GOGAT operates in combination with GS2, while the NADH dependent-GOGAT is mostly correlated with GS1 (Coschigano et al., 1998). As evidenced by mutants defective in Fd-GOGAT that show reversible lethal phenotypes, Fd-independent GOGAT was found to supply the only source of glutamate as an amino donor (Ferrario-Mery et al., 2002). On the other hand, NADH-GOGAT is found to be essential for the re-use of glutamine that has been transported in developing organs as suggested by overexpression patterns with a specific promoter. The increased expression levels of NADH-GOGAT and GS1 were identified throughout leaf senescence of various plant species, including Arabidopsis and populus species such as aspen, wheat, potato, and cotton (Buchanan-Wollaston et al., 2003, 2005; Lin and Wu, 2004; Bhalerao et al., 2003; Teixeira et al., 2005; Andersson el: al., 2004; Kichey et al., 2005; Gregersen and Holm, 2007; Shen et al., 2006). Moreover, in the context of N-recycling 2-oxoglutarate acts as a key enzyme engaged in a variety of reactions, particularly transaminating reactions that involve re-filling of glutamate depleted by nitrogen assimilation reactions and translocation to sink organs. Aconitate hydratase, citrate synthase, and isocitrate dehydrogenase (NADP) can supply 2-oxoglutarate for the assimilation of N through GS. In this model, multiple gene transcripts involved in enzymatic processes have been reported to be upregulated in various senescent plant leaves, such as Arabidopsis, cotton, species of populus, potatoes, and wheat (Buchanan-Wollaston et al., 2003, 2005; Lin and Wu, 2004; Andersson et al., 2004; Kichey et al., 2005; Teixeira et al., 2005; Gregersen and Holm, 2007). It should be noted that in Arabidopsis leaves transcriptional upregulation of NADP was not reported but the peroxisomal citrate synthase transcript is known to be upregulated (Buchanan-Wollaston et al., 2005).

8.7.2 THE PPDK-GS/GOGAT PATHWAY

In the PPDK-GS/GOGAT pathway (Figure 8.1), phosphoenolpyruvate (PEP) is produced using ATP and pyruvic acid by the action of pyruvate orthophosphate dikinase (PPDK; I-. C.2.7.9.1). PPDK was isolated from C4 plants for the first time but has also been found in C3 plants. Microarray studies on expression profile in dark-induced Arabidopsis senescing leaves have shown substantial upregulation of cytosolic PPDK (Lin and Wu, 2004). PEP is then converted to oxaloacetic acid (OAA) through the action of PEP carboxykinase (PEPC; EC 4.1.1.31), which offers carbon skeletons for the reconstruction of amino acids. Contrary to GS-GOGAT, glutamate supplies the amino donor for the amination of OAA rather than free ammonia, and this reaction is catalyzed by aspartate aminotransferase (AspAT; E.C. 2.6.1.1), which produces aspartate for further reactions. The 2-oxoglutarate, along with glutamine (generated from glutamate amination), by offering additional outgrowth of this amination process, act as a fusion between GS/GOGAT and PPDK-GS/COGAT. For instance, the glutamine form PPDK-GS/GOGAT interacts reversibly with 2-oxoglutarate, giving glutamate by the action of GOGAT, rendering glutamate accessible again for the next cycle of N-resorption. Moreover, the primary nitrogen-remobilization N carrier has been asparagine, which is produced by the action of asparagine synthase (Taylor et al., 2012). Several studies have demonstrated that the percentage of available asparagine molecules and the functionality of asparagine synthetase increase dramatically during leaf senescence (Lin and Wu, 2004; Shen et al., 2006; Gregersen and Holm, 2007). In summary, the role of degrading processes for protein breakdown and N-remobilization particularly in Arabidopsis has been well characterized, which has facilitated our acquisition of information. However, in other plant species more useful knowledge on functional gene regulation is still needed to validate the results derived from the study of N-resorption during leaf senescence in Arabidopsis. The increasingly evolving technologies for the functional analysis of possible resorption and regulatory genes and their extension to the N-metabolization research are a priority for the next decade. The application of such research would be very significant for crop plants, including the manipulation of genetic diversity.

8.8 TRANSLOCATION OF NITROGEN COMPOUNDS TO THE SINK TISSUES

Amino acids and small peptides produced by improved plastidial proteolysis and other proteins experience a sequence of phases of transportation between the protein degradation site and the reuse or storage site of the remobilized nitrogen. Depending upon the organelle in which an amino acid is identified after proteolysis, membrane transport proteins are necessary for export from vacuoles or chloroplasts. Amino acids must be transferred to the vascular bundles within senescing mesophyll tissues. A particularly interesting case can be found in soybeans and some similar plants, where a specific single-cell layer, termed the paraveinal mesophyll (PVM), spans the area between the palisade parenchyma, the spongy mesophyll, and the minor veins (Lansing and Franceschi, 2000; Tegeder, 2014). This cell layer is believed to serve as a symplastic conduit for photosynthetic assimilates, including N compounds from mesophyll cells towards the minor veins. In addition, an important dataset reveals that the PVM cell layer also serves as a transitory store for N and (possibly) carbon compounds before their export from leaves to sinks (Bunker et al., 1995; Fan et al., 2009; Tegeder, 2014). Various experiments in microscopy, physiology, and biochemistry have shown that phloem loadings occur from the apoplast in most species, with the possible exception of plants that utilize raffinose-family oligosaccharides for long-distance carbohydrate transport (Turgeon and Medville, 2004). Consequently, metabolites must be released at some stage before phloem loading from the symplasm (cytosol of leaf cells) into the cell wall area (apoplast). It was found that nitrogen and other metabolites are primarily absorbed into a sieve tube/companion cell complex through a specialized transport protein, by secondary active transport of amino acids combined with an electrochemical (H+) gradient (Lalonde et al., 2004). The metabolites are transported by bulk flow from source to sink tissues once they enter the phloem. Both at the cellular level and tissues such as developing endosperms or cotyledons the transport processes involved with phloem unloading to the actual sink tissues are complex. In developing seeds the nitrogen metabolites move symplastically from the phloem until they reach the inner epidermis of the seed coat and are then released into the apoplast surrounding the embryo (Debeaujon et al., 2000). Nutrients are then acquired by the embryo through carrier proteins present in its outer epidermal cells (Distelfeld et al., 2014). Molecular approaches have contributed to the discovery of various membrane transporters. For example, the best-studied plant membrane transporters involved include sucrose transport, potassium transport, and oligopeptide/amino acid transport (Lalonde et al., 2004). There are 53 and 59 putative amino acid transporter genes found to be present in the genomes of Arabidopsis and rice, respectively (Lalonde et al., 2004). These genes belong to three "gene families": (a) ATF1 (amino acid transporter superfamily 1; (b) APC (amino acid-polyamine-choline transporter superfamily; and (c) MFS (main facilitator superfamily) (Lalonde et al., 2004). The best studies of these transporters are the amino acid permeases of the "subfamily" ATF1, which are located in vascular tissues and instigate the H+-coupled absorption of a vast array of amino acids. The contribution of these genes to the transport of amino acids is shown by evidence that the amino acid quantity of StAAP1 potato antisense line tuber is reduced and that other associated genes of this family are found in seeds, indicating the role in metabolite translocation to the seed (Lalonde et al., 2004; Koch et al., 2003). Moreover, T-DNA insertion in AtOPT3 and anti-sense suppression of AtPTR2 (members of peptide transporter families) decreased the metabolic supply of seeds that ultimately led to the failure of embryo development in Arabidopsis (Stacey et al., 2002). However, the characterization of peptides transported selectively by these protein transporters is challenging, because of the significant variety of amino acid compositions in small peptides. Overall, we are still at the beginning of our current understanding of the physiology, molecular biology, and biochemistry of proteolysis and transport processes necessary for the efficient recycling of N. In both situations, the ambiguity of plant genetics leads to difficulties in identifying specific roles to key genes. Although some important insights have been gained, there are still some knowledge gaps that persist. However, investigating the underlying mechanisms of N-resorption in different plant species offers great potential benefits throughout the food chain, either through the manipulation of the genome or environmental triggers.

8.9 TARGETS TO IMPROVE NITROGEN FERTILIZATION

Nitrogen is the most critical and commonly used agricultural input after water. The unjudicial use of these fertilizers creates both economic and environmental problems. Therefore, scientists must look for an alternative to reduce the environmental issues without compromising the yield. To enhance the plant's ability to use nitrogen most effectively, the processes of nitrate absorption, transport, assimilation, and remobilization within plants should be understood. The remobilization of N is known to be one of the essential steps in the improvement of NUE of plants (Mascluaux Daubreasse et al., 2010). As evident from the cereals, the N metabolites that are remobilized from leaves act as the main source of N for grain filling. Nitrogen accumulated in grains has been shown to account for 60–90% of the remobilized metabolites, and the rate of remobilization depends on the N availability and the efficiency of remobilization (Qiao et al., 2012). However, apart from the source/sink demands, it has been acknowledged that environmental factors along with genotype are crucial for the NUE by plants (McAllister et al., 2012). The genes involved in remobilization and translocation are thus promising priorities for enhancing NUE (McAllister et al., 2012). The genes aimed primarily for the NUE are usually those that code for N remobilization pathway components, particularly those participating in the rate-limiting steps. For example, studies have reported that transgenic overexpression of Asparagine synthase genes that are enthusiastically engaged in amino acid remobilization and translocation leads to an increase in the protein content of the seed and overall plant (Masclaux-Daubresse et al., 2010; Pathak et al., 2011). In addition, genetically modified rice (Oryza sativa) was created by the introduction of a complementary DNA (cDNA) of barley alanine aminotransferase induced by the OsAnt1 (tissue-specific) promoter (Shrawat et al.,

2008; Pathak et al., 2011). These plants exhibited higher biomass and grain yield along with major improvements in nitrate content and key metabolites indicating improved NUE. In addition, the overexpression of AlaAT (alanine aminotransferase) under the control of tissue-specific promoters exhibiting a strong NUE phenotype was shown by the root and shoot transcriptome evaluation of engineered rice (Beatty et al., 2013). The function of GDH in N remobilization is said to be controversial, but genetically engineered plants have been shown to overexpress the gdhA gene, thereby improving amino acid content and yields in maize and wheat (Lightfoot, 2009). Moreover, the cytosolic GS1, which is responsible for the reassimilation of ammonia released from the proteolysis, offers the potential target for increasing the NUE (Brugiere et al., 2000). The justification for this statement comes from the work on maize, in which overexpression of GS1 isoenzyme under the control of the CsVMV promoter amplified kernel size and number by many folds, thereby improving yield (Masclaux Daubresse et al., 2010). Although transcriptome analyzes during leaf senescence have been performed in many crop plants (Table. 8.2), many of the genes assumed to be involved in the efficient use of nitrogen remain functionally uncharacterized. The CRISPR/Cas9-based gene manipulation and high-throughput screening method with a non-invasive phenotyping platform in mutants, transgenic lines, or varieties are likely to enable the identification and alteration of essential genes (Ito et al., 2015). Although there are several interesting genetic targets to be explored to maximize the NUE of plants, the final yield depends on the nature of the gene, allele, and promoter used. With the discovery of more new genes and signaling factors together with the advancement of new techniques, it may be possible to reduce the use of nitrogen fertilizers for crop production shortly.

TABLE 8.2
List of Transgenic Plants with Observed NUE Phenotypes

Enzyme/Gene product	Gene source	Promoter used	Target plant	Phenotypes Acquired
GS2-Clp glutamine synthetase	*Rice*	CaMV35S	*Rice*	Improved photorespiration, salinity tolerance
GS1-Cystolic glutamine synthetase	*Common bean*	Rubisco	*Wheat*	Enhanced capacity to accrue nitrogen
NADH–GOGAT–NADH-dependent glutamate synthetase	*Rice*	O. Sativa	*Rice*	Enhanced grain filling & grain weight
GDH–glutamate dehydrogenase	*E. coli*	OsUBI	*Maize*	Improved N assimilation, herbicide tolerance, biomass, Seed amino acid content
GDH–glutamate dehydrogenase	E. coli	CaMV 35S	Tobacco	Increased dry weight, ammonium assimilation, and sugar content
GS1–cytosolic glutamine G. max synthetase	Alfalfa	CaMV 35S	Lotus	Higher biomass and leaf proteins
OsENOD93-1	nodulin gene	35sC4PDK	*Rice*	Increased shoot mass and seed yield
ASN1–glutamine-dependent asparagine synthetase	*Arabidopsis*	CaMV 35S	Arabidopsis	Enhanced seed protein
ASNI–asparagine synthetase	Pea	CaMV 35S	Tobacco	Reduced biomass and increased level of free asparagine
AlaAT–alanine aminotransferase	barley	btg26	Rapeseed	Increased yields even with 50% less N fertilizer

Source: Pathak et al., 2011.

8.10 CONCLUSION

The senescence process can be described as a phenological stage that marks the end of leaf development with the remobilization and orderly transition of reserves to growing tissues. This chapter summarized current knowledge on various facets of protein degradation as part of the leaf senescence. Protein degradation is the main molecular and cellular process that typically occurs during senescence and remobilization of nitrogen. The chloroplast protein (about 2/3 of leaf protein) meltdown in senescing leaves is accomplished by the concerted action of at least three pathways, executing their function in different subcellular compartments. Therefore, it can be concluded that preliminary steps in the degradation of chlorophyll pigment and chloroplast protein take place by the chloroplast degradation pathway in the intact organelles. The breakdown products are further catabolized employing vesicles that emerge from chloroplast and end up in central lytic vacuole. Every vesicle can be asserted to have its chloroplast protein array, giving the cell the versatility needed to independently/separately deplete stromal and thylakoid proteins with various time-courses depending on environmental conditions. Moreover, the scientific data suggests that senescence entails the involvement of proteases of all catalytic types virtually from every family that are upregulated during natural or stress-induced senescence. Furthermore, the ammonia generated by the aminotransferase reactions is reassimilated by the action of GS-GOGAT and PPDK/GS-GOGAT into amino acids, which are readily translocated to the sink tissues through the phloem. As far as the regulatory mechanism of senescence is concerned, the identification and availability of senescence-related genes, promoters, and mutants not only allow to obtain a clear picture but also open new perspectives on the processes that occur during senescence. Finally, the overarching purpose of increasing our knowledge of nitrogen remobilization processes is to boost the NUE of plants. Exploiting transgenic technology to overexpress genes that are actively engaged in the remobilization of nitrogen metabolites such as AS, GS1, and GS-GOGAT pose attractive targets for increasing the NUE of plants. Instead of altering one or several genes at the same time, the most reasonable way to boost NUE is to exploit transcription factors that facilitate a sequence of genes functioning in a network of nitrogen remobilization. This theory would be expedited by the advances in the genome sequences of crop varieties along with the recent advent of advanced genetic techniques that can be used in any species. Thus, the exploitation of natural variation and quantitative genetic methods with the application of agronomic, breeding, and biotechnological techniques will contribute to attaining crops with higher NUE.

Acknowledgments: The authors appreciate the facilities offered by the Department of Botany, University of Kashmir, Srinagar, J&K, India, to conduct this work. Also thankful to University Grants Commission for providing fellowship to authors through the JRF scheme. We are also thankful to the anonymous referees for their helpful comments.

REFERENCES

Adam, Z. (2001). Chloroplast proteases and their role in photosynthesis regulation. In E.M. Aro and B. Andersson (eds.), *Regulation of Photosynthesis* (pp. 265–276). Springer, Dordrecht.

Adam, Z., & Clarke, A. K. (2002). Cutting edge of chloroplast proteolysis. Trends in *Plant Science, 7*(10), 451–456.

Andersson, A., Keskitalo, J., Sjödin, A., Bhalerao, R., Sterky, F., Wissel, K., Tandre, K., Aspeborg, H., Moyle, R., Ohmiya, Y., & Bhalerao, R. (2004). A transcriptional timetable of autumn senescence. *Genome Biology, 5*(4), R24.

Aubert, S., Bligny, R., Douce, R., Gout, E., Ratcliffe, R. G., & Roberts, J. K. M. (2001). Contribution of glutamate dehydrogenase to mitochondrial glutamate metabolism studied by 13 C and 31 P nuclear magnetic resonance. *Journal of Experimental Botany, 52*(354), 37–45.

Avila-Ospina L., Moison M., Yoshimoto K., & Masclaux-Daubresse C. (2014) Autophagy, plant senescence, and nutrient recycling. *Journal of Experimental Botany, 65*(14), 3799–3811.

Bachmann, A., Fernández-López, J. O. S. E., Ginsburg, S., Thomas, H., Bouwkamp, J. C., Solomos, T., & Matile, P. (1994). Stay-green genotypes of *Phaseolus vulgaris* L.: chloroplast proteins and chlorophyll catabolites during foliar senescence. *New Phytologist*, *126*(4), 593–600.

Beatty, P. H., Carroll, R. T., Shrawat, A. K., Guevara, D., & Good, A. G. (2013). Physiological analysis of nitrogen-efficient rice overexpressing alanine aminotransferase under different N regimes. *Botany*, *91*(12), 866–883.

Ben-Porath, I., & Weinberg, R. A. (2005). The signals and pathways activating cellular senescence. *The International Journal of Biochemistry & Cell Biology*, *37*(5), 961–976.

Bernard, S. M., & Habash, D. Z. (2009). The importance of cytosolic glutamine synthetase in nitrogen assimilation and recycling. *New Phytologist*, *182*(3), 608–620.

Bhalerao, R., Keskitalo, J., Sterky, F., Erlandsson, R., Björkbacka, H., Birve, S. J., Karlsson, J., Gardeström, P., Gustafsson, P., Lundeberg, J., & Jansson, S. (2003). Gene expression in autumn leaves. *Plant Physiology*, *131*(2), 430–442.

Brouquisse, R., Masclaux, C., Feller, U., & Raymond, P. (2001). Protein hydrolysis and nitrogen remobilisation in plant life and senescence. In P. Lea and J.F. Morot-Gaudry (Eds.), *Plant Nitrogen* (pp. 275–293). Springer, Berlin, Heidelberg.

Brugière, N., Dubois, F., Masclaux, C., Sangwan, R. S., & Hirel, B. (2000). Immunolocalization of glutamine synthetase in senescing tobacco (*Nicotiana tabacum* L.) leaves suggests that ammonia assimilation is progressively shifted to the mesophyll cytosol. *Planta*, *211*(4), 519–527.

Buchanan-Wollaston, V. (1994). Isolation of cDNA clones for genes that are expressed during leaf senescence in Brassica napus (identification of a gene encoding a senescence-specific metallothionein-like protein). *Plant Physiology*, *105*(3), 839–846.

Buchanan-Wollaston, V., Earl, S., Harrison, E., Mathas, E., Navabpour, S., Page, T., & Pink, D. (2003). The molecular analysis of leaf senescence–a genomics approach. *Plant Biotechnology Journal*, *1*(1), 3–22.

Buchanan-Wollaston, V., Earl, S., Harrison, E., Mathas, E., Navabpour, S., Page, T., & Pink, D. (2003). The molecular analysis of leaf senescence–A genomics approach. *Plant Biotechnology Journal*, *1*(1), 3–22.

Buchanan-Wollaston, V., Page, T., Harrison, E., Breeze, E., Lim, P. O., Nam, H. G., & Leaver, C. J. (2005). Comparative transcriptome analysis reveals significant differences in gene expression and signalling pathways between developmental and dark/starvation-induced senescence in Arabidopsis. *The Plant Journal*, *42*(4), 567–585.

Buchanan-Wollaston, V., Page, T., Harrison, E., Breeze, E., Lim, P. O., Nam, H. G., Lin, J.F., Wu, S.H., Swidzinski, J., Ishizaki, K., & Leaver, C. J. (2005). Comparative transcriptome analysis reveals significant differences in gene expression and signalling pathways between developmental and dark/starvation-induced senescence in Arabidopsis. *The Plant Journal*, *42*(4), 567–585.

Bunker, T. W., Koetje, D. S., Stephenson, L. C., Creelman, R. A., Mullet, J. E., & Grimes, H. D. (1995). Sink limitation induces the expression of multiple soybean vegetative lipoxygenase mRNAs while the endogenous jasmonic acid level remains low. *The Plant Cell*, *7*(8), 1319–1331.

Chang, X., Donnelly, L., Sun, D., Rao, J., Reid, M. S., & Jiang, C. Z. (2014). A petunia homeodomain-leucine zipper protein, PhHD-Zip, plays an important role in flower senescence. *PLoS One*, *9*(2), e88320.

Chiba, A., Ishida, H., Nishizawa, N. K., Makino, A., & Mae, T. (2003). Exclusion of ribulose-1, 5-bisphosphate carboxylase/oxygenase from chloroplasts by specific bodies in naturally senescing leaves of wheat. *Plant and Cell Physiology*, *44*(9), 914–921.

Coschigano, K. T., Melo-Oliveira, R., Lim, J., & Coruzzi, G. M. (1998). Arabidopsis gls mutants and distinct Fd-GOGAT genes: implications for photorespiration and primary nitrogen assimilation. *The Plant Cell*, *10*(5), 741–752.

Cuervo, A. M., Gomes, A. V., Barnes, J. A., & Dice, J. F. (2000). Selective degradation of annexins by chaperone-mediated autophagy. *Journal of Biological Chemistry*, *275*(43), 33329–33335.

Curty, C., & Engel, N. (1996). Detection, isolation and structure elucidation of a chlorophyll a catabolite from autumnal senescent leaves of *Cercidiphyllum japonicum*. *Phytochemistry*, *42*(6), 1531–1536.

Debeaujon, I., Léon-Kloosterziel, K. M., & Koornneef, M. (2000). Influence of the testa on seed dormancy, germination, and longevity in Arabidopsis. *Plant Physiology*, *122*(2), 403–414.

Deschênes-Simard, X., Lessard, F., Gaumont-Leclerc, M. F., Bardeesy, N., & Ferbeyre, G. (2014). Cellular senescence and protein degradation: breaking down cancer. *Cell Cycle*, *13*(12), 1840–1858.

Diab, H., & Limami, A. M. (2016). Reconfiguration of N metabolism upon hypoxia stress and recovery: roles of alanine aminotransferase (AlaAT) and glutamate dehydrogenase (GDH). *Plants*, *5*(2), 25.

Diaz, C., Lemaître, T., Christ, A., Azzopardi, M., Kato, Y., Sato, F.,.... & Masclaux-Daubresse, C. (2008). Nitrogen recycling and remobilization are differentially controlled by leaf senescence and development stage in Arabidopsis under low nitrogen nutrition. *Plant Physiology*, *147*(3), 1437–1449.

Díaz-Mendoza, M., Velasco-Arroyo, B., González-Melendi, P., Martínez, M., & Díaz, I. (2014). C1A cysteine protease–cystatin interactions in leaf senescence. *Journal of Experimental Botany*, *65*(14), 3825–3833.

Diaz-Mendoza, M., Velasco-Arroyo, B., Santamaria, M. E., González-Melendi, P., Martinez, M., & Diaz, I. (2016). Plant senescence and proteolysis: two processes with one destiny. *Genetics and Molecular Biology*, *39*(3), 329–338.

Distelfeld, A., Avni, R., & Fischer, A. M. (2014). Senescence, nutrient remobilization, and yield in wheat and barley. *Journal of Experimental Botany*, *65*(14), 3783–3798.

Doelling, J. H., Walker, J. M., Friedman, E. M., Thompson, A. R., & Vierstra, R. D. (2002). The APG8/12-activating enzyme APG7 is required for proper nutrient recycling and senescence in Arabidopsis thaliana. *Journal of Biological Chemistry*, *277*(36), 33105–33114.

Eckhardt, U., Grimm, B., & Hörtensteiner, S. (2004). Recent advances in chlorophyll biosynthesis and breakdown in higher plants. *Plant Molecular Biology*, *56*(1), 1–14.

Fan, S. C., Lin, C. S., Hsu, P. K., Lin, S. H., & Tsay, Y. F. (2009). The Arabidopsis nitrate transporter NRT1. 7, expressed in phloem, is responsible for source-to-sink remobilization of nitrate. *The Plant Cell*, *21*(9), 2750–2761.

Feller, U., Anders, I., & Mae, T. (2008). Rubiscolytics: fate of Rubisco after its enzymatic function in a cell is terminated. *Journal of Experimental Botany*, *59*(7), 1615–1624.

Ferrario-Méry, S., Valadier, M. H., Godefroy, N., Miallier, D., Hirel, B., Foyer, C. H., & Suzuki, A. (2002). Diurnal changes in ammonia assimilation in transformed tobacco plants expressing ferredoxin-dependent glutamate synthase mRNA in the antisense orientation. *Plant Science*, *163*(1), 59–67.

Fischer, A. M. (2012). The complex regulation of senescence. *Critical Reviews in Plant Sciences*, *31*(2), 124–147.

Frank, S., Hollmann, J., Mulisch, M., Matros, A., Carrión, C. C., Mock, H. P., Hensel, G., & Krupinska, K. (2019). Barley cysteine protease PAP14 plays a role in degradation of chloroplast proteins. *Journal of Experimental Botany*, *70*(21), 6057–6069.

Gan, S. (2003). Mitotic and postmitotic senescence in plants. *Science of Aging Knowledge Environment*, *2003*(38), 7.

Gepstein, S., Sabehi, G., Carp, M. J., Hajouj, T., Nesher, M. F. O., Yariv, I., Dor, C., & Bassani, M. (2003). Large-scale identification of leaf senescence-associated genes. *The Plant Journal*, *36*(5), 629–642.

Good, A. G., Shrawat, A. K., & Muench, D. G. (2004). Can less yield more? Is reducing nutrient input into the environment compatible with maintaining crop production?. *Trends in Plant Science*, *9*(12), 597–605.

Gregersen, P. L., & Holm, P. B. (2007). Transcriptome analysis of senescence in the flag leaf of wheat (*Triticum aestivum* L.). *Plant Biotechnology Journal*, *5*(1), 192–206.

Gregersen, P. L., Culetic, A., Boschian, L., & Krupinska, K. (2013). Plant senescence and crop productivity. *Plant Molecular Biology*, *82*(6), 603–622.

Gregersen, P. L., Holm, P. B., & Krupinska, K. (2008). Leaf senescence and nutrient remobilisation in barley and wheat. *Plant Biology*, *10*, 37–49.

Guiboileau, A., Sormani, R., Meyer, C., & Masclaux-Daubresse, C. (2010). Senescence and death of plant organs: Nutrient recycling and developmental regulation. *Comptes Rendus Biologies*, *333*(4), 382–391.

Hara-Nishimura, I., Hatsugai, N., Nakaune, S., Kuroyanagi, M., & Nishimura, M. (2005). Vacuolar processing enzyme: an executor of plant cell death. *Current Opinion in Plant Biology*, *8*(4), 404–408.

Himelblau, E., & Amasino, R. M. (2001). Nutrients mobilized from leaves of *Arabidopsis thaliana* during leaf senescence. *Journal of Plant Physiology*, *158*(10), 1317–1323.

Hörtensteiner, S., & Feller, U. (2002). Nitrogen metabolism and remobilization during senescence. *Journal of Experimental Botany*, *53*(370), 927–937.

Hortensteiner, S., Krautler, B. (2011). Chlorophyll breakdown in higher plants. *Biochimica et Biophysica Acta*, *1807*, 977–88.

Ichimura, Y., Kirisako, T., Takao, T., Satomi, Y., Shimonishi, Y., Ishihara, N., Mizushima, N., Tanida, I., Kominami, E., Ohsumi, M., & Noda, T. (2000). A ubiquitin-like system mediates protein lipidation. *Nature*, *408*(6811), 488–492.

Ishida, H., Anzawa, D., Kokubun, N., Makino, A., & Mae, T. (2002). Direct evidence for non-enzymatic fragmentation of chloroplastic glutamine synthetase by a reactive oxygen species. *Plant, Cell & Environment*, *25*(5), 625–631.

Ishida, H., Nishimori, Y., Sugisawa, M., Makino, A., & Mae, T. (1997). The large subunit of ribulose-1, 5-bisphosphate carboxylase/oxygenase is fragmented into 37-kDa and 16-kDa polypeptides by active oxygen in the lysates of chloroplasts from primary leaves of wheat. *Plant and Cell Physiology*, *38*(4), 471–479.

Ishida, H., Yoshimoto, K., Izumi, M., Reisen, D., Yano, Y., Makino, A., Ohsumi, Y., Hanson, M.R., & Mae, T. (2008). Mobilization of rubisco and stroma-localized fluorescent proteins of chloroplasts to the vacuole by an ATG gene-dependent autophagic process. *Plant Physiology*, *148*(1), 142–155.

Ito, Y., Nishizawa-Yokoi, A., Endo, M., Mikami, M., & Toki, S. (2015). CRISPR/Cas9-mediated mutagenesis of the RIN locus that regulates tomato fruit ripening. *Biochemical and Biophysical Research Communications*, *467*(1), 76–82.

Jarvis, P., & López-Juez, E. (2013). Biogenesis and homeostasis of chloroplasts and other plastids. *Nature Reviews Molecular Cell Biology*, *14*(12), 787–802.

Jin, Z., Zhu, Y., Li, X., Dong, Y., & An, Z. (2015). Soil N retention and nitrate leaching in three types of dunes in the Mu Us desert of China. *Scientific Reports*, *5*, 14222.

Kamachi, K., Yamaya, T., Hayakawa, T., Mae, T., & Ojima, K. (1992). Vascular bundle-specific localization of cytosolic glutamine synthetase in rice leaves. *Plant Physiology*, *99*(4), 1481–1486.

Kichey, T., Hirel, B., Heumez, E., Dubois, F., & Le Gouis, J. (2007). In winter wheat (*Triticum aestivum* L.), post-anthesis nitrogen uptake and remobilisation to the grain correlates with agronomic traits and nitrogen physiological markers. *Field Crops Research*, *102*(1), 22–32.

Kichey, T., Le Gouis, J., Sangwan, B., Hirel, B., & Dubois, F. (2005). Changes in the cellular and subcellular localization of glutamine synthetase and glutamate dehydrogenase during flag leaf senescence in wheat (*Triticum aestivum* L.). *Plant and Cell Physiology*, *46*(6), 964–974.

Killingbeck, K. T. (2004). Nutrient resorption. In: L.D. Noodén (ed.), *Plant Cell Death Processes* (pp. 215–226). Elsevier Academic Press, San Diego.

Kim, J., & Klionsky, D. J. (2000). Autophagy, cytoplasm-to-vacuole targeting pathway, and pexophagy in yeast and mammalian cells. *Annual Review of Biochemistry*, *69*(1), 303–342.

Koch, W., Kwart, M., Laubner, M., Heineke, D., Stransky, H., Frommer, W. B., & Tegeder, M. (2003). Reduced amino acid content in transgenic potato tubers due to antisense inhibition of the leaf H+/amino acid symporter StAAP1. *The Plant Journal*, *33*(2), 211–220.

Komenda, J., Sobotka, R., & Nixon, P. J. (2012). Assembling and maintaining the Photosystem II complex in chloroplasts and cyanobacteria. *Current Opinion in Plant Biology*, *15*(3), 245–251.

Kräutler, B., Jaun, B., Matile, P., Bortlik, K., & Schellenberg, M. (1991). On the enigma of chlorophyll degradation: the constitution of a secoporphinoid catabolite. *Angewandte Chemie International Edition in English*, *30*(10), 1315–1318.

Krupinska, K. (2007). Fate and activities of plastids during leaf senescence. In *The Structure and Function of Plastids* (pp. 433–449). Springer, Dordrecht.

Lalonde, S., Wipf, D., & Frommer, W. B. (2004). Transport mechanisms for organic forms of carbon and nitrogen between source and sink. *Annual Review of Plant Biology*, *55*, 341–372.

Lansing, A. J., & Franceschi, V. R. (2000). The paraveinal mesophyll: A specialized path for intermediary transfer of assimilates in legume leaves. *Functional Plant Biology*, *27*(9), 757–767.

Levine, B., & Klionsky, D. J. (2004). Development by self-digestion: Molecular mechanisms and biological functions of autophagy. *Developmental Cell*, *6*(4), 463–477.

Li, F., & Vierstra, R. D. (2012). Autophagy: A multifaceted intracellular system for bulk and selective recycling. *Trends in Plant Science*, *17*(9), 526–537.

Lightfoot, D. A. (2009). 7 Genes for use in improving nitrate use efficiency in crops. *Genes for Plant Abiotic Stress*, 167.

Lim, P. O., Kim, H. J., & Gil Nam, H. (2007). Leaf senescence. *Annual Review of Plant Biology*, *58*, 115–136.

Lim, P. O., Lee, I. C., Kim, J., Kim, H. J., Ryu, J. S., Woo, H. R., & Nam, H. G. (2010). Auxin response factor 2 (ARF2) plays a major role in regulating auxin-mediated leaf longevity. *Journal of Experimental Botany*, *61*(5), 1419–1430.

Lin, J. F., & Wu, S. H. (2004). Molecular events in senescing Arabidopsis leaves. *The Plant Journal*, *39*(4), 612–628.

Luciński, R., Misztal, L., Samardakiewicz, S., & Jackowski, G. (2011). The thylakoid protease Deg2 is involved in stress-related degradation of the photosystem II light-harvesting protein Lhcb6 in Arabidopsis thaliana. *New Phytologist*, *192*(1), 74–86.

Lundquist, P. K., Poliakov, A., Bhuiyan, N. H., Zybailov, B., Sun, Q., & van Wijk, K. J. (2012). The functional network of the Arabidopsis plastoglobule proteome based on quantitative proteomics and genome-wide coexpression analysis. *Plant Physiology*, *158*(3), 1172–1192.

Maathuis, F. J. (2009). Physiological functions of mineral macronutrients. *Current Opinion in Plant Biology*, *12*(3), 250–258.

Mae, T., Makino, A., & Ohira, K. (1983). Changes in the amounts of ribulose bisphosphate carboxylase synthesized and degraded during the life span of rice leaf (*Oryza sativa* L.). *Plant and Cell Physiology*, *24*(6), 1079–1086.

Marschner, H. (1995). *Mineral Nutrition of Higher Plants*. Academic Press, London p. 889.

Marshall, R. S., McLoughlin, F., & Vierstra, R. D. (2016). Autophagic turnover of inactive 26S proteasomes in yeast is directed by the ubiquitin receptor Cue5 and the Hsp42 chaperone. *Cell Reports*, *16*(6), 1717–1732.

Martin, A., Belastegui-Macadam, X., Quilleré, I., Floriot, M., Valadier, M. H., Pommel, B., Andrieu, B., Donnison, I., & Hirel, B. (2005). Nitrogen management and senescence in two maize hybrids differing in the persistence of leaf greenness: Agronomic, physiological and molecular aspects. *New Phytologist*, *167*(2), 483–492.

Martínez, D. E., Bartoli, C. G., Grbic, V., & Guiamet, J. J. (2007). Vacuolar cysteine proteases of wheat (*Triticum aestivum* L.) are common to leaf senescence induced by different factors. *Journal of Experimental Botany*, *58*(5), 1099–1107.

Masclaux, C., Quillere, I., GALLAIS, A., & Hirel, B. (2001). The challenge of remobilisation in plant nitrogen economy. A survey of physio-agronomic and molecular approaches. *Annals of Applied Biology*, *138*(1), 69–81.

Masclaux, C., Valadier, M. H., Brugière, N., Morot-Gaudry, J. F., & Hirel, B. (2000). Characterization of the sink/source transition in tobacco (*Nicotiana tabacum* L.) shoots in relation to nitrogen management and leaf senescence. *Planta*, *211*(4), 510–518.

Masclaux-Daubresse, C., Chen, Q., & Havé, M. (2017). Regulation of nutrient recycling via autophagy. *Current Opinion in Plant Biology*, *39*, 8–17.

Masclaux-Daubresse, C., Daniel-Vedele, F., Dechorgnat, J., Chardon, F., Gaufichon, L., & Suzuki, A. (2010). Nitrogen uptake, assimilation and remobilization in plants: challenges for sustainable and productive agriculture. *Annals of Botany*, *105*(7), 1141–1157.

Masclaux-Daubresse, C., Daniel-Vedele, F., Dechorgnat, J., Chardon, F., Gaufichon, L., & Suzuki, A. (2010). Nitrogen uptake, assimilation and remobilization in plants: challenges for sustainable and productive agriculture. *Annals of Botany*, *105*(7), 1141–1157.

Matile, P. H. I. L. I. P. P. E. (1992). Chloroplast senescence. *Crop photosynthesis: Spatial and Temporal Determinants*, 413–440.

McAllister, C. H., Beatty, P. H., & Good, A. G. (2012). Engineering nitrogen use efficient crop plants: the current status. *Plant Biotechnology Journal*, *10*(9), 1011–1025.

Mitsuhashi, W., & Feller, U. (1992). Effects of light and external solutes on the catabolism of nuclear-encoded stromal proteins in intact chloroplasts isolated from pea leaves. *Plant Physiology*, *100*(4), 2100–2105.

Mitsuhashi, W., Crafts-Brandner, S. J., & Feller, U. (1992). Ribulose-1, 5-bis-phosphate carboxylase/oxygenase degradation in isolated pea chloroplasts incubated in the light or in the dark. *Journal of Plant Physiology*, *139*(6), 653–658.

Mizushima, N., Noda, T., Yoshimori, T., Tanaka, Y., Ishii, T., George, M. D., Klionsky, D. J., Ohsumi, M., & Ohsumi, Y. (1998). A protein conjugation system essential for autophagy. *Nature*, *395*(6700), 395–398.

Mühlecker, W., & Kräutler, B. (1996). Breakdown of chlorophyll: constituent of nonfluorescing chlorophyll-catabolites from senescent cotyledons of the dicot rape. *Plant Physiology and Biochemistry (Paris)*, *34*(1), 61–75.

Nie, H., Zhao, C., Wu, G., Wu, Y., Chen, Y., & Tang, D. (2012). SR1, a calmodulin-binding transcription factor, modulates plant defense and ethylene-induced senescence by directly regulating NDR1 and EIN3. *Plant Physiology*, *158*(4), 1847–1859.

Noquet, C., Avice, J. C., Rossato, L., Beauclair, P., Henry, M. P., & Ourry, A. (2004). Effects of altered source–sink relationships on N allocation and vegetative storage protein accumulation in *Brassica napus* L. *Plant Science*, *166*(4), 1007–1018.

Orsel, M., Moison, M., Clouet, V., Thomas, J., Leprince, F., Canoy, A. S.,... & Masclaux-Daubresse, C. (2014). Sixteen cytosolic glutamine synthetase genes identified in the *Brassica napus* L. genome are differentially regulated depending on nitrogen regimes and leaf senescence. *Journal of Experimental Botany*, 65(14), 3927–3947.

Ougham, H. J., Morris, P., & Thomas, H. (2005). 4 The colors of autumn leaves as symptoms of cellular recycling and defenses against environmental stresses. *Current Topics in Developmental Biology*, 66(1), 135–160.

Pageau, K., Reisdorf-Cren, M., Morot-Gaudry, J. F., & Masclaux-Daubresse, C. (2006). The two senescence-related markers, GS1 (cytosolic glutamine synthetase) and GDH (glutamate dehydrogenase), involved in nitrogen mobilization, are differentially regulated during pathogen attack and by stress hormones and reactive oxygen species in *Nicotiana tabacum* L. leaves. *Journal of Experimental Botany*, 57(3), 547–557.

Park, J. H., Oh, S. A., Kim, Y. H., Woo, H. R., & Nam, H. G. (1998). Differential expression of senescence-associated mRNAs during leaf senescence induced by different senescence-inducing factors in Arabidopsis. *Plant Molecular Biology*, 37(3), 445–454.

Pathak, R. R., Lochab, S., & Raghuram, N. (2011). Plant systems | improving plant nitrogen-use efficiency. In: M. Moo-Young (ed.), *Comprehensive Biotechnology* (pp. 209–218), 2nd Edn., Elsevier, New York.

Peerzada Y.Y., & Iqbal M. (2021) Leaf senescence and ethylene signaling. In: T. Aftab and K.R. Hakeem (eds.), *Plant Growth Regulators*. Springer, Cham. https://doi.org/10.1007/978-3-030-61153-8_7

Peoples, M. B., & Dalling, M. J. (1988). The interplay between proteolysis and amino acid metabolism during senescence and nitrogen reallocation. In: L.D. Noodén and A.C. Leopold (eds.), *Senescence and Aging in Plants* (pp. 181–217). Academic Press, San Diego.

Qiao, J., Yang, L., Yan, T., Xue, F., & Zhao, D. (2012). Nitrogen fertilizer reduction in rice production for two consecutive years in the Taihu Lake area. *Agriculture, Ecosystems & Environment*, 146(1), 103–112.

Robatzek, S., & Somssich, I. E. (2001). A new member of the Arabidopsis WRKY transcription factor family, AtWRKY6, is associated with both senescence-and defence-related processes. *The Plant Journal*, 28(2), 123–133.

Roberts, I. N., Caputo, C., Criado, M. V., and Funk, C. (2012). Senescence associated proteases in plants. *Physiologia Plantarum*, 145, 130–139.

Salon, C., Munier-Jolain, N., Duc, G., Voisin, A. S., Grandgirard, D., Larmure, A., & Ney, B. (2001). Grain legume seed filling in relation to nitrogen acquisition: a review and prospects with particular reference to pea. *Agronomie*, 21, 539–552.

Schippers, J. H., Schmidt, R., Wagstaff, C., & Jing, H. C. (2015). Living to die and dying to live: The survival strategy behind leaf senescence. *Plant Physiology*, 169(2), 914–930.

Shen, F., Yu, S., Xie, Q., Han, X., & Fan, S. (2006). Identification of genes associated with cotyledon senescence in upland cotton. *Chinese Science Bulletin*, 51(9), 1085–1094.

Shrawat, A. K., & Good, A. G. (2008). Genetic engineering approaches to improving nitrogen use efficiency. *ISB News Report*, 1–5.

Smalle, J., & Vierstra, R. D. (2004). The ubiquitin 26S proteasome proteolytic pathway. *Annual Review of Plant Biology*, 55, 555–590.

Stacey, G., Koh, S., Granger, C., & Becker, J. M. (2002). Peptide transport in plants. *Trends in Plant Science*, 7(6), 257–263.

Sullivan, J. A., Shirasu, K., & Deng, X. W. (2003). The diverse roles of ubiquitin and the 26S proteasome in the life of plants. *Nature Reviews Genetics*, 4(12), 948–958.

Suzuki, N. N., Yoshimoto, K., Fujioka, Y., Ohsumi, Y., & Inagaki, F. (2005). The crystal structure of plant ATG12 and its biological implication in autophagy. *Autophagy*, 1(2), 119–126.

Tabuchi, M., Abiko, T., & Yamaya, T. (2007). Assimilation of ammonium ions and reutilization of nitrogen in rice (*Oryza sativa* L.). *Journal of Experimental Botany*, 58(9), 2319–2327.

Tamary, E., Nevo, R., Naveh, L., Levin-Zaidman, S., Kiss, V., Savidor, A., & Adam, Z. (2019). Chlorophyll catabolism precedes changes in chloroplast structure and proteome during leaf senescence. *Plant Direct*, 3(3), e00127.

Taylor, S. H., Parker, W. E., & Douglas, A. E. (2012). Patterns in aphid honeydew production parallel diurnal shifts in phloem sap composition. *Entomologia experimentalis et applicata*, 142(2), 121–129.

Tegeder, M. (2014). Transporters involved in source to sink partitioning of amino acids and ureides: Opportunities for crop improvement. *Journal of Experimental Botany*, 65(7), 1865–1878.

Teixeira, J., Pereira, S., Cánovas, F., & Salema, R. (2005). Glutamine synthetase of potato (*Solanum tuberosum* L. cv. Desiree) plants: Cell-and organ-specific expression and differential developmental regulation reveal specific roles in nitrogen assimilation and mobilization. *Journal of Experimental Botany*, *56*(412), 663–671.

Tercé-Laforgue, T., Dubois, F., Ferrario-Méry, S., de Crecenzo, M. A. P., Sangwan, R., & Hirel, B. (2004). Glutamate dehydrogenase of tobacco is mainly induced in the cytosol of phloem companion cells when ammonia is provided either externally or released during photorespiration. *Plant Physiology*, *136*(4), 4308–4317.

Thoenen, M., & Feller, U. (1998). Degradation of glutamine synthetase in intact chloroplasts isolated from pea (*Pisum sativum*) leaves. *Functional Plant Biology*, *25*(3), 279–286.

Thomas, H., & Donnison, I. (2000). Back from the brink: plant senescence and its reversibility. In: *Symposia of the Society for Experimental Biology* (Vol. 52, p. 149).

Thompson, A. R., Doelling, J. H., Suttangkakul, A., & Vierstra, R. D. (2005). Autophagic nutrient recycling in Arabidopsis directed by the ATG8 and ATG12 conjugation pathways. *Plant Physiology*, *138*(4), 20 97–2110.

Thompson, J. E., Froese, C. D., Madey, E., Smith, M. D., & Hong, Y. (1998). Lipid metabolism during plant senescence. *Progress in Lipid Research*, *37*(2–3), 119–141.

Tsay, Y. F., Chiu, C. C., Tsai, C. B., Ho, C. H., & Hsu, P. K. (2007). Nitrate transporters and peptide transporters. *FEBS Letters*, *581*(12), 2290–2300.

Turgeon, R., & Medville, R. (2004). Phloem loading. A reevaluation of the relationship between plasmodesmatal frequencies and loading strategies. *Plant Physiology*, *136*(3), 3795–3803.

Wada, S., Ishida, H., Izumi, M., Yoshimoto, K., Ohsumi, Y., Mae, T., & Makino, A. (2009). Autophagy plays a role in chloroplast degradation during senescence in individually darkened leaves. *Plant Physiology*, *149*(2), 885–893

Waditee-Sirisattha, R., Shibato, J., Rakwal, R., Sirisattha, S., Hattori, A., Nakano, T., & Tsujimoto, M. (2011). The Arabidopsis aminopeptidase LAP2 regulates plant growth, leaf longevity and stress response. *New Phytologist*, *191*(4), 958–969.

Wang, P., Mugume, Y., & Bassham, D. C. (2018, August). New advances in autophagy in plants: regulation, selectivity and function. *Seminars in Cell & Developmental Biology*, *80*, 113–122.

Watanabe, M., Balazadeh, S., Tohge, T., Erban, A., Giavalisco, P., Kopka, J., Mueller-Roeber, B., Fernie, A.R., & Hoefgen, R. (2013). Comprehensive dissection of spatiotemporal metabolic shifts in primary, secondary, and lipid metabolism during developmental senescence in Arabidopsis. *Plant Physiology*, *162*(3), 1290–1310.

Woo, H. R., Chung, K. M., Park, J. H., Oh, S. A., Ahn, T., Hong, S. H., & Nam, H. G. (2001). ORE9, an F-box protein that regulates leaf senescence in Arabidopsis. *The Plant Cell*, *13*(8), 1779–1790.

Woo, H. R., Kim, H. J., Lim, P. O., & Nam, H. G. (2019). Leaf senescence: Systems and dynamics aspects. *Annual Review of Plant Biology*, *70*, 347–376.

Xiong, Y., Contento, A. L., & Bassham, D. C. (2005). AtATG18a is required for the formation of autophagosomes during nutrient stress and senescence in Arabidopsis thaliana. *The Plant Journal*, *42*(4), 535–546.

Xu, F., Meng, T., Li, P., Yu, Y., Cui, Y., Wang, Y., Gong, Q., & Wang, N. N. (2011). A soybean dual-specificity kinase, GmSARK, and its Arabidopsis homolog, AtSARK, regulate leaf senescence through synergistic actions of auxin and ethylene. *Plant Physiology*, *157*(4), 2131–2153.

Xu, G., Fan, X., & Miller, A. J. (2012). Plant nitrogen assimilation and use efficiency. *Annual Review of Plant Biology*, *63*, 153–182.

Yoneyama, T., Ito, O., & Engelaar, W. M. (2003). Uptake, metabolism and distribution of nitrogen in crop plants traced by enriched and natural 15 N: progress over the last 30 years. *Phytochemistry Reviews*, *2*(1–2), 121–132.

Yoshida, S., Ito, M., Nishida, I., & Watanabe, A. (2001). Isolation and RNA gel blot analysis of genes that could serve as potential molecular markers for leaf senescence in Arabidopsis thaliana. *Plant and Cell Physiology*, *42*(2), 170–178.

Yoshimoto, K., Hanaoka, H., Sato, S., Kato, T., Tabata, S., Noda, T., & Ohsumi, Y. (2004). Processing of ATG8s, ubiquitin-like proteins, and their deconjugation by ATG4s are essential for plant autophagy. *The Plant Cell*, *16*(11), 2967–2983.

Yoshioka, M., Uchida, S., Mori, H., Komayama, K., Ohira, S., Morita, N., Nakanishi, T., & Yamamoto, Y. (2006). Quality control of photosystem II cleavage of reaction center D1 protein in spinach thylakoids by FtsH protease under moderate heat stress. *Journal of Biological Chemistry*, *281*(31), 21660–21669.

Żelisko, A., García-Lorenzo, M., Jackowski, G., Jansson, S., & Funk, C. (2005). AtFtsH6 is involved in the degradation of the light-harvesting complex II during high-light acclimation and senescence. *Proceedings of the National Academy of Sciences*, *102*(38), 13699–13704.

Zienkiewicz, M., Ferenc, A., Wasilewska, W., & Romanowska, E. (2012). High light stimulates Deg1-dependent cleavage of the minor LHCII antenna proteins CP26 and CP29 and the PsbS protein in Arabidopsis thaliana. *Planta*, *235*(2), 279–288.

9 Role of Phytohormones in Nitrogen Metabolism

Samina Mazahar and Ruchi Raina
Department of Botany, Dyal Singh College, University of Delhi,
New Delhi, India
Corresponding author: Email: samina.thaimee@gmail.com

CONTENTS

9.1 Introduction 119
9.2 Phytohormones and Nitrogen Regulation 121
    9.2.1 Auxin 121
    9.2.2 Cytokinin 122
        9.2.2.1 Nitrate Uptake Regulation through Cytokinin 122
    9.2.3 Abscisic Acid 123
    9.2.4 Ethylene 123
    9.2.4 Jasmonic Acid 123
9.3 Conclusion 123
References 124

9.1 INTRODUCTION

Phytohormones are a diverse group of plant growth regulators and are needed in very low concentrations in higher plants. They are the most important endogenous substance that modulates physiological, biochemical, and molecular processes and hence are necessary for plant survival as a sessile organism (Fahad et al., 2015a). Despite regulating the vital plant activities phytohormones are also responsive to different environmental factors such as nutrient, temperature, light, salt, drought, and microorganisms (Kiba et al., 2011). Hence, the phytohormones work as chemical messengers and can regulate various cellular activities in higher plants and maintain plants homeostasis. They are also known to regulate internal as well as external stimuli (Kazan, 2015) and hence interact to control growth. Therefore, their central role in promoting plant adaptation to ever-changing environmental conditions mediating growth, development, source/sink transitions, and nutrient allocation are well-known (Fahad et al., 2015a). The classical plant hormones are auxin, ethylene, cytokinin (CK), abscisic acid (ABA), and gibberellins (GA3), and recently there have been new additions to the family of growth regulators, including the brassinosteroids, strigolactones, salicylic acids, and jasmonic acids (Santner and Estelle, 2009).

Various studies have revealed the function of phytohormones concerning environmental signals and endogenous plant growth as well as development (Halliday et al., 2009, Kazan and Manners, 2009, Patel and Franklin, 2009). There is a well-coordinated pathway between hormone signaling and nitrogen metabolism, which in turn promotes plant growth and development through a range of complex interactions (Yeshitila and Gobena, 2020). According to various studies, there are three phytohormones, auxin, CK, and ABA, that coordinate the demand and acquisition of nitrogen (Signora et al., 2001; Wilkinson and Davies, 2002; Walch-Liu et al., 2006; Argueso et al., 2009).

DOI: 10.1201/9781003248361-9

The most important macronutrient, nitrogen, is available to plants as nitrate or ammonia (inorganic forms) or as free amino acids (organic forms). Nitrate being the most important N source is involved in various cellular activities (Konishi and Yanagisawa, 2014; Medici and Krouk, 2014). The nitrate uptake system constitutes the high- and low-affinity nitrate transporters encoded as the NRT1 and NRT2 gene family (Gojan et al., 2009), and their expression is regulated by numerous signals. One of the major high-affinity nitrate transport systems is the AtNRT2.1, which is repressed by the product of nitrogen assimilation and CKs and induced by nitrate and sugars (Brenner et al., 2005).

Hence, for the improvement of plant growth and crop yield, it's important to understand the plant signaling pathways in relevance to the N sources and phytohormones. However, nitrogen acquisition and its regulation involve the modulation of nitrate uptake system and lateral root proliferation (Forde and Walch-Liu, 2009). The outgrowth of lateral growth is a complex developmental process regulated by various signals such as nitrate, N-assimilation products, and growth promoters, particularly auxin, CK, and ABA (Walch-Liu et al., 2006, Zhang et al., 2007, Fukaki and Tasaka, 2009). In this chapter, we have summarized the role of different phytohormones in regulating plant metabolic activities and overall growth and development as shown in Table 9.1. Further, recent progress related to the role of phytohormones in regulating the nitrate uptake system, lateral root proliferation in response to N availability, and hence regulation of nitrogen metabolism is also discussed.

TABLE 9.1
Brief Outline of Phytohormones and Their Functions

Plant Hormone	Functions	References
Auxin IAA (most abundant)	Cell division, cell enlargement, organ development, defense, and stress response	(Kopittke, 2016; Egamberdieva et al., 2017)
	Vascular tissue development and apical dominance	(Wang et al., 2006)
	Embryonic and post-embryonic development	(Davies, 2004)
	Lateral root development enhances virulence in insects	(Liao et al., 2017; Yeshitila and Gobena, 2020)
Cytokinin (CK)	Photosynthesis and growth regulation, maintaining cell proliferation, cell differentiation, and senescence retardation	(Schmulling, 2002; Boivin et al., 2016) (Yeshitila and Gobena, 2020)
	Development of lateral roots, axillary branching in shoots.	Mohapatra et al. (2011)
	Improves grain filling	(Davies 2004; Fahad et al. 2015a)
	Nutrient mobilization, apical dominance, chloroplast biogenesis, anthocyanin production, vascular differentiation, and photo-morphogenic development	(Barciszewski et al. 2000; Iqbal et al. 2006, 2014, Fahad et al. 2015a)
	Alleviates salinity stress on plant growth	
Abscisic acid (ABA)	Maintains dormancy of buds, inhibits germination	(Lymperopoulos et al., 2018)
	Improves stress response, regulates growth and adaptation	(Egamberdieva et al., 2017)
	Controls downstream response under stress condition	(Wilkinson et al., 2012)
	Regulates the expression of stress-responsive genes	(Sah et al., 2016)
	Regulates root growth by allowing deeper rooting, modulates water and nutrient uptake under drought stress	(Vysotskaya et al., 2009, Cutler et al., 2010)
	Acts as antitranspirant and hence regulates stomatal closure	(Wilkinson and Davies, 2002, 2010)
	Controls hydraulic conductivities of shoots and roots and hence maintains tissue turgor potential	(Chaves et al., 2003)
	Improves antioxidant enzymes activity	(Guajardo et al., 2016)

TABLE 9.1 (Continued)
Brief Outline of Phytohormones and Their Functions

Plant Hormone	Functions	References
Gibberellic Acid	Promotes seed germination, breaks seed dormancy, shoot elongation, fruit and flower maturation	(Lymperopoulos et al., 2018) (Olszewski et al., 2002)
	Lateral shoot growth	(Ahmad, 2010; Iqbal and Ashraf, 2013)
	Promotes plant growth and metabolism under normal and stressful conditions	(Maggio et al., 2010)
	Enhances nutrient and water uptake reduces stomatal resistance	(Egamberdieva et al., 2017) (Ahmad, 2010)
	Increases yield, leaf area, mineral nutrition, and nitrogen metabolism	(Manjili et al., 2012)
	Maintains osmotic stress and tissue water content	
	Lowers the level of reactive oxygen species (ROS) enhances antioxidant enzyme activity and hence promotes better growth under stress	
Ethylene	Regulates plant growth and development, stress responses, flower senescence, leaf and petal abscission, ripening of fruits	(Groen and Whiteman, 2014; Shi et al., 2012) (Gamalero and Glick, 2012)
	Provides defense to plants against environmental stress	
Brassinosteroids	Photosynthesis, seed germination, pollen tube growth, activation of the proton pump, epinasty and leaf bending, nucleic acids and protein biosynthesis, reproductive growth, production of flowers and fruits	(Hayat et al. 2010; Fahad et al. 2015b) (Khripach et al. 2000, Zhang et al. 2007)
	Regulates ion uptake, enhances dry mass accumulation and antioxidant enzymes under salinity	(Sharma et al. 2013)
	Enhances proline contents in salt-stressed plants	

9.2 PHYTOHORMONES AND NITROGEN REGULATION

Extensive physiological, genetic, and molecular studies describe the nitrogen response to various phytohormones and their underlying mechanisms (Ueda et al., 2017). Nitrate is known to induce various cellular responses and hence function as a signaling molecule (Konishi and Yanagisawa, 2014; Medici and Krouk, 2014).

9.2.1 Auxin

Auxin is the main hormone involved in root development and N-mediated control of root architecture, which has been studied extensively in Arabidopsis (Ueda et al., 2017). The nitrate transporters 1.1 (NRT1.1) are involved in the dual function of taking up nitrate and transport of auxin and hence the changes in the auxin level as mediated by NRT1.1 in the lateral root primordium play a fundamental role in nitrate-induced lateral root growth (Krouk et al., 2010; Tsay et al., 2011; Mounier et al., 2014; Bouguyon et al., 2016; Krouk, 2016). The formation of the lateral root is known to be influenced by various factors, but auxin and nitrogen are considered the most important ones (Hu et al., 2020). According to Zhang et al. (2007) auxin is not only the factor regulating the lateral root development as the inhibitory effect of high nitrate on lateral root growth is not enhanced by exogenous auxin application.

During auxin signaling, the AUXIN RESPONSE FACTOR 8 (ARF8), which encodes a transcription factor of auxin signaling machinery, has been reported as the N-responsive gene in the pericycle (Gifford et al., 2008). Studies in Arabidopsis seedlings suggest that seedling grown on low nitrogen (LN) possess a higher level of auxin in roots than seedlings grown on high nitrogen concentration (Kiba et al., 2011). Various studies suggest that ammonium transporters (AMTs) and GDP-mannose pyrophosphate are involved in ammonium-induced changes in the root morphology, although the underlying mechanism is not well-known (Lima et al. 2010; Tsay et al. 2011). Further, the auxin biosynthesis, transport, and accumulation vary in response to different N availability in Arabidopsis (Ma et al., 2014; Krouk, 2016) and maize (Chen et al., 2013). Auxin-related regulatory genes such as TAR2 respond to N availability in Arabidopsis. Further, TAR2 is involved in auxin biosynthesis and transport such as PIN-FORMED1 (PIN1), PIN2, PIN4, and PIN7, which regulates the subcellular trafficking (Krouk et al., 2010). However, under the condition of N starvation the ammonia transporter gene AtAmt1.1 plays a key role in regulating the structure of lateral roots (Engineer and Kranz, 2007). Hence, the study of root architecture modification in response to nitrogen is a very complex phenomenon and requires a lot more research to fully understand it.

9.2.2 Cytokinin

CK, another important phytohormone, is involved in nitrogen signaling and a strong correlation between the two is observed in tobacco (Singh et al., 1992), barley (Samuelson and Larsson, 1993), and Urtica dioica (Wagner and Beck, 1993) and also in Plantago major where the growth-limiting effect caused by low nitrogen availability is partially overcome by exogenous treatment with cytokinin (Kuiper, 1988). CK functions as a long-distance root-to-shoot signal for nitrogen supplement (Sakakibara et al., 2008) as nitrogen supplementation causes an increase in CK content in the xylem sap and roots and shoots of maize (Takei et al., 2001). A similar study has been reported in Arabidopsis (Takei et al., 2004). A higher level of CK is reported in Arabidopsis seedlings grown on the high concentration of nitrate (HN, 10 Mm) and low CK level was observed in those grown on low nitrate (LN, 0.1 mM) (Kiba et al., 2011). Hence, according to this study, the CKs are considered as "nitrogen status signal" reflecting the nitrogen status along with the nitrogen supplement signal. According to various studies, CKs act as a local signal or as a shoot-to-root long-distance signal (Matsumoto-Kitano et al., 2008, Hirose et al., 2008; Kudo et al., 2010).

9.2.2.1 Nitrate Uptake Regulation through Cytokinin

It has been reported that the AtIPT3 gene among the differentially expressed seven genes in different tissues (encoding and regulating the CK biosynthesis catalyzed by isopentenyl transferase IPT) is nitrate inducible and is mainly involved in nitrate-dependent CK biosynthesis (Miyawaki et al., 2004; Takei et al., 2004). The nitrate-inducible expression of AtIPT3 is also expressed in detached shoots in phloem throughout the plant (Miyawaki et al., 2004, Takei et al., 2004), and this expression overlaps with NRT1.1 in roots, although in shoots, the expression of NRT1.1 is detected only in younger leaves (Guo et al., 2001). Similar studies are reported in detached sunflower and tobacco leaves where nitrogen supplementation induces CK accumulation (Singh et al., 1992). CKs are known to upregulate many shoot-expressed AtNRTs under both high and low nitrate conditions (Kiba et al., 2011). Wang et al. (2009) and Ho et al. (2009) reported that the nitrate-inducible expression of AtIPT3 is partly mediated by a protein named NRT1.1/CHL1, which serves as nitrate sensor and dual-affinity nitrate transporter.

The CKs may signal the availability of nitrogen through their translocation in the shoots or being produced locally in the roots. Hence, CK is involved in the negative regulation of nitrogen uptake-related genes and also controls the root architecture in response to nitrogen availability (Kiba et al., 2011). According to microarray analysis, the exogenous application of CK represses two AtNRT2 genes (AtNRT2.1 and AtNRT2.3), a urea transporter gene, three ammonium transporter genes, and three amino acid transporter genes in Arabidopsis (Brenner et al., 2005; Kiba et al., 2005; Sakakibara et al., 2006; Yokoyama et al., 2007). Hence, the repression of AtNRT genes leads to reduced nitrate uptake activity. The main components of the nitrate uptake and xylem-loading system are AtNRT2.1, AtNRT2.2, AtNRT1.1, and AtNRT1.5 amongst the root type genes (Li et al., 2007; Lin et al., 2008). Therefore, it is concluded that CKs are closely correlated with nitrogen, acting as a satiety signal of nitrogen and inhibiting the nitrate uptake in the roots, and on the other hand, CKs positively regulate a few AtNRT genes (Kiba et al., 2011).

9.2.3 ABSCISIC ACID

ABA is also reported to be involved in nitrogen signaling in several plant species (Wilkinson and Davies, 2002), however, the correlation between the two seems not very statistically significant. For instance, Arabidopsis seedling grown under high nitrogen (HN) and low nitrogen (LN) did not show any significant difference in the ABA levels (Kiba et al., 2011). Hence, there is still a question of whether ABA is involved in nitrogen signaling, which would be an interesting study. Various genetic studies suggest the involvement of ABA in lateral root development in Arabidopsis in response to high nitrate supply (Kiba et al., 2011).

Studies suggest that ABA controls the lateral root development after the emergence of lateral root primordium (De Smet et al., 2003). According to Zhang et al. (2007), a set of mutants (labi mutants), which produces lateral roots in the presence of ABA, showed reduced sensitivity to the inhibitory effect of high nitrate. The identification of the mutant labi gene offered a better understanding of the mechanism underlying the inhibition effect. The correlation between nitrogen signaling and ABA was studied in Medicago tranculata through a mutant latd (Yendrek et al., 2010), which is characterized by many defects in the maintenance of root growth and root meristem and is overcome by ABA application exogenously (Liang et al., 2007). According to Yendrek et al. (2010) the LATD gene encodes a transporter belonging to the NRT1 family and the primary root growth of the latd mutant is insensitive to nitrate.

9.2.4 ETHYLENE

Ethylene promotes physiological and morphological responses to nitrogen deficiency, and the nitrate transporters (NRT1.1 and NRT2.1) are responsive to ethylene (Tian et al., 2009). Further, the signaling activities and ethylene biosynthesis are induced and promoted by NRT2.1 when the external nitrate concentration is low (Zheng et al., 2013).

9.2.4 JASMONIC ACID

Jasmonic acid also plays an important role in regulating nitrogen metabolism. The jasmonate responsive gene (JR1) and D-AMINO ACID RASEMASE2 (DAAR2) promote lateral root growth under low nitrate levels (0.03mM KNO_3) (Gifford et al., 2013). According to Roses et al. (2013), the ratio of primary root to lateral root in seedling grown in 1mM nitrate medium is determined by ROOT SYSTEM ARCHITECTURE1 (RSA1). JR1 encodes a mannose-binding lectin superfamily protein involved in flowering time and heat stress tolerance (Xiao et al., 2015, Echevarría-Zomeño et al. 2016), and RSA1 and DAAR2 encode uncharacterized proteins such as tyrosine transaminase family protein and phenazine biosynthesis like domain-containing protein (Ueda et al., 2017).

9.3 CONCLUSION

Various research has concluded that there is a strong significant association between phytohormone and nitrogen signaling pathways. The phytohormones auxin, cytokinin, ABA, ethylene, and jasmonic acid are involved in nitrogen signaling. It has been reported that phytohormones interact with nitrogen and also promote sulfur (Maruyama-Nakashita et al., 2004), phosphorous (Franco-Zorrilla et al., 2002), and iron signaling (Seguela et al., 2008). Hence, the phytohormones have a complex signaling network regulating the plant metabolic activities as well as the nutritional status. Hence, the omics-based approach is much desirable as it involves the study of genomics, proteomics, metagenomics, and metabolomics and also will open avenues for future research.

REFERENCES

Ahmad, P. Growth and antioxidant responses in mustard (*Brassica juncea* L.) plant subjected to the combined effect of gibberellic acid and salinity. *Arch. Agron. Soil Sci.* 2010, 56, 575–588.

Argueso, C.T.; Ferreira, F.J.; Kieber, J.J. Environmental perception avenues: the interaction of cytokinin and environmental response pathways. *Plant Cell Environ.* 2009, 32, 1147–1160.

Barciszewski, J.; Siboska, G.; Rattan, S. I. S.; Clark, B. F. C. Occurrence, biosynthesis and properties of kinetin (N6-furfuryladenine). *Plant Growth Regul.* 2000, 32, 257–265.

Boivin, S.; Fonouni-Farde, C.; Frugier, F. How auxin and cytokinin phytohormones modulate root microbe interactions. *Front. Plant Sci.* 2016, 7, 1240.

Bouguyon, E.; Perrine-Walker, F.; Pervent, M. *et al*. Nitrate controls root development through post transcriptional regulation of the NRT1.1/NPF6.3 transporter/sensor. *Plant Physiol.* 2016, 172, 1237–1248.

Brenner, W.G.; Romanov, G.A.; Kollmer, I., Burkle, L.; Schmulling, T. Immediate-early and delayed cytokinin response genes of Arabidopsis thaliana identified by genome-wide expression profiling reveal novel cytokinin-sensitive processes and suggest cytokinin action through transcriptional cascades. *The Plant J.* 2005, 44, 314–333.

Brenner, W.G.; Romanov, G.A.; Kollmer, I., Burkle, L.; Schmulling, T. Immediate-early and delayed cytokinin response genes of Arabidopsis thaliana identified by genome-wide expression profiling reveal novel cytokinin-sensitive processes and suggest cytokinin action through transcriptional cascades. *The Plant J.* 2005, 44, 314–333.

Chaves, M. M.; Maroco, J. P.; Pereira, J. S. Understanding plant responses to drought – from genes to the whole plant. *Funct. Plant Biol.* 2003, 30, 239–264.

Chen, F.; Fang, Z.; Gao, Q., Ye, Y.; Jia, L.; Yuan, L.; Mi, G.; Zhang, F. Evaluation of the yield and nitrogen use efficiency of the dominant maize hybrids grown in North and Northeast China. *Sci. China Life Sci.* 2013, 56, 552–560.

Cutler, S. R.; Rodriguez, P. L.; Finkelstein, R. R.; Abrams, S. R. Abscisic acid: emergence of a core signaling network. *Annu. Rev. Plant Biol.* 2010, 61, 651–679.

Davies, P. J.; *Plant Hormones: Biosynthesis, Signal Transduction Action.* Kluwer Academic Publisher. 2004.

De Smet, I.; Signora, L.; Beeckman, T.; Inze, D.; Foyer, C. H.; Zhang, H. M. An abscisic acid-sensitive checkpoint in lateral root development of *Arabidopsis*. *Plant J.* 2003, 33, 543–55.

Echevarría-Zomeño, S.; Fernández-Calvino, L.; Castro-Sanz, A. B.; López, J. A.; Vázquez, J.; Castellano, M. M. Dissecting the proteome dynamics of the early heat stress response leading to plant survival or death in *Arabidopsis*. *Plant, Cell Environ.* 2016, 39, 1264–1278.

Egamberdieva, D.; Wirth, S. J.; Alqarawi, A. A.; Abd Allah, E. F.; Hashem, A. Phytohormones and beneficial microbes: Essential components for plants to balance stress and fitness. *Front. Microbiol.* 2017, 8, 2104.

Engineer, C.B.; Kranz, R.G. Reciprocal leaf and root expression of AtAmt1.1 and root architectural changes in response to nitrogen starvation. *Plant Physiol.* 2007, 143, 236–50.

Fahad, S.; Hussain, S.; Bano, A.; Saud, S.; Hassan, S.; Shan, D. *et al*. Potential role of phytohormones and plant growth-promoting rhizobacteria in abiotic stresses: consequences for changing environment. *Environ. Sci. Pollut. Res.* 2015a, 22, 4907–4921.

Fahad, S.; Nie, L.; Chen, Y.; Wu, C.; Xiong, D.; Saud, S. *et al*. Crop plant hormones and environmental stress. *Sustain Agric Rev.* 2015b, 15, 371–400.

Forde, B.G.; Walch-Liu, P. Nitrate and glutamate as environmental cues for behavioural responses in plant roots. *Plant Cell Environ.* 2009, 32, 682–693.

Franco-Zorrilla, J. M., Martin, A. C.; Solano, R.; Rubio, V.; Leyva, A.; Paz-Ares, J. Mutations at CRE1 impair cytokinin-induced repression of phosphate starvation responses in *Arabidopsis*. *The Plant J.* 2002, 32, 353–360.

Fukaki, H.; Tasaka, M. Hormone interactions during lateral root formation. *Plant Mol. Bio.* 2009, 69, 437–449.

Gamalero, E.; Glick, B.R. Ethylene and abiotic stress tolerance in plants, in: P. Ahmed, M.N.V. Prasad (Eds.), *Environmental Adaptations and Stress Tolerance of Plants in the Era of Climate Change.* Springer, New York, 2012, pp. 395–412.

Gifford, M. L.; Banta, J. A.; Katari, M. S, Hulsmans, J.; Chen, L.; Ristova, D. *et al.* Plasticity regulators modulate specific root traits in discrete nitrogen environments. *PLoS Genet.* 2013 9, e1003760.

Gifford, M.L.; Dean, A.; Gutierrez, R.A.; Coruzzi, G.M.; Birnbaum, K.D. Cell-specific nitrogen responses mediate developmental plasticity. *Proceedings of the National Academy of Sciences, USA*, 2008, 105, 803–808.

Gojon, A.; Nacry, P.; Davidian, J.C. Root uptake regulation: A central process for NPS homeostasis in plants. *Curr. Opi. Plant Bio.* 2009. 12, 328–338.

Groen, S.C.; Whiteman, N.K. The evolution of ethylene signaling in plant chemical ecology. *J. Chem. Ecol.* 2014, 40, 700–716.

Guajardo, E.; Correa, J. A.; Contreras-Porcia, L. Role of abscisic acid (ABA) in activating antioxidant tolerance responses to desiccation stress in intertidal seaweed species. *Planta.* 2016, 243, 767–781.

Guo, F.Q.; Wang, R.; Chen, M.; Crawford, N.M. The Arabidopsis dual-affinity nitrate transporter gene AtNRT1.1 (CHL1) is activated and functions in nascent organ development during vegetative and reproductive growth. *The Plant Cell.* 2001, 13, 1761–1777.

Halliday, K.J.; Martinez-Garcia, J.F.; Josse, E.M. Integration of light and auxin signaling. *Cold Spring Harbor Perspectives in Biology*, 2009, 1, a001586.

Hayat, S.; Mori, M.; Fariduddin, Q.; Bajguz, A.; Ahmad, A. Physiological role of brassinosteroids: an update. *Indian J. Plant Physiol.* 2010, 15, 99–109.

Hirose, N.; Takei, K.; Kuroha, T.; Kamada-Nobusada, T.; Hayashi, H.; Sakakibara, H. Regulation of cytokinin biosynthesis, compartmentalization, and translocation. *J. Exp. Bot.* 2008, 59, 75–83.

Ho, C.H.; Lin, S.H.; Hu, H.C.; Tsay, Y.F. CHL1 functions as a nitrate sensor in plants. *Cell.* 2009, 138, 1184–1194.

Hu, S.; Zhang, M.; Yang, Y.; Xuan, W.; Zou, Z.; Arkorful, E. *et al.* A novel insight into nitrogen and auxin signaling in lateral root formation in tea plant [*Camellia sinensis* (L.) O. Kuntze]. *BMC Plant Bio.* 2020, 20, 232.

Iqbal, M.; Ashraf, M. Gibberellic acid mediated induction of salt tolerance in wheat plants: growth, ionic partitioning, photosynthesis, yield and hormonal homeostasis. *Environ. Exp. Bot.* 2013, 86, 76–85.

Iqbal, M.; Ashraf, M.; Jamil, A. Seed enhancement with cytokinins: Changes in growth and grain yield in salt stressed wheat plants. *Plant Growth Regul.* 2006, 50, 29–39.

Iqbal, N.; Umar, S.; Khan, N. A.; Khan, M.I.R. A new perspective of phytohormones in salinity tolerance: regulation of proline metabolism. *Environ. Exp. Bot.* 2014, 100, 34–42.

Kazan, K. Diverse roles of jasmonates and ethylene in abiotic stress tolerance. *Trends Plant Sci.* 2015, 20, 219–229.

Kazan, K.; Manners, J.M. Linking development to defense: Auxin in plant–pathogen interactions. *Trends Plant Sci.* 2009, 14, 373–382.

Khripach, V.; Zhabinskii, V. N.; deGroot, A. E. Twenty years of brassinosteroids: steroidal plant hormones warrant better crops for the XXI century. Ann. Bot. 2000, 86, 441–447.

Kiba, T.; Kudo, T.; Kojima, M.; Sakakibara, H. Hormonal control of nitrogen acquisition: Roles of auxin, abscisic acid, and cytokinin. *J. Exp. Bot.* 2011, 62, 1399–1409.

Kiba, T.; Naitou, T.; Koizumi, N.; Yamashino, T.; Sakakibara, H.; Mizuno, T. Combinatorial microarray analysis revealing Arabidopsis genes implicated in cytokinin responses through the His/Asp phosphorelay circuitry. *Plant Cell Physiol.* 2005, 46, 339–355.

Konishi, M.; Yanagisawa, S. Emergence of a new step towards understanding the molecular mechanisms underlying nitrate-regulated gene expression. *J. Exp. Bot.* 2014, 65, 5589–5600.

Kopittke, P. M. Role of phytohormones in aluminium rhizotoxicity. *Plant Cell Environ.* 2016, 33, 2319–2328.

Krouk, G. Hormones and nitrate: a two-way connection. *Plant Mol. Biol.* 2016, 91, 599–606.

Krouk, G.; Lacombe, B.; Bielach, A. *et al.* Nitrate-regulated auxin transport by NRT1.1 defines a mechanism for nutrient sensing in plants. *Dev. Cell.* 2010, 18, 927–937.

Krouk, G.; Lacombe, B.; Bielach, A.; Perrine-Walker, F.; Malinska, K.; Mounier, E. et al. Nitrate-regulated auxin transport by NRT1.1 defines a mechanism for nutrient sensing in plants. *Dev. Cell.* 2010, 18, 927–937.

Kudo, T.; Kiba, T.; Sakakibara, H. Metabolism and long-distance translocation of cytokinins. *J. Integ. Plant Biol.* 2010, 52, 53–60.

Kuiper, D. Growth responses of Plantago major L. ssp. pleiosperma (Pilger) to changes in mineral supply: evidence for regulation by cytokinins. *Plant Physiol.* 1988, 87, 555–557.

Li, W.; Wang, Y.; Okamoto, M.; Crawford, N.M.; Siddiqi, M.Y.; Glass, A.D. Dissection of the AtNRT2.1:AtNRT2.2 inducible high-affinity nitrate transporter gene cluster. *Plant Physiol.* 2007, 143, 425–433.

Liang, Y., Mitchell, D.M.; Harris, J.M. Abscisic acid rescues the root meristem defects of the *Medicago truncatula* latd mutant. *Develop. Biol.* 2007, 304, 297–307.

Liao, X.; Lovett, B.; Fang, W.; St Leger, R. J. Metarhizium robertsii produces indole-3-acetic acid, which promotes root growth in Arabidopsis and enhances virulence to insects. *Microbiol.* 2017,163, 980–991.

Lima, J.E.; Kojima, S.; Takahashi, H.; von Wirén, N. Ammonium triggers lateral root branching in Arabidopsis in an Ammonium Transporter 1; 3-dependent manner. *Plant Cell*, 2010, 22, 3621–3633.

Lin, S.H.; Kuo, H.F.; Canivenc, G. *et al.* Mutation of the Arabidopsis NRT1.5 nitrate transporter causes defective root-to-shoot nitrate transport. *The Plant Cell* 2008, 20, 2514–2528.

Lymperopoulos, P.; Msanne, J.; Rabara, R. Phytochrome and phytohormones: working in tandem for plant growth and development. *Front. Plant Sci.* 2018, 9, 1037.

Ma, W.; Li, J.; Qu, B.; He, X.; Zhao, X.; Li, B.; Fu, X.; Tong, Y. Auxin biosynthetic gene TAR2 is involved in low nitrogen-mediated reprogramming of root architecture in Arabidopsis. *Plant J Cell Mol. Biol.* 2014, 78, 70–79.

Maggio, A.; Barbieri, G.; Raimondi, G.; De Pascale, S. Contrasting effects of ga3 treatments on tomato plants exposed to increasing salinity. *J. Plant Growth Regul.* 2010, 29, 63–72.

Manjili, F. A.; Sedghi, M.; Pessarakli, M. Effects of phytohormones on proline content and antioxidant enzymes of various wheat cultivars under salinity stress. *J. Plant Nutr.* 2012, 35, 1098–1111.

Maruyama-Nakashita, A.; Nakamura, Y.; Yamaya, T., Takahashi, H. A novel regulatory pathway of sulfate uptake in Arabidopsis roots: Implication of CRE1/WOL/AHK4-mediated cytokinin-dependent regulation. *The Plant J.* 2004, 38, 779–789.

Matsumoto-Kitano, M.; Kusumoto, T.; Tarkowski, P.; Kinoshita-Tsujimura, K.; Vaclavikova, K.; Miyawaki, K.; Kakimoto, T. Cytokinins are central regulators of cambial activity. *Proceedings of the National Academy of Sciences*, USA 2008, 105, 20027–20031.

Medici, A.; Krouk, G. The primary nitrate response: a multifaceted signalling pathway. *J. Exp. Bot*, 2014, 65, 5567–5576.

Miyawaki, K.; Matsumoto-Kitano, M.; Kakimoto, T. Expression of cytokinin biosynthetic isopentenyltransferase genes in Arabidopsis: tissue specificity and regulation by auxin, cytokinin, and nitrate. *The Plant J.* 2004, 37, 128–138.

Mohapatra, P. K.; Panigrahi, R.; Turner, N. C. Physiology of spikelet development on the rice panicle: is manipulation of apical dominance crucial for grain yield improvement? *Adv. Agron.* 2011, 110, 333–360.

Mounier, E.; Pervent, M.; Ljung, K.; Gojon, A.; Nacry, P. Auxin-mediated nitrate signalling by NRT1.1 participates in the adaptive response of Arabidopsis root architecture to the spatial heterogeneity of nitrate availability. *Plant Cell Environ.* 2014, 37, 162–174.

Olszewski, N.; Sun, T. P.; Gubler, F. Gibberellin signaling: biosynthesis, catabolism, and response pathways. *Plant Cell.* 2002, 14(Suppl.), S61–S80.

Patel, D.; Franklin, K.A. Temperature-regulation of plant architecture. *Plant Signal. Beh.* 2009, 4, 577–579.

Rosas, U.; Cibrian-Jaramillo, A.; Banta, J. A *et al.* Integration of responses within and across Arabidopsis natural accessions uncovers loci controlling root systems architecture. *Proc. Natl. Acad. Sci. USA*, 2013, 110, 15133–15138.

Sah, S. K.; Reddy, K. R.; Li, J. Abscisic acid and abiotic stress tolerance in crop plants. *Front. Plant Sci.* 2016, https://doi.org/10.3389/fpls.2016.00571

Sakakibara H. 2008. Regulation of cytokinin biosynthesis, compartmentalization, and translocation. *J Exp. Bot.* 59, 75–83.

Sakakibara, H.; Takei, K.; Hirose, N. Interactions between nitrogen and cytokinin in the regulation of metabolism and development. *Trends Plant Sci.* 2006, 11, 440–448.

Samuelson, M.E.; Larsson, C.M. Nitrate regulation of zeatin riboside levels in barley roots. Effects of inhibitors of N-assimilation and comparison with ammonium. *Plant Sci.* 1993, 93, 77–84.

Santner, A.; Estelle, M. Recent advances and emerging trends in plant hormone signaling. *Nature.* 2009, 459(7250), 1071–8. doi: 10.1038, PMID: 19553990

Schmulling, T. New insights into the functions of cytokinins in plant development. *J. Plant Growth Regul.* 2002, 21, 40–49.

Seguela, M.; Briat, J. F.; Vert, G.; Curie, C. Cytokinins negatively regulate the root iron uptake machinery in Arabidopsis through a growth-dependent pathway. *The Plant J.* 2008, 55, 289–300.

Sharma, I.; Chin, I.; Saini, S.; Bhardwaj, R.; Pati, P. K. Exogenous application of brassinosteroid offers tolerance to salinity by altering stress responses in rice variety Pusa Basmati-1. *Plant Physiol. Biochem.* 2013, 69, 17–26.

Shi, Y.; Tian, S.; Hou, L.; Huang, X.; Zhang, X.; Guo, H.; Yang, S. Ethylene signaling negatively regulates freezing tolerance by repressing expression of CBF and Type-A ARR genes in Arabidopsis. *Plant Cell.* 2012, 24, 2578–2595.

Signora, L.; De Smet, I.; Foyer, C. H.; Zhang, H. ABA plays a central role in mediating the regulatory effects of nitrate on root branching in Arabidopsis. *The Plant J.* 2001, 28, 655–662.

Singh, S.; Letham, D.S.; Zhang, R.; Palni, L.M.S. Cytokinin biochemistry in relation to leaf senescence: IV. Effect of nitrogenous nutrient on cytokinin levels and senescence of tobacco leaves. *Physiol. Plantar.* 1992, 84, 262–268.

Takei, K.; Sakakibara, H.; Taniguchi, M.; Sugiyama, T. Nitrogen dependent accumulation of cytokinins in root and the translocation to leaf: implication of cytokinin species that induces gene expression of maize response regulator. *Plant Cell Physiol.* 2001, 42, 85–93.

Takei, K.; Ueda, N.; Aoki, K.; Kuromori, T.; Hirayama, T.; Shinozaki, K. AtIPT3 is a key determinant of nitrate-dependent cytokinin biosynthesis in Arabidopsis. *Plant Cell Physio.* 2004. 45, 1053–1062.

Tian, Q.Y.; Sun, P.; Zhang, W.H. Ethylene is involved in nitrate-dependent root growth and branching in Arabidopsis thaliana. *New Phytol.* 2009, 184, 918–31.

Tsay, Y.F.; Ho, C.H.; Chen, H.Y.; Lin, S.H. Integration of nitrogen and potassium signaling. *Annu. Rev. Plant Biol.* 2011, 62, 207–226.

Ueda, Y.; Konishi, M.; Yanagisawa, S. Molecular basis of the nitrogen response in plants. *Soil Sci. Plant Nut.* 2017, 63, 329–341.

Vysotskaya, L. B.; Korobova, A. V.; Veselov, S. Y.; Dodd, I. C.; Kudoyarova, G. R. ABA mediation of shoot cytokinin oxidase activity: assessing its impacts on cytokinin status and biomass allocation of nutrient deprived durum wheat. *Funct. Plant Biol.* 2009, 36, 66–72.

Wagner, B.M.; Beck, E. Cytokinins in the perennial herb *Urtica dioica* L. as influenced by its nitrogen status. *Planta.* 1993, 4, 511–518.

Walch-Liu, P.; Ivanov, I.I.; Filleur, S.; Gan, Y.; Remans, T.; Forde, B.G. Nitrogen regulation of root branching. *Annals Bot.* 2006, 97, 875–881.

Wang, R.; Xing, X.; Wang, Y.; Tran, A.; Crawford, N.M. A genetic screen for nitrate regulatory mutants captures the nitrate transporter gene NRT1.1. *Plant Physiol.* 2009,151, 472–478.

Wang,Y.; Mopper, S.; Hasentein, K. H. Effects of salinity on endogenous ABA, IAA, JA, and SA in Iris hexagona. *J. Chem. Ecol.* 2001, 27, 327–342.

Wilkinson, S.; Davies, W. J. ABA-based chemical signalling: The coordination of responses to stress in plants. *Plant Cell Environ.* 2002, 25, 195–210.

Wilkinson, S.; Davies, W. J. Drought, ozone, ABA and ethylene: new insights from cell to plant to community. *Plant Cell Environ.* 2010, 33, 510–525.

Wilkinson, S.; Davies, W.J. ABA-based chemical signalling: The co-ordination of responses to stress in plants. *Plant Cell Environ.* 2002, 25, 195–210.

Wilkinson, S.; Kudoyarova, G. R.; Veselov, D. S.; Arkhipova, T. N.; Davies, W. J. Plant hormone interactions: innovative targets for crop breeding and management. *J. Exp. Bot.* 2012, 63, 3499–3509.

Xiao, J.; Li, C.; Xu, S.; Xing, L.; Xu, Y.; Chong, K. JACALIN-LECTIN LIKE1 regulates the nuclear accumulation of GLYCINE-RICH RNA-BINDING PROTEIN7, influencing the RNA processing of FLOWERING LOCUS Cantisense transcripts and flowering time in Arabidopsis. *Plant Physiol.* 2015, 169, 2102–2117.

Yendrek, C.R.; Lee, Y.C.; Morris, V. *et al.* A putative transporter is essential for integrating nutrient and hormone signaling with lateral root growth and nodule development in *Medicago truncatula*. *The Plant J.* 2010, 62, 100–112.

Yeshitila, M.; Gobena, A. Review on Phytohormones signaling cross-talk: to control plant growth and development. Int. J. Adv. Res. Biol. Sci. 2020, 7(3), 54–71. DOI: http://dx.doi.org/10.22192/ijarbs.2020.07.03.008

Yeshitila, M.; Gobena, A. Review On Phytohormones signaling cross-talk: to control plant growth and development. *Int. J. Adv. Res. Biol. Sci.* 2020, 7(3), 54–71.

Yokoyama, A.; Yamashino, T.; Amano, Y, Tajima Y, Imamura A, Sakakibara H, Mizuno T. 2007. Type-B ARR transcription factors, ARR10 and ARR12, are implicated in cytokinin-mediated regulation of protoxylem differentiation in roots of Arabidopsis thaliana. *Plant Cell Physiol.* 48, 84–96.

Zhang, H.; Rong, H.; Pilbeam, D. Signalling mechanisms underlying the morphological responses of the root system to nitrogen in Arabidopsis thaliana. *J. Exp. Bot.* 2007, 58, 2329–2338.

Zhang, S.; Hu, J.; Zhang, Y.; Xie, X. J.; Knapp, A. Seed priming with brassinolide improves lucerne (*Medicago sativa* L.) seed germination and seedling growth in relation to physiological changes under salinity stress. *Aust. J Agric. Res.* 2007, 58, 811–815.

Zheng, D. C.; Han, X.; An, Y.; Guo, H.W.; Xia, X. L.; Yin, W. L. The nitrate transporter NRT2.1 functions in the ethylene response to nitrate deficiency in Arabidopsis. *Plant Cell Environ.* 2013, 36, 1328–37.

10 Biological Nitrogen Fixation
An Overview

Hanan Javid, Rouf ul Qadir, Sufiya Rashid, Kausar Rashid, Junaid Ahmad Magray, Tajamul Islam, Bilal Ahmad Wani, Shabana Gulzar, and Irshad Ahmad Nawchoo

¹Plant Reproductive Biology, Genetic Diversity and Phytochemistry Research Laboratory Department of Botany, University of Kashmir, Srinagar, J & K, India
Corresponding author e-mail: Hanan Javid bhathanan123@gmail.com

CONTENTS

10.1 Introduction	129
10.2 Microbiology Involved in Nitrogen Fixation	130
10.3 Process of Biological Nitrogen Fixation	130
10.4 Genetic Regulation of Biological Nitrogen Fixation	131
10.5 Role of Flavonoids in Biological Nitrogen Fixation	133
10.5.1 Expression of Nod Genes	133
10.5.2 Regulation of Genes Involved in the T3SS Strategy	134
10.5.3 Growth Stimulants and Chemoattractants for Rhizobium Spp.	135
10.5.4 Act as Phytoalexins during Nodulation	135
10.5.5 Nitrogen Fixation of Actinorhizal Plant	135
10.5.6 Establishment of Actinorhizal Symbioses	135
10.5.7 Host Specificity	136
10.5.8 Development and Function Of Nodule	136
10.5.9 The Prospective of Manipulating Flavonoids	136
10.6 Role of Mineral Nutrients in Biological Nitrogen Fixation	137
10.7 Conclusion	137
References	138

10.1 INTRODUCTION

Fixed sources of nitrogen (N) are needed for all life forms and their availability frequently confines productivity in natural ecosystems (Falkowski et al., 1997). The bulk of the nitrogen on earth exists as dinitrogen (N_2) that is not available for various biological processes. Abiotic activities include electrical (lightning)-based N oxidation into nitric oxide (Yung et al., 1979). The reduction of N_2 can also be mineral (e.g., ferrous sulfide) based (Schoonen et al., 2001), nitrous oxide, or nitrite (NO_2)/nitrate (NO) (Summers et al., 2012) to NH_3. "Abiotic sources of fixed N (e.g., NO, NO_3, NH_3) are assumed to have become limiting to an expanding global biome which may have hastened the innovation of biological mechanisms to reduce N_2" (Summers et al., 2005).

Nitrogen fixation is the transformation of inactive N_2 gas into a metabolically active form (i.e., ammonia). Fixation of nitrogen in biology is extremely sensitive to oxygen and is limited to a small group of different microorganisms, also cooperatively known as diazotrophs or "nitrogen eaters." Historically, the non-availability of biologically active, fixed nitrogen has limited global food production (Long et al., 1987; Heytler et al., 1985; Ludwig et al., 1986; Marschner et al., 1995). However, around 1920, conditions were greatly improved by the use of fertilizer manufactured by industry. The Haber-Bosch method that is used for the production of industrial fertilizers is being promoted as a technological breakthrough that is having the greatest impact on the modern world. Therefore, it has played a significant role in driving the last century's green revolution and accelerating exponential population growth. Nevertheless, the usage of fertilizers manufactured by industry to increase agriculture leads to significant agronomic, economic, and environmental losses. These penalties include the use of non-renewable fossil fuel, tremendous creation of greenhouse gases, costs linked with fertilizer application and distribution, cossetting of watersheds as a result of fertilizer run-off, and the social and political problems linked to unregulated population development ((Long et al., 1987; Heytler et al., 1985; Ludwig et al., 1986; Marschner et al., 1995). Hence, there is mounting interest in enhancing the involvement of biological nitrogen fixation for the growth of crop plants in agriculture (Mus et al., 2016).

10.2 MICROBIOLOGY INVOLVED IN NITROGEN FIXATION

Nitrogen is one of the major constituents of all biomolecules and is very important for life. Atmosphere constitutes a vast amount of dinitrogen, but due to the less reactive nature of N_2, this dinitrogen is not directly available to most of the species. To show proper growth and development nitrogen is required by every species, being one of the important constituents of biomolecules. Prokaryotes play a very important role in reducing this dinitrogen to usable ammonia and keeping it available for those that can use this nitrogen directly. All this is made possible by the nitrogenase enzyme, which plays a very important role in fixing this atmospheric nitrogen.

The group of prokaryotes responsible for fixing dinitrogen to usable form is known as diazotrophs. This group comprises some blue-green algae and bacteria. These diazotrophs can be divided into two types depending upon the available carbon source. One group is known as autotrophs while the other group is known as heterotrophs. These organisms can be further divided as free-living or symbionts depending upon the environment in which these organisms are found during fixing this dinitrogen into ammonia. Symbiotic heterotrophs belonging to the genera Frankia (actinomycetes) and Rhizobium, numerous free-living heterotrophs that include azotobacters, and many free-living autotrophs such as rhodospirillum are examples of bacterial diazotrophs. In both symbiotic and free-living forms, the blue-green algae, which are mainly autotrophic, have diazotrophic members. Plant forms that are found to be in symbiotic association with blue-green algae are fungi, liverworts, ferns, gymnosperms, and angiosperms. Diazotrophs in their original habitats are found in a free-living state; however, during their symbiotic contacts, colonization of diazotrophs inside and around plants is seen making an alternative favorable microenvironment for biological nitrogen fixation. This process also requires a heavy expenditure of energy and is very sensitive to oxygen and different forms of nitrogen (N). Keeping the above consideration in view, fixation is preferred under the presence of sufficient carbon content and low partial pressure of oxygen. All of these conditions are fulfilled in the root nodule microenvironment with regard to rhizosphere and phyllosphere created by diazotrophs to altering degrees. The extent of mutuality among microorganisms and the symbiotic plant varies greatly. The capacity of some plant species to fund some of their requirement for nitrogen by biological fixation gives them a considerable competitive advantage over those that lack this capacity (Dong et al., 2020).

10.3 PROCESS OF BIOLOGICAL NITROGEN FIXATION

During biological nitrogen fixation, atmospheric nitrogen is utilized to produce ammonia. The reaction taking place during this process is

$$N_2 + 8\ e^- + 8\ H^+ + 16\ MgATP \rightarrow 2\ NH_3 + H_2 + 16\ MgADP + 16\ Pi$$

The reduction of N_2 to $2NH_3$, a shift of six electrons, is linked with the reduction of two protons for H_2 to develop. The enzyme complex responsible for catalyzing this reaction is known as the nitrogenase enzyme. This enzyme complex consists of two parts, one known as Fe protein and another as MoFe protein. Both of these parts are important to carry out the catalytic activity. Among these two protein components of the enzyme complex, the Fe protein is the smaller one possessing two subunits of 30–70 Kda depending on the organism.

During the conversion of N_2 to NH_3, S2-iron-sulfur (4 Fe-4 S2) plays an important role, being a component of each subunit. O_2 inactivates Fe irreversibly (Fe protein having a half-life period of 30–45 seconds) (Dixon and wheeler, 1986). MoFe protein has four subunits possessing a molecular mass of 180–235 Kda, with each cluster having two Mo-Fe-S subunits. Oxygen is responsible for inactivating MoFe protein, having a half-life decay time of about 10 minutes in the atmosphere.

Ferredoxin functions as a donor of electrons to the Fe protein in the process of nitrogen reduction reaction. In this process, ATP is hydrolyzed and MoFe protein is reduced. The protein MoFe will further reduce various substrates, while H^+ reacts only with N_2 under normal conditions. The reaction in which acetylene is reduced to ethylene catalyzed by nitrogenase is required for the calculation of nitrogenase function. The energetics involved in nitrogen fixation is nuanced. During the formation of NH_3 from N_2 and H_2 a change in free energy of -27 KJ Mol^{-1} occurs, therefore making the reaction exergonic. Moreover, the industrial formation of NH_3 from H_2 and N_2 is endergonic, utilizing enormous energy to sever the bonds present in N_2 due to the activation energy needed. For the same reason, the reduction of dinitrogen by enzyme nitrogenase often involves enormous energy expenditure but the precise changes that occur in free energy remain uncertain.

Depending on calculations, the metabolism of carbohydrate legumes reveals that plants utilize organic carbon of around 12 g per gram of fixed N_2 (Heytler et al., 1984). Depending on the above equation the $G^{0'}$ is about -200 kJ mol^{-1} for the overall reaction of biological nitrogen fixation. As the reaction is exceedingly exergonic, the production of ammonia is restricted by the slow action of the nitrogenase complex (i.e., the number of N2 is reduced per unit time) (Ludwig and de Vries, 1986).

Ample quantities of H^+ are converted to H_2 under normal conditions. This process will interfere with the reduction of N_2 nitrogenase electrons. In fact, 30 to 60% of the nitrogenase energy available is exhausted as H_2, thus reducing the effectiveness of fixation of nitrogen. However, few rhizobia produce hydrogenase enzymes that can break H_2 generated and produce electrons for the reduction of dinitrogen (N_2). This process leads to an upsurge in the effectiveness of nitrogen fixation (Marschner et al.,1995). Prokaryotic nitrogen-fixing organisms that are symbiotic emit ammonia. This ammonia produced in the rood nodules must be quickly transformed into organic forms to prevent contamination before being transferred by the xylem to the shoot. Based on the structure of xylem sap, legumes that fix nitrogen can be classified based on the exported product (i.e., ureide or amide). Amides (primarily glutamine or asparagine amino acids) are transported by legumes of the temperate zone, such as clover, pea, lentils, and large beans. Ureides are transported in tropical legumes that include kidney beans, soybeans, southern peas, and peanuts. Allantoin, allantoic acid, and citrulline are the three main ureides. Allantoin is formed in peroxisomes from uric acid and allantoin, and allantoic acid is formed in the endoplasmic reticulum. Citrulline is synthesized from ornithine but the site has not yet been revealed. Ultimately, the three compounds formed are emitted into the xylem and from there transported to the shoot. After that, these are promptly broken into ammonium, which then joins the path of assimilation.

10.4 GENETIC REGULATION OF BIOLOGICAL NITROGEN FIXATION

Nitrogenases are multifaceted metalloenzymes with preserved structural and mechanistic characteristics that have the potential of catalyzing the reduction of dinitrogen to ammonia (Rees et al., 2000; Lawson et al., 2002). Nitrogenase comprises two components. These components are named based on their metal composition. One component is smaller and is dimeric called the iron (Fe) protein. Iron (Fe) protein acts as an electron donor to the larger component and utilizes ATP. The larger component, the molybdenum–iron (MoFe) protein, is heterotetrameric and comprises the enzyme catalytic site. Most diazotrophs contain a molybdenum-iron nitrogenase mechanism, but some species such as *Rhodobacter capsulatus* and *Azotobacter* encourage the making of alternate nitrogenases that contain iron-iron or vanadium-iron co-factors under conditions of molybdenum depletion. Both of the proteins in the nitrogenase are highly sensitive to oxygen. The Fe protein's oxygen sensitivity is convened by a surface-exposed [4Fe-4S] cluster. This cluster bridges the dimer's two subunits. Two types of metal centers are found in the MoFe protein. The first one is the P cluster (a [8Fe-7S] cluster) and the second is the FeMo co-factor ($MoFe_7S_9$·homocitrate). The second metallic center is the site of substrate reduction (Seefeldt et al., 2004). The complete stoichiometry of the reduction of dinitrogen under optimum settings is

$$N_2 + 8\ e^- + 8\ H^+ + 16\ MgATP \rightarrow 2\ NH_3 + H_2 + 16\ MgADP + 16\ Pi$$

This enzymatic reaction includes the reduction of the Fe protein. The main electron donors in this reaction are flavodoxin and ferredoxin. Transferal of electrons from the Fe protein to the MoFe protein is reliant on the hydrolysis of MgATP. Eventually, the transfer of internal electrons to the MoFe protein with the help of the P cluster to the substrate-binding site of the FeMo co-factor occurs. A mandatory period of connection amongst the Fe and MoFe proteins is involved in each electron-transfer step, leading to the formation of a complex. The two components dissociate after that (Hageman et al., 1978). Nitrogenase is a comparatively less effective enzyme, having a turnover time of $\sim 5\ s^{-1}$. The rate-limiting step is the dissociation of the complex (Thorneley et al., 1985). The creation of the complex has a vital role in the process of the enzyme as it is significant for linking electron transfer to ATP hydrolysis. The MoFe protein, Fe protein, and stabilized complexes of the two proteins X-ray crystallographic structures have been attained. The Fe protein's inclusive architecture is comparable to other nucleotide-binding proteins such as Ras p21. However, it has exclusive structural homology with the other two ATPases, MinD and ArsA, that act in spatial cell division regulation and oxyanion extrusion, respectively (Lutkenhaus et al., 2003). While these proteins play different roles, their similarity in structure is expected to represent a basic requisite for cell division. Benzoyl-CoA reductase and 2-hydroxyglutaryl-CoA dehydratase also catalyze electron transfer in an ATP-dependent manner, and hence are enzymes with a functional resemblance to nitrogenase (Locher et al., 2001; Möbitz et al., 2004). Investigation of nitrogenase complexes stabilized with ATP hydrolysis ADP-AlF-4 reveals TRANSITION-STATE ANALOGUE that the conformational changes of Fe protein are coupled with repositioning of the [4Fe-4S] cluster upon ATP turnover. This takes the cluster in close vicinity to the MoFe protein and thus promotes the electron inter-protein (Schindelin et al., 1997). Enzymes that are accountable for the action of nitrogenase are prone to oxygen damage, contributing to the "Oxygen Paradox." Whereas respiration and oxygen are normally required in the processing of a huge amount of ATP, this ATP is essential for the operation of nitrogenase. Oxygen is terminally inactivating the enzyme. To cope with the Oxygen Paradox, the following methods are used:

1. The easiest technique is oxygen avoidance. Some diazotrophs, for instance, are obligate anaerobes. Other bacteria are optional anaerobes: they have the capability of aerobic or anaerobic development. Diazotrophic species of this specific group normally fix nitrogen under anaerobic conditions (Postgate, 1998).
2. Microaerobic diazotrophy is demonstrated by certain strains of rhizobia. Oxygen exposure to microaerophilia is demonstrated by many other forms of nitrogen-fixing aerobes (Postgate, 1998).

3. Another mechanism for enabling (micro) aerobic fixation of nitrogen is respiration. High respiration levels are an effective method of generating ATP (as it is significant for the functioning of nitrogenase) in aerobes while shielding nitrogenase from damage due to oxygen (respiratory protection). Nitrogenase "conformational protection" also exists (Postgate, 1998).
4. Oxygenic phototrophic prokaryotes that include cyanobacteria express the machinery for BNF, which is widespread. To permit the two mismatched procedures (i.e., BNF and oxygen photosynthesis to occur in one organism), two key strategies have evolved:
    a) sequential expression in diel cycles
    b) creation of specialized structures namely heterocysts, where BNF occurs (Flores et al., 2015).
5. Highly specialized structures occur in leguminous plant stems and roots where bacteroids and various types of inducible rhizobia help in nitrogen fixation for the host. The plant cells, which bind and transfer oxygen, develop leghemoglobin in these nitrogen-fixing nodules. Leghemoglobin has a great attraction towards oxygen. Hence, nitrogenase remains comparatively harmless (Postgate, 1998).

10.5 ROLE OF FLAVONOIDS IN BIOLOGICAL NITROGEN FIXATION

Flavonoids are a collection of secondary metabolites formed by the phenylpropanoid pathway. There are 9000 diverse compounds that have been studied to date (Ferrer et al., 2008). Flavonoids are vital signaling molecules in the symbiosis between legumes and their nitrogen-fixing symbionts, the rhizobia (Liu et al., 2016). The main known subgroups include flavones, aurones, flavonols, proanthocyanidins, and anthocyanins (Winkel-Shirley et al., 2001). Flavonoids are used in a varied plant process such as deterrence from harmful radiation, protection against pests and pathogens, sexual reproduction, and coloration of tissue. A variety of distinct enzymatic steps (Winkel-Shirley et al., 2002) are involved in their synthesis. For the development of nodules in legumes, flavonoids are known to be necessary and therefore expected to be essential in AM and actinorhizal symbiosis.

Fabaceae is one of the largest flowering plant families, having 18,000 species scattered across the globe with members of more than 650 genera. Symbiosis of Fabaceae with Rhizobium species helps plants to endure soils that are lacking in nitrogen, which is part of their evolutionary progress. It has been reported that 28–84 kg per hectare per year of nitrogen is fixed by legumes in natural environments, while this may increase to a few hundred kilograms in a cropping area (Frankow-Lindberg et al., 2013). To accomplish biological nitrogen fixation, legume symbionts can follow several strategies that include the Nod strategy, T3SS strategy, and the non-T3SS strategy (non-Nod) (Figure 10.1). It has been well recognized that the Nod strategy (Abdel-Lateif et al., 2012) is controlled by flavonoids produced by the legume species roots. In addition, flavonoids are involved in many other phases of the nodulation mechanism, such as Rhizobium chemoattraction, T3SS strategy, nodule growth, selection of symbionts, and so on.

The following are the processes regulated by flavonoids.

10.5.1 Expression of Nod Genes

Genetic, biochemical, and molecular approaches have led to the documentation of several genes involved in the regulation of nodulation (Desbrosses and Stougaard, 2011). The regulation of nod genes by flavonoids exuded by roots in Rhizobium spp. has been studied in detail (Masson-Boivin et al., 2018). This flavonoid function was discovered as luteolin was found in alfalfa (Cooper et al., 2007) and 7.4' dihydroxyflavone was found in white clover (Peters et al., 1986) that function as inducers of nod genes. Usually, the flavonoid concentration required for the induction of nod genes is within the range of nanomoles to micromoles. However, combinations of different flavonoids are more effective than solitary compounds (Redmond et al., 1986). A variety of products from nod

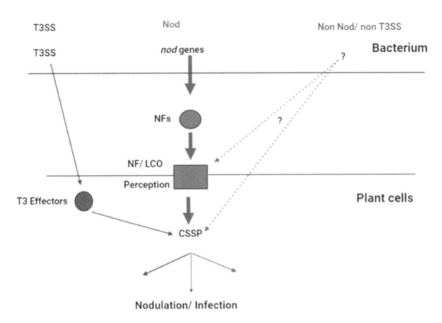

FIGURE 10.1 Nodulation strategies in legume symbionts.

genes cooperate in the synthesis of nod factors that are required during the formation of a nodule in a diversity of alpha- and β-proteobacteria rhizobial organisms (Hassan et al., 2012). A variety of products from nod gene products work together in the production of nod factors required for the establishment of nodules. NodD, which is a transcription factor belonging to the LysR family, regulates these nod genes. It is suspected that the attachment of a sufficient flavonoid with NodD promotes the contact of RNA polymerase, therefore improving nod gene transcription. Described as a nod box, the NodD-flavonoid complex adheres to the DNA sequences of the target. Flavonoid sensitivity by Rhizobium spp. is related to an increase in calcium concentration inside cells, due to which NodD expression is induced (Wang et al., 2011). Plant LysM receptor-like kinases recognize secreted nod factors, activating the specific curling of the tip of the root hair, due to which the symbiont is trapped inside a pouch-like structure. From this structure, the symbiont is picked into the root by an infection thread (Horvath et al., 1993). Symbionts finally get endocytosed in the cells of the nodule until they enter the inner root and continue to fix nitrogen. In addition to this nod factors have the potential to promote cell division and gene expression that initiates nodule development (Cullimore et al., 2001).

10.5.2 Regulation of Genes Involved in the T3SS Strategy

In certain rhizobial species, flavonoids can stimulate a variety of genes accountable for T3SS machinery (Figure 10.1). In *B. elkanii* USDA61 this mechanism has been found (Guldan et al., 1996; Staehelin et al., 2015). The promoter region of the majority of these genes contains a tts box cis-element. This gene expression is often determined by the manifestation of the transcription factor of TtsI. This transcription factor can bind to the tts box. TtsI expression (which is controlled by the fusion constructs of the promoter-lacZ reporter gene) gets highly prompted by flavonoids. Since the ttsI promoter, similar to the nod gene, consists of the nod box, secretions of T3SS proteins, particularly certain Nops, are induced in the host by flavonoids. These induced proteins serve to inhibit the response of pathogen defense by the host plant and also facilitate multiple methods essential to construct a symbiosis. It is suspected that certain *Nops* encourage symbiosis further specifically by interacting with the signaling machinery of nodulation in the host (Lee et al., 2011). Nops are transmitted through the lumen of structures that are needle-like from the Rhizobium spp. cytoplasm. These structures can be seen as appendages in electron micrographs referred to as T3 pili (Gazi et al., 2012). In the case of mutant strains of Rhizobium spp., which are deficient in the synthesis or exudation potential of 13 pili (Michiels et al., 1995; Okazaki et al., 2009; Zhang et al., 2011), their ability for the formation of a stable symbiosis is negotiated. The expression of Nops genes depends on the existence of NodD and unique flavonoids that are host-derived, including those genes convoluted in T3SS.

10.5.3 Growth Stimulants and Chemoattractants for Rhizobium Spp.

For Rhizobium spp., flavonoids can serve as chemoattractants. It was concluded from several studies that the abundance of symbionts is elevated in the proximity of root tip (Graham et al., 1991; Zuanazzi, 1998) and especially in the proximity of root hairs that are emerging, where Rhizobium spp. infection (Aguilar et al., 1988) is initiated. To facilitate their migration towards the roots of their host, Sinorhizobium meliloti cells use flavonoids (Phillips et al., 1992). The effective amount of the flavonoid varies from 1 μM to 0.1 nM. The development of Rhizobium spp. can also be controlled by flavonoids. The growth of two symbionts, *S. meliloti* and *B. japonicum*, was augmented with the endowment of luteolin-7xO-glucoside or quercetin-3-O-galactoside, daidzein produced by *Medicago sativa* (Hartwig et al., 1991; Nouwen et al., 2019). The accumulation of the flavonoids, namely naringenin and apigenin, when used in vitro in the medium to grow *Bradyrhizobium* sp. or the host plant exudate, considerably augments cell proliferation (Cesco et al., 2010). A comparable effect is implemented by a variety of naïve phenolic acids (caffeic, p-coumaric, phenyl lactic acids, protocatechuic, and p hydroxybenzoic acid) that exist in the rhizosphere, as the products of breakdown from flavonoid (Bennett et al., 1998).

10.5.4 Act as Phytoalexins during Nodulation

Phytoalexins are antimicrobial compounds of low molecular weight that plants produce as a reaction to a different type of stress, which include stress from both biotic and abiotic factors. Flavonoids suppress the microbiota of the rhizosphere battling with Rhizobium spp. towards establishment. Both soybeans and *M. truncatula* have been identified with several primary flavonoids, as they are necessary for initiation and development of infection, and act as a phytoalexin to enhance specificity (Yuan et al., 2016).

10.5.5 Nitrogen Fixation of Actinorhizal Plant

A distinct family of plants (Fabaceae) that acts as host for various symbionts has prospered in progressing a relationship with Rhizobium spp. that is symbiotic. However, symbiosis amongst actinobacterium Frankia spp. comprises eight families of plants, which include Myricaceae, Betulaceae, Elaeagnaceae, Datiscaceae, Rosaceae, Coriariaceae, Rhamnaceae, and Casuarinaceae (Dawson, 2007). Some herbaceous plants of the genus Datisca are the only exceptions as all other "actinorhizal" species are mostly woody trees or woody shrubs (Santi et al., 2013).

10.5.6 Establishment of Actinorhizal Symbioses

Although the genetics behind flavonoid activity in actinorhizal symbiosis is not known evidently, the consensus is that stable interaction between actinorhizal plant and *Frankia* spp. depends on an interchange of signals amongst the two partners, or during the early stage of the relationship, flavonoids are involved in the signaling process (Redmond et al., 1986). Flavonoids gather within the actinorhizal nodule. Development of nodule by Frankia on the roots of Red alder is endorsed by treatment with flavonoid-containing seed washes. These seed washes are made from the host species (Benoit et al., 1997); these findings have been strengthened by treatment with the exudates of kaempferol and quercetin contained in Black alder root exudates (Hughes et al., 1999).

The coiling of root hairs, which is an important incident in the formation of symbiosis, is stimulated by revealing the cells of Frankia spp. to the filtrate that is made from the roots of Black alder (Ghelue et al., 1997). Clear evidence has been given for the role of flavonoids through the preliminary phases of actinorhizal nodulation by demonstrating that the abolition of chalcone synthase (flavonoid pathway's first enzyme) function in Scaly oak greatly impedes nodulation (Abdel-Lateif et al., 2013). The transcription of 22 Frankia spp. is affected by the fruit of *Myrica gale* exudate. When *M. gale* is inoculated with Frankia spp. the transcription level of certain genes involved in the pathway of flavonoid formation is altered (Popovici et al., 2010). A study for a *C. gluca* nodule and root-expressed sequence tag (EST) database revealed the distinctiveness of some genes that encode for enzymes that have a role in flavonoid synthesis (Hocher et al., 2006).

10.5.7 HOST SPECIFICITY

Extracted flavonoids from *M. gale* (fruit) increase both nitrogen fixation productivity and growth that is obtained by a corresponding Frankia spp. strain, but has a detrimental impact on a discordant strain (Popovici et al., 2010). This suggests the role of flavonoids in the development of symbiont selection by the host. Extracts of flavonoids from the root of River oak have been found to change some components of the surface of a compliant Frankia spp. strain concerning infectivity (Beauchemin et al., 2012). Several laboratory data confirms the idea that in either process of chemoattraction and multiplication of the cells Frankia spp. cells are supported by flavonoids in the rhizosphere (Perrine-Walker et al., 2011).

10.5.8 DEVELOPMENT AND FUNCTION OF NODULE

In Frankia spp. beneficial interactions, a specific class of flavonoids called flavan are gathered explicitly in the lobes of nodules (Laplaze et al., 1999). Even though similar compounds can occur in the nodulated and non-nodulated roots, flavan quantity is greater in the nodulated roots. The technique of in situ hybridization has set up those gene transcripts that encode chalcone synthase amass in cells containing flavin that are found at the zenith of the nodule. The hugeness of this compartmentalization isn't clear; however, its growth essentially needs the trading of signs amongst the symbiont and the host. In Autumn olive the plenitude of a transcript created from the gene that encodes chalcone isomerase is especially elevated in nodules, aggregating through the promotion of nodule (Kim et al., 2003).

10.5.9 THE PROSPECTIVE OF MANIPULATING FLAVONOIDS

Augmentation of inoculants with flavonoids is used commercially for the promotion of Rhizobium-legume symbiosis (Smith et al., 1999). For instance, the produce SoyaSignal™ comprises the supplementary daidzein and genistein. In *B. Japonicum*, this helps in the induction of nod genes. The prospect of controlling the rhizosphere and root relationship (particularly biological nitrogen fixation) by changing the structure or quality of flavonoids in the plant is being proposed as a promising area of research. However, for the time being, the related expertise is yet to be established, which requires a clearer understanding of the type of interactions that exist amongst the host plants, applicable soil micro-organisms, and flavonoids. Because the rhizosphere is biologically complex, the functioning of symbiotic or non-symbiotic soil microorganisms may be affected by the change in the content of a particular flavonoid. Such unpredicted effects need to be taken into consideration and studied. Even though some flavonoids can improve nodulation, they might affect the quorum sensing in bacteria and other interactions between plants. As has been found, flavonoids can alter the arrangement of the microbial population in the rhizosphere, and the application of DNA-based approaches will be a significant path for future study to monitor the changes in species composition of rhizosphere concerning the alteration of flavonoids. These data may also increase the degree of understanding regarding the involvement of the different flavonoids for the functioning of the microbiota of the rhizosphere.

10.6 ROLE OF MINERAL NUTRIENTS IN BIOLOGICAL NITROGEN FIXATION

Several functions are carried out by mineral nutrients. They are active in various plant metabolic processes, including nucleic acids, proteins, synthesis of cell walls, management of cell sap osmotic concentrations, electron transport systems, chlorophyll molecule components, and enzymatic actions, and function as important components of co-enzymes, macromolecules, co-enzymes, and fixation of nitrogen. Amongst the most significant biological processes, biological nitrogen fixation is one of them, and it would be crucial to continue developing the knowledge regarding the relationship between mineral nutrients and biological nitrogen fixation to preserve the food supplies for its dwellers. The occurrence of mineral N in the soil prevents two processes: the production of nodules and the activity of nitrogenase (Sprent et al., 1988). One study reported (Voisin et al., 2002) symbiotic nitrogen fixation is inhibited by mineral N present in the soil; these findings were true when taken relative to the initiation of nodulation and N_2 fixation at low concentrations during initial vegetative growth. Mineral N has an inhibiting effect on N_2 fixation and nodulation of soybean evident at high concentrations (>5 mM); however, it has a less effect at low concentrations. Nevertheless, nitrogen fertilization impacts the nodulation of bean plants and therefore it has been recommended that at 40–60 kg N ha-1 nitrogen fixation be suppressed. Only a few studies of beneficial effects of low nitrate concentration on N_2 fixation in legumes such as soybean (Streeter et al., 1982) have been done. In one study (Olsson et al., 2005) it was shown that plants when grown in high nitrogen agar media minimize carbon distribution towards arbuscular mycorrhizae. A serious constraint of nitrogen fixation and symbiotic relationships occurs because of the lack of phosphorous supply and availability (Leung et al., 1987). At the molecular level, calcium plays a central role in symbiotic interactions. Cell-surface association and growth of pea nodules are affected by Boron (Bolanos et al., 1996). In the subterranean clover-reduced nitrogen fixation occurred due to deficiency of copper (Snowball et al., 1980). Nitrogenase, the nitrogen-fixing enzyme, consists of iron and molybdenum; additionally, nitrogen fixation does not occur without sufficient amounts of these elements. Restricted concentrations of nickel are important for the development of root nodules and activation of hydrogenase in certain legumes. For much nitrogen-fixing microbiota (cyanobacteria) cobalt is significant.

10.7 CONCLUSION

The fixation of N_2 by the biological process is restrained to a comparatively minor but varied cluster of prokaryotic species commonly referred to as the diazotrophs. A large range of bacteria and a few blue-green algae are found in the diazotrophs. Biological nitrogen fixation is mediated by the nitrogenases, which are multifaceted metalloenzymes with preserved structural and mechanistic characteristics that help in catalyzing the biological reductional reaction of converting dinitrogen to ammonia. Nitrogenase comprises two components, and their name is based on the composition of metals. One component is dimeric, smaller, and is called the iron (Fe) protein. It acts as an electron donor to the larger component that is heterotetrameric and is ATP- dependent. The larger component is called the molybdenum–iron (MoFe) protein and it holds the catalytic site of the enzyme. Nitrogenases are prone to oxygen destruction, contributing to the "Oxygen

Paradox." Nitrogen-fixing bacteria have acquired various means to cope with this matter in several ways such as certain diazotrophs respire anaerobically thus obligating anaerobic mode. Numerous other bacteria are optional anaerobes, which implies that they have the capacity for anaerobic or aerobic development. Diazotrophic species of this particular group normally fix nitrogen under anaerobic conditions. Cyanobacteria fix nitrogen in specialized structures called heterocysts, which can maintain low oxygen levels. Flavonoids belong to a class of secondary metabolites in plants that are formed by the phenylpropanoid pathway. Flavonoids play a significant part in varied plant processes including defense against pests and pathogens and prevention from detrimental radiation. Flavonoids are known to be crucial and are therefore likely to be essential for symbiosis by acting as chemoattractants, phytoalexins, and regulators of nod genes. Flavonoids can be further modulated to enhance the symbiotic association between the nitrogen-fixing bacteria and the plant concerned. Mineral nutrients, most importantly N, P, B, and Co, also impact the course of biological nitrogen fixation by affecting nodule formation and nitrogenase activity.

REFERENCES

Abdel-Lateif, K.; Bogusz, D.; Hocher, V. The role of flavonoids in the establishment of plant roots endosymbioses with arbuscular mycorrhiza fungi, rhizobia and Frankia bacteria. *Plant Signal. Behav.* 2012, *7*(6), 636–41.

Abdel-Lateif, K.; Vaissayre, V.; Gherbi, H.; Verries, C.; Meudec, E.; Perrine-Walker, F.; Cheynier, V.; Svistoonoff, S.; Franche, C.; Bogusz., D.; Hocher, V. Silencing of the chalcone synthase gene in C asuarina glauca highlights the important role of flavonoids during nodulation. *New Phytol.* 2013, *199*(4), 1012–21.

Aguilar, J.M.; Ashby A.M.; Richards, A. J.; Loake, G. J.; Watson, M. D.; Shaw, C.H. Chemotaxis of Rhizobium leguminosarum biovar phaseoli towards flavonoid inducers of the symbiotic nodulation genes. *Microbiol.* 1988, *134*(10), 2741–6.

Beauchemin, N. J.; Furnholm, T.; Lavenus, J.; Svistoonoff, S.; Doumas, P.; Bogusz, D.; Laplaze, L.; Tisa, L.S. Casuarina root exudates alter the physiology, surface properties, and plant infectivity of Frankia sp. strain CcI3. *Appl. Environ. Microbiol.* 2012, *78*(2), 575–80.

Bennett, M.J.; Marchant, A.; May, S.T.; Swarup, R. Going the distance with auxin: unravelling the molecular basis of auxin transport. *Philos. Trans. R. Soc. Lond., B, Biol. Sci.* 1998, *353*(1374), 1511–5.

Benoit, L. F.; Berry, A. M. Flavonoid-like compounds from seeds of red alder (Alnus rubra) influence host nodulation by Frankia (Actinomycetales). *Physiol. Plant.* 1997, *99*(4), 588–93.

Bolanos, L.; Brewin, N. J.; Bonilla, I. Effects of boron on Rhizobium-legume cell-surface interactions and nodule development. *Plant Physiol.* 1996, *110*(4), 1249–56.

Cesco, S.; Neumann, G.; Tomasi, N.; Pinton, R.; Weisskopf, L. Release of plant-borne flavonoids into the rhizosphere and their role in plant nutrition. *Plant and Soil*, 2010, *329*(1), 1–25.

Cooper, J.E. Early interactions between legumes and rhizobia: disclosing complexity in a molecular dialogue. *J. Appl. Microbiol.* 2007, *103*(5), 1355–65.

Cullimore, J.V.; Ranjeva, R.; Bono, J. J. Perception of lipo-chitooligosaccharidic Nod factors in legumes. *Trends Plant Sci.* 2001, *6*(1):24–30.

Dawson, J. O. Ecology of actinorhizal plants In: Pawlowski, K. and Newton, W.E. (eds.), *Nitrogen-Fixing Actinorhizal Symbioses, Nitrogen Fixation Research: Origins, Applications, and Research Progress*, 2007, Vol. 6. Springer, Dordrecht, pp. 199–233. https://doi.org/10.1007/978-1-4020-3547-0_8

Desbrosses, G.J.; Stougaard, J. Root nodulation: A paradigm for how plant-microbe symbiosis influences host developmental pathways. 2011, *Cell Host & Microbe*. 2011, *10*(4), 348–58.

Dixon, R. O. D.; Wheeler, C. T. *Nitrogen Fixation in Plants*. Chapman and Hall, New York, 1986.

Dong, W.; Song, Y. The significance of flavonoids in the process of biological nitrogen fixation. *Int. J. Mol. Sci.* 2020, *21*(16), 5926.

Falkowski, P.G. Evolution of the nitrogen cycle and its influence on the biological sequestration of CO_2 in the ocean. *Nature*. 1997, *387*(6630), 272–5.

Ferrer, J. L.; Austin, M. B.; Stewart, Jr. C.; Noel, J. P. Structure and function of enzymes involved in the biosynthesis of phenylpropanoids. *Plant Physiol. Biochem.* 2008, *46*(3), 356–70.

Flores, E.; López-Lozano, A.; Herrero, A. Nitrogen fixation in the oxygenic phototrophic prokaryotes (cyanobacteria):The fight against oxygen. In: F. J. De Bruijn (Ed.), *Biological Nitrogen Fixation*, Vol. 2, Wiley, Hoboken, NJ, pp. 879–889.

Frankow-Lindberg, B. E.; Dahlin, A. S. N_2 fixation, N transfer, and yield in grassland communities including a deep-rooted legume or non-legume species. *Plant and Soil*. 2013, *370*(1), 567–81.

Gazi, A.D.; Sarris, P.F.; Fadouloglou, V.E.; Charova, S. N.; Mathioudakis, N.; Panopoulos, N. J.; Kokkinidis, M. Phylogenetic analysis of a gene cluster encoding an additional, rhizobial-like type III secretion system that is narrowly distributed among Pseudomonas syringae strains. *BMC Microbiol*. 2012, *12*(1), 1–5.

Ghelue, M.V.; Løvaas, E.; Ringø, E.; Solheim, B. Early interactions between Alnus glutinosa and Frankia strain ArI3. Production and specificity of root hair deformation factor (s). *Plant Physiol*. 1997, *99*(4):579–87.

Graham, T. L. Flavonoid and isoflavonoid distribution in developing soybean seedling tissues and in seed and root exudates. *Plant Physiol*. 1991, *95*(2), 594–603.

Guldan, G. S. Obstacles to community health promotion. *Soc. Sci. Med*. 1996, 43(5), 689–695.

Hageman, R.V.; Burris, R.H. Nitrogenase and nitrogenase reductase associate and dissociate with each catalytic cycle. *Proc. Natl. Acad. Sci*. 1978, 75(6):2699–702.

Hartwig, U.A.; Joseph, C.M.; Phillips, D.A. Flavonoids released naturally from alfalfa seeds enhance growth rate of Rhizobium meliloti. *Plant Physiol*. 1991, *95*(3), 797–803.

Hassan, S.; Mathesius, U. The role of flavonoids in root–rhizosphere signalling: opportunities and challenges for improving plant–microbe interactions. *J. Exp. Bot*. 2012, *63*(9), 3429–3444.

Heytler, P. G.; Reddy, G. S.; Hardy, R. W. F. *In Vivo Energetics of Symbiotic Nitrogen Fixation in Soybeans*. Elsevier, New York, 1985, pp. 283–292.

Hocher, V.; Auguy, F.; Argout, X.; Laplaze, L.; Franche, C.; Bogusz, D.; Expressed sequence-tag analysis in Casuarina glauca actinorhizal nodule and root. *New Phytol*. 2006, *169*(4), 681–8.

Horvath, B.; Heidstra, R.; Lados, M.; Moerman, M.; Spaink, H.P.; Promé, J. C.; Van Kammen, A.; Bisseling, T. Lipo-oligosaccharides of Rhizobium induce infection-related early nodulin gene expression in pea root hairs. *Plant J*. 1993, *4*(4), 727–33.

Hughes, M.; Donnelly, C.; Crozier, A.; Wheeler, C. T. Effects of the exposure of roots of Alnus glutinosa to light on flavonoids and nodulation. *Can. J. Bot*. 1999, 77(9), 1311–1315.

Kim, H. B.; Oh, C. J.; Lee, H.; An, C. S. A Type-A chalcone isomerase mRNA is highly expressed in the root nodules of Elaeagnus umbellate. *J. Plant Biol*. 2003, *46*(4):263–70.

Laplaze, L.; Gherbi, H.; Frutz, T.; Pawlowski, K.; Franche, C.; Macheix, J. J.; Auguy, F.; Bogusz, D.; Duhoux, E. Flavan-containing cells delimit Frankia-infected compartments in Casuarina glauca nodules. *Plant Physiol*. 1999, *121*(1):113–22.

Lawson, D. M.; Smith, B.E. Department of Biological Chemistry, John Innes Centre, Norwich NR4 7UH, UK. *Metals Ions in Biological System: Volume 39: Molybdenum and Tungsten: Their Roles in Biological Processes*. 2002; pp. 53.

Lee, W. K.; Jeong, N.; Indrasumunar, A.; Gresshoff, P.M.; Jeong, S.C. Glycine max non-nodulation locus rj1: a recombinogenic region encompassing a SNP in a lysine motif receptor-like kinase (GmNFR1α). *Theor. Appl. Genet*. 2011, *122*(5), 875–84.

Leung, K.; Bottomley, P. J. Influence of phosphate on the growth and nodulation characteristics of Rhizobium trifolii. *Appl. Environ. Microbiol*. 1987, *53*(9), 2098–105.

Liu, C. W.; Murray, J. D. The role of flavonoids in nodulation host-range specificity: an update. *Plants*. 2016, *5*(3),33.

Locher, K. P.; Hans, M.; Yeh, A. P.; Schmid, B.; Buckel, W.; Rees, D.C. Crystal structure of the Acidaminococcus fermentans 2-hydroxyglutaryl-CoA dehydratase component A. *J. Mol. Biol*. 2001, *307*(1), 297–308.

Ludwig, R. A., De Vries, G. E. Biochemical physiology of Rhizobium dinitrogen fixation. *Nitrogen Fixation*. 1986, *4*, 50–69.

Lutkenhaus, J.; Sundaramoorthy, M.; MinD and role of the deviant Walker A motif, dimerization and membrane binding in oscillation. *Mol. Microbiol*. 2003, *48*(2), 295–303.

Marschner, H. *Mineral Nutrition of Higher Plants, 2nd ed*. Academic Press, London, 1995.

Masson-Boivin, C.; Sachs, J. L. Symbiotic nitrogen fixation by rhizobia—the roots of a success story. *Curr. Opin. Plant Biol*. 2018, 44, 7–15.

Michiels, J.; Pelemans, H.; Vlassak, K.; Verreth, C.; Vanderleyden, J. Identification and characterization of a Rhizobium leguminosarum bv. phaseoli gene that is important for nodulation competitiveness and shows structural homology to a Rhizobium fredii host-inducible gene. *Mol Plant Microbe Interact.* 1995, *8*(3), 468–72.

Möbitz, H.; Friedrich, T.; Boll, M. Substrate binding and reduction of benzoyl-CoA reductase: Evidence for nucleotide-dependent conformational changes. *Biochem.* 2004, *43*(5), 1376–85.

Mus, F.; Crook, M. B.; Garcia, K.; Garcia Costas, A.; Geddes, B. A.; Kouri, E. D.; Paramasivan, P.; Ryu, M. H.; Oldroyd, G. E.; Poole, P. S.; Udvardi, M. K. Symbiotic nitrogen fixation and the challenges to its extension to nonlegumes. *Applied and Environmental Microbiology.* 2016, *82*(13), 3698–710.

Nouwen, N.; Gargani, D.; Giraud, E. The modification of the flavonoid naringenin by Bradyrhizobium sp. Strain ORS285 changes the nod genes inducer function to a growth stimulator. *Appl. Environ. Microbiol.* 2019, *32*(11), 1517–25.

Okazaki, S.; Zehner, S.; Hempel, J.; Lang, K,; Göttfert, M. Genetic organization and functional analysis of the type III secretion system of Bradyrhizobium elkanii. *FEMS Microbiol. Lett..* 2009, *295*(1), 88–95.

Olsson, P. A.; Burleigh, S. H.; Van Aarle, I. M.; The influence of external nitrogen on carbon allocation to Glomus intraradices in monoxenic arbuscular mycorrhiza. *New Phytol.* 2005, *168*(3), 677–86.

Perrine-Walker, F.; Gherbi, H.; Imanishi, L.; Hocher, V.; Ghodhbane-Gtari, F.; Lavenus, J.; Meriem Benabdoun, F.; Nambiar-Veetil, M.; Svistoonoff, S.; Laplaze, L. Symbiotic signaling in actinorhizal symbioses. *Curr. Protein Pept. Sci.* 2011, *12*(2), 156–64.

Peters, N. K.; Frost, J. W, Long, S. R. A plant flavone, luteolin, induces expression of Rhizobium meliloti nodulation genes. *Science.* 1986, 233(4767), 977–80.

Phillips, D. A.; Tsai, S. M. Flavonoids as plant signals to rhizosphere microbes. *Mycorrhiza.* 1992, *1*(2), 55–8.

Popovici, J.; Comte, G.; Bagnarol, É.; Alloisio, N.; Fournier, P.; Bellvert, F.; Bertrand, C.; Fernandez, M. P. Differential effects of rare specific flavonoids on compatible and incompatible strains in the Myrica gale-Frankia actinorhizal symbiosis. *Appl. Environ. Microbiol.* 2010, 76(8):2451–60.

Postgate, J. *Nitrogen Fixation.* Cambridge University Press, Cambridge, 1998, p. 112.

Redmond, J.W.; Batley, M.; Djordjevic, M.A.; Innes, R. W.; Kuempel, P. L.; Rolfe, B.G. Flavones induce expression of nodulation genes in Rhizobium. *Nature.* 1986, *323*(6089), 632–5.

Rees, D.C.; Howard, J.B. Nitrogenase: Standing at the crossroads. *Curr Opin Chem Biol.* 2000, *4*(5), 559–66.

Santi, C.; Bogusz, D.; Franche, C. Biological nitrogen fixation in non-legume plants. Ann. Bot. 2013, *111*(5), 743–67.

Schindelin, H.; Kisker, C.; Schlessman, J. L.; Howard, J. B.; Rees, D.C. Structure of ADP· AIF 4⁻-stabilized nitrogenase complex and its implications for signal transduction. *Nature.* 1997, *387*(6631), 370–6.

Schoonen, M.A.; Xu, Y. Nitrogen reduction under hydrothermal vent conditions: Implications for the prebiotic synthesis of CHON compounds. *Astrobiology.* 2001, *1*(2),133–42.

Seefeldt, L. C.; Dance, I. G.; Dean, D. R.; Substrate interactions with nitrogenase: Fe versus Mo. *Biochem* 2004, *43*(6), 1401–9.

Smith, D. L., Zhang, F.. Washington, DC: U.S. patent and trademark office; 1999. *U.S. Patent No. 5,922,316.*

Snowball, K.; Robson, A. D.; Loneragan, J. F. The effect of copper on nitrogen fixation in subterranean clover (*Trifolium subterraneum*). *New Phytol.* 1980, *85*(1), 63–72.

Sprent, J. I.; Stephens, J. H.; Rupela, O. P. Environmental effects on nitrogen fixation. In: *World Crops: Cool Season Food Legumes.* Springer, Dordrecht, 1988, pp. 801–810.

Staehelin, C.; Krishnan, H. B. Nodulation outer proteins: double-edged swords of symbiotic rhizobia. *Biochem. J.* 2015, *470*(3), 263–74.

Streeter, J.G. Enzymes of sucrose, maltose, and α, α-trehalose catabolism in soybean root nodules. *Planta.* 1982, *155*,112–5.

Summers, D. P. Ammonia formation by the reduction of nitrite/nitrate by FeS: ammonia formation under acidic conditions. *Orig. Life Evol. Biosph.* 2005, *35*(4), 299–312.

Summers, D. P.; Basa, R. C.; Khare, B.; Rodoni, D. Abiotic nitrogen fixation on terrestrial planets: reduction of NO to ammonia by FeS. *Astrobiology.* 2012, 2,107–14.

Thorneley, R. N.; Lowe, D. J. Kinetics and mechanism of the nitrogenase enzyme system. *Molybdenum Enzymes*, 7, 89–116.

Voisin, A. S.; Salon, C.; Munier-Jolain N.G.; Ney, B. Effect of mineral nitrogen on nitrogen nutrition and biomass partitioning between the shoot and roots of pea (Pisum sativum L.). *Plant and Soil.* 2002, *242*(2), 251–62.

Wang, Y.; Chen, S.; Yu, O. Metabolic engineering of flavonoids in plants and microorganisms. *Appl. Microbiol. Biotechnol.* 2011, *91*(4), 949–56.

Winkel-Shirley, B. Biosynthesis of flavonoids and effects of stress. *Curr. Opin. Plant Biol.* 2002, *5*(3), 218–23.

Winkel-Shirley, B. Flavonoid biosynthesis. A colorful model for genetics, biochemistry, cell biology, and biotechnology. *Plant Physiol.* 2001, *126*(2), 485–93.

Yuan, S.; Li, R.; Chen, S.; Chen, H.; Zhang, C.; Chen, L.; Hao, Q.; Shan, Z.; Yang, Z.; Qiu, D.; Zhang, X. RNA-Seq analysis of differential gene expression responding to different rhizobium strains in soybean (Glycine max) roots. *Front. Plant Sci.* 2016, *30*, 7:721.

Yung, Y. L.; McElroy, M. B. Fixation of nitrogen in the prebiotic atmosphere. *Science.* 1979, *203*(4384),1002–4.

Zhang, L.; Chen, X. J.; Lu, H. B.; Xie, Z. P.; Staehelin, C. Functional analysis of the type 3 effector nodulation outer protein L (NopL) from Rhizobium sp. NGR234: Symbiotic effects, phosphorylation, and interference with mitogen-activated protein kinase signaling. *J. Biol. Chem.* 2011, *286*(37), 32178–87.

Zuanazzi, J. A. S.; Clergeot, P. H.; Quirion, J.C.; Husson, H.P.; Kondorosi, A.; Ratet, P. Production of Sinorhizobium meliloti nod gene activator and repressor flavonoids from Medicago sativa roots. *Mol Plant Microbe Interact.* 1998, *11*(8), 784–94.

11 Nitrogenase Enzyme Complex
Functions, Regulation, and Biotechnological Applications

Anandkumar Naorem[1], Jyotsana Tilgam[2],
Parichita Priyadarshini[3], Yamini Tak[4], Alka Bharati[5],
and Abhishek Patel[1]*

[1]ICAR-Central Arid Zone Research Institute, RRS-Bhuj, Gujarat, India
[2]ICAR- National Bureau of Agriculturally Important Microorganisms, MaunathBhanjan, U.P., India
[3]ICAR-Crop Improvement Division, Indian Grassland and Fodder Research Institute, Jhansi, U.P., India
[4]Agricultural Research Station, Agriculture University, Kota, Rajasthan, India
[5]ICAR-Central Agroforestry Research Institute, Jhansi, U.P., India
*Corresponding author: E-mail: Anandkumar.naorem@icar.gov.in

CONTENTS

11.1	Introduction	143
11.2	Functions of Nitrogenase	144
11.3	Nitrogenase Regulation	145
	11.3.1 ADP Ribosylation	145
	11.3.2 Nif Genes	146
	11.3.3 Environmental Signals	147
	11.3.4 Post-translation Control	148
11.4	Biotechnological Application of Nitrogenase	148
	11.4.1 Enzymatic Bioelectrocatalysis	149
	11.4.2 Engineering Biological Nitrogen Fixation	149
11.5	Way Forward	151
References		151

11.1 INTRODUCTION

The ever-rising human population has increased food demand and affected the ecosystem functioning owing to several anthropogenic activities. Crop production under a climate-changing scenario has been a focal point of discussion for ages since several constraints are limiting crop productivity, thereby threatening global food security. Hundreds of tonnes of fertilizers are applied to replenish the nutrient deficiencies in the soil system when mismanaged, which could lead to environmental pollution and poor human health. Nitrogen is one of the crucial major essential elements required for the proper growth and productivity of crops. Nitrogen fertilizers are generally produced through the Haber-Bosch process and have been substantially contributing to enhancing global food production. However, the use and burning of huge amounts of fossil fuel in nitrogen fertilizer production

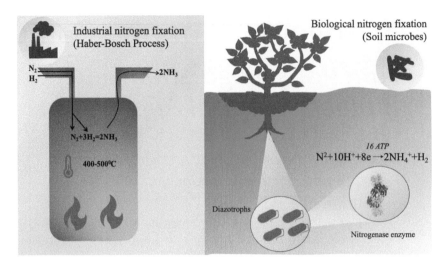

FIGURE 11.1 Simplified representation of industrial nitrogen fixation (left) and biological nitrogen fixation through nitrogenase (right).

is a big environmental concern. Moreover, while some parts of nitrogen fertilizer applied to the soil are taken up and assimilated by plants, much of the parts are lost through leaching or volatilization, which might lead to the greenhouse gas effect and global warming. In addition, 78% of the atmosphere contains nitrogen. Therefore, a more sustainable approach needs to be identified so that the huge reservoir of nitrogen in the atmosphere can be utilized without burning fossil fuels.

With all these issues in mind, scientists proposed an alternative to industrial nitrogen fixation, which is referred to as "biological nitrogen fixation." Biological nitrogen fixation is a microbial process through which the nitrogen in the atmosphere is transformed into ammonia with the help of soil microbes. The nitrogen in the atmosphere is present in non-available form and therefore an intervention from soil microbes is necessary to break down the triple bond and convert it into more available forms for living organisms (Figure 11.1). The conversion of atmospheric nitrogen to useable ammonia is carried out with the production of the nitrogenase enzyme. Such soil microorganisms that can fix the atmospheric nitrogen to ammonia are referred to as nitrogen-fixing microbes or diazotrophs. However, it must be added that atmospheric nitrogen can be converted through lightning. However, due to unpredicted duration and its short-term effect, it is relatively less addressed and highlighted. The diazotrophs can be either symbiotic or free living. Some examples of diazotrophs include *Azospirillum, Rhizobium, Acetobacter, Burkholderia, Gluconacetobacteria diazotrophicus*, etc.

11.2 FUNCTIONS OF NITROGENASE

The early 1960s marked the investigation and development of methods to extract nitrogenase inactive form. These studies revealed some important functions of nitrogenase such as:

(i) Nitrogenase is not a single unit, rather it is a double component system. It includes two distinct parts: dinitrogenase and dinitrogenase reductase.
(ii) The dinitrogenase component is composed of MoFe protein and is also referred to as component I.
(iii) The dinitrogenase reductase is composed of the electron transfer Fe protein and is also referred to as component II.

Nitrogenase Enzyme Complex

(iv) For successful catalysis during the nitrogen fixation process through nitrogenase, the presence of a reducing protein and MgATP are crucial.

(iv) The dinitrogenase reductase and dinitrogenase associate and dissociate in a catalytic cycle.

(v) The dinitrogenase contains a Fe-Mo cofactor and a P cluster; the former serves the active site for binding and reducing the substrate and the latter participates in electron transfer from Fe protein to Fe-Mo cofactor.

Four types of nitrogenase enzymes differ in the set of genes and the presence of metals in their active sites. Mo-dependent nitrogenase is the most widely studied and abundant nitrogenase enzyme that has a FeMo-cofactor. The fixation of nitrogen through Mo-dependent nitrogenase enzyme can be represented by

$$N_2 + 8H^+ + 16\text{MgATP} + 8e^- \rightarrow 2NH_3 + H_2 + 16\text{MgADP} + 16\text{Pi} \qquad (11.1)$$

11.3 NITROGENASE REGULATION

11.3.1 ADP Ribosylation

Nitrogenase inhibition has long been thought to be caused by the covalent moderation in the subunit of the Fe protein by a nucleotide factor; this nucleotide has recently been identified by various researchers as ADP-ribose (Halbleib et al., 2000). The Fe protein is homodimeric and is linked to a Fe_4S_4 cluster between the two subunits via cysteine residues. Furthermore, each subunit has an ATP/ADP binding site. Fe protein donates the protein electron MoFe, and for each electron transferred, ATP molecules (two) are hydrolyzed. Later electrons are shifted to the FeMo cofactor, where hydrogen is produced by the reduction of the molecular nitrogen (Rutledge and Tezcan, 2020). To catalyze the whole reaction, the nitrogenase enzyme needs at least 16 ATP molecules, and this process is strictly regulated at the transcriptional level. The α-proteobacteria *Azospirillum brasilense* and *Rhodospirillum rubrum* have been extensively studied for Fe protein ADP ribosylation (Moure et al., 2014).

Dinitrogenase reductase ADP-ribosyltransferase (DraT) catalyzes Fe protein ADP-ribosylation. DraT is a monomer with 30 kDa molecular weight and catalysis ADP-ribosylation reaction utilizing NAD^+ as a contributor for ADP-ribose moiety. In vitro, as an acceptor, DraT is distinct for Fe protein (Moure et al., 2013). DraT's biochemical activity is close to that of various ADP-ribosylating bacterial toxins whose structures are known, like diphtheria toxin, iota toxin, and certhrax toxin (Koch-Nolte et al., 2002; Suge et al., 2003). ADP-ribosylation deactivates the Fe protein, most likely through congesting the binding site for the Fe and the MoFe proteins, halting the Fe protein from interacting with the MoFe protein and preventing the transfer of electrons (Chen et al., 2019a). DraG, a dinitrogenase reductase-activating glycohydrolase, detaches the ADP-ribosyl moiety fastened to the Fe protein, promoting nitrogenase complex reactivation. DraG is monomeric with a 32 kDa of molecular weight, which catalyzes the hydrolysis of the Fe protein's ADP-ribosyl moiety, enabling the Fe protein to freely interconnect with and shifts electrons to the MoFe protein. The reaction of hydrolysis necessitates the use of ATP and, with Mn^{2+} and Fe^{2+} divalent cation, both are equally effective. DraG accepts both synthetic and denatured ADP-ribosylated Fe protein substrates (Nordlund and Högbom, 2013).

In vivo, the DraT and DraG functions are controlled in opposite directions. DraT and DraG enzymes were discovered in *Rhodospirillum rubrum*, where the draTG operon encodes them, which comes before the nitrogenase structural genes *nifHDK*(Oetjen and Reinhold-Hurek, 2009). By negative factors like the existence of ammonium or diminished inaccessible cell energy, DraT is turned on, while DraG is turned off, encouraging nitrogenase demission via Fe protein ADP-ribosylation. DraT is deactivated when the negative impulse is removed. While DraG is activated, ADP-ribose elimination from the Fe protein and nitrogenase activity is increased (Klassen et al., 2005). According to structural comparisons, DraG in bacteria is related to ADP-ribosylhydrolases in Archaea and mammalian ADP-ribosylhydrolases (ARH). P_{II} protein is known as a nitrogen biosensor, influencing nitrogenase regulation in almost all aspects. The P_{II} protein is a trimeric protein that regulates different processes in prokaryotic cells related to DraT/DraG regulation. P_{II} proteins attach to ATP/ADP and 2–oxoglutarate, modifying the uridylylation of a conserved tyrosine residue (Truan et al., 2010). When there is less nitrogen available in the cell, GlnD, a bifunctional uridyltransferase/uridylyl-removing enzyme, catalyzes the uridylylation. 2-oxoglutarate, ATP, and divalent cations (Mg^{+2} or Mn^{+2}) are bound to P_{II} proteins, which are uridylylated. When there is a lack of energy, ADP hitches to P_{II} proteins and encourages the unuridylylated state (Watzer et al., 2019). *R. rubrum* has three P_{II} paralogues (GlnB, GlnJ, and GlnK) whereas *A. brasilense* only has two (GlnB and GlnZ). GlnZ interrelates with the solvent-disclosed α-helix five on the DraG surface. This helix is may be found in other DraG homologs, making it a possible place for the docking of assumed ARH transformers in some creatures (Zhang et al., 2004). GlnB and GlnZ are fully uridylylated during biological nitrogen fixation, and the nitrogenase enzyme is active. Nitrogenase is inactivated by Fe protein ADP-ribosylation when ammonium levels are high. Concurrently, GlnB and GlnZ are deuridylylated and interrelate with DraT and DraG (Moure et al., 2013). By interconnecting with the AmtB ammonium transporter, the DraG-GlnZ complex reaches the membrane, where it inactivates DraG. Ternary complex (AmtB-GlnZ-DraG) genesis on the membrane may increase the harmony of GlnZ and DraG, preventing the ADP-ribosylateddinitrogenase reductase substrate from accessing the DraG mobile site (Moure et al., 2019).

11.3.2 Nif Genes

Nitrogenase biosynthesis and activity require a huge number of nitrogen fixation (*nif*) genes, which vary with the host organism. MoFe protein constitutes two clusters, namely Mo-7Fe-9S-C-homocitrate and a P-cluster (8Fe-7S). Mo-7Fe-9S-C-homocitrate has an active site for binding the substrate and P-cluster (8Fe-7S) transports electrons to FeMo-cofactor (encoded by *nif*D and *nif*K). *nif*H encodes the Fe protein, which transports electrons to MoFe protein (Burén et al., 2020). In addition to the structural subunits encoded by *nif*H, *nif*D, and *nif*K, genes, including *nif*E, *nif*Q,*nif*N, *nif*B,*nif*X, *nif*V, *nif*Y, and *nif*H, aid in the biosynthesis and incorporation of FeMo-co into the nitrogenase enzyme complex. The genes *nif*U, *nif*S, and *nif*Z are important in the biosynthesis of metalloclusters, and *nif*M plays an important function in a formal folder of the Fe protein (Poza-Carrión et al., 2014). NifI1 and NifI2 are two different P_{II} protein subfamily other than encoded by the *nif* operon genes. The joining of the NifI1/NifI2 heteromer to NifDK stops the interaction with NifHv protein, resulting in the disruption of electron transfer to the nitrogenase complex. When the intracellular concentration of 2-OG is high during nitrogen fixation, it attaches to NifI and forms NifI1/NifI2 oligomers that are unable to bind to NifDK. When ammonia concentration is low, intracellular 2-OG concentration is low, and the NifI1/NifI2 heteromer reverts to a hexamer, which binds to NifDK and inactivates nitrogenase (Dodsworth et al., 2006).

The second level is formed by transcriptional control of the *nif*A gene acting as a regulatory cascade of nitrogenase activity and control of the NifA protein itself. Nitrogenase is very sensitive to oxygen, so an excess of NH_4^+ and O_2 inhibits Nif A activity (Wang et al., 2018). The O_2 sensitive NifA proteins present in *Bradyrhizobium japonicum* are distinguished by a cysteine-rich interdomain bridge between the central and C-terminal domains, which is thought to have O_2 reactivity. O_2 bearable Nif A proteins are found in *K. pneumoniae* that lack an interdomain bridge, and O_2-dependent regulation of this Nif A protein is complemented by Nif L regulatory protein, which commands NifA activity by protein-protein interaction, ligand binding, and interrelation with GlnK (Grabbe et al., 2001).

PII proteins are found in many diazotrophic bacteria and regulate the NH4-dependent modulation of NifA. In A.*vinelandii* and K. *pneumonia*, GlnK is needed to control NifL-mediated inhibition of NifAfunction (Huergo et al., 2013). Under excess nitrogen conditions, GlnK promotes NifL-NifA interaction in *A. vinelandii*, whereas under nitrogen deficiency conditions GlnK inhibits NifL-NifA interaction in *K. pneumonia* (Martinez-Argudo et al., 2004). In addition, under high nitrogen conditions the presence of GlnB is required for NifA protein function in some bacteria, such as *Rhodospirillum rubrum, Herbaspirillumseropedicae*, and *Rhodospirillum brasilense*. Under high nitrogen concentrations, GlnB and GlnK inhibit NifA protein activity in *Azorhizobiumcaulinodans* and *RhodobacterCapsulatus* (Schnabel et al., 2021).

11.3.3 Environmental Signals

Nitrogen fixation is not a constitutive process; instead, the gene responsible for the functioning of the nitrogenase complex to convert atmospheric nitrogen into ammonia is dependent on various environmental signals. Environmental signals such as the availability of external oxygen concentration, nitrogen, and carbon can control the process of biological nitrogen fixation (Dixon and Kahn, 2004). *NTr, Nif A, Dra T, and Dra G* are key regulatory genes that are responsible for sensing nitrogen, ATP, and oxygen. The nitrogen availability is sensed by *the Ntr* gene, which then activates the Ntr C gene. Further, the oxygen and high nitrogen sensing mechanism is controlled by Nif A. Various other environmental signals and nitrogen availability sensing is controlled by Dra T and Dra G. Apart from these factors, the nitrogenase complex is also governed by light, temperature, ammonium, molybdenum, and iron availability (Rao, 2014; Masepohl & Forchhammer, 2007). To control nitrogen assimilation and fixation, various signaling proteins like GlnB and GlnK sense the nitrogen status and work accordingly (Dixon and Kahn, 2004). The 2-oxoglutarate takes a readout of the status of carbon and nitrogen availability because it confers to the carbon and nitrogen metabolism. The metabolism of nitrogen and carbon and their coordination play important roles in bacterial adaptation in response to nutritional status in the external environment. In photosynthetic bacteria ribulose, 1,5 bisphosphate carboxylase/oxygenase (RuBisCO) is a bifunctional enzyme having both carboxylase and oxygenase activity in response to CO_2. In the presence of low CO2 concentration, oxygenase activity leads to the production of 2-phosphoglycolate. 2-oxoglutarate (2-OG) and 2-phosphoglycolate (accumulates when carbon level is higher than nitrogen level) are reversibly correlated. Among various enzymes controlling carbon metabolism, NAD(P)H dehydrogenase regulator (NdhR) is the key enzyme. The 2-PG acts as an inducer whereas 2-OG acts as a corepressor of the NdhR. The NdhR can sense 2-PG and 2-OG (both are an indicator of carbon and nitrogen status inside the cell) to fine-tune the C/N metabolism in photosynthetic bacteria (Jiang et. al., 2018).

To conserve the energy level of the cell, microbes are highly thermodynamically regulated. They initiate the anabolic or catabolic reactions in the way to save the energy level. For example, when the ammonium is present and the energy level in the form of ATP is low, there is no need to convert dinitrogen into ammonia and therefore they activate DraT in response to the external environment and inactivate NifH linked to DraT. But when the ammonium levels decrease and energy level increases, they reactivate the NifH leading to activation of nitrogenase (Huergo et. al., 2012). In the same way, in response to external oxygen and carbon monoxide concentration, CoWn and FdxD genes become activated and protect the nitrogenase (Hoffmann et. al., 2014). The cowNndFdxD genes are responsible for the Mo-nitrogenase-dependent growth of nitrogen-fixing bacteria at different carbon monoxide concentrations. In cyanobacteria, heterocysts are the cells responsible for the fixation of environmental nitrogen into ammonia. The enzyme nitrogenase is very sensitive towards oxygen, which inactivates the nitrogenase activity, and therefore heterocysts keep oxygen concentration very low by controlling the influx of oxygen (Stal, 2017). In this way, it may be concluded that various environmental factors affect the biological nitrogen fixation via controlling nitrogenase complex activity. To prevent the adverse effects of environmental factors, there are different strategies present in diazotrophic organisms at the gene expression level. In the presence of oxygen, to prevent any negative consequences some organisms fix atmospheric nitrogen to ammonia whereas some others have mechanisms to fix nitrogen at night when photolysis and oxygen generation is negligible. Among them, the most common is the presence of some specific cells like heterocysts, which have multiple layers of cell wall along with lack of photosystem II to keep oxygen concentration very low inside the cell. These are a few mechanisms by which diazotrophs keep themselves safe and do nitrogen fixation in response to environmental stimuli.

11.3.4 Post-translation Control

Apart from various modes of nitrogenase regulation by environmental signals, ADP ribosylations, and *nif* genes, the enzyme dinitrogenase is also regulated by transcriptional and posttranslation mechanisms. Although the posttranslational mechanism also targets various common *nif* genes and ADP ribosylation genes to regulate the conversion of dinitrogen into ammonia few unique genes also have been reported in various literature. DraG and DraT are the key enzymes whose activity is regulated in response to reversible ADP ribosylations (Huergo et. al., 2005). Grunwald et al. (1995) overexpressed the two key enzymes DraT and DraG in *Rhodospirillum rubrum* and concluded that proper balance of these enzymes levels is required for appropriate regulation of DraG. In response to the presence of ammonium, DraT bind to the arginine residue of nitrogenase and switch off its activity whereas in lack of ammonium DraG restores the activity of nitrogenase (Moure et al., 2014). In *Rhodopseudomonas palustris*, it has been found that mutation of *the nifA* gene leads to nitrogenase resistance to posttranslational modification. In these mutants, even in presence of ammonium, the low levels of GlnK2 and DraT2 lead to the production of hydrogen (Heiniger et. al., 2012). The posttranslational control, which is independent of ADP ribosylations, in mutant strains carrying mutant *nifH* or lack *DraF* and *DraT* is not dependent on changes in Arg101 of nitrogenase in *R. capsulatus* but how independently this system function remains unclear (Masepohl and Hallenbeck, 2010). The regulation of nitrogenase independent of DraT and DraG was observed in *glnB* and *glnK* double mutant. Zhang et. al. (2005) concluded that *glnD* mutants, which have intermediate uridylyltransferase activity, are defective in the expression of the *nif* genes in the photosynthetic bacterium *Rhodospirillum rubtrum*. In the same way, it has been proved that mutant *ntrBC* has normal nitrogenase function even in *nif* genes-depressing environments proving that ntrBC is not essential for *nif* genes in a photosynthetic bacterium. In this mutant posttranslational control by ADP ribosylations in ammonium response and darkness was affected, which indicates that there is a link between *ntr* and energy signal transduction in a photosynthetic bacterium (Zhang et. al., 1995). In various other mutation studies, it has been found that nitrogenase regulation by posttranslational mechanisms involves *nif*, *ntr*, and *DraG/T* genes to convert dinitrogen into ammonium in response to various factors. Although the process is highly coordinated and cross-linked at various genetic levels, the study of other key genes in different organisms is in progress to elucidate the identification and functional characterization of novel genes.

11.4 BIOTECHNOLOGICAL APPLICATION OF NITROGENASE

Nitrogen is the most abundant element in the atmosphere (78%) but is present in kinetically inert and biologically unavailable dinitrogen form. Most organisms can only assimilate reactive forms of nitrogen (i.e., in oxidized (NOx, NO^{3-}, HNO_3) and reduced form (NH_3, NH^{4+}, and amines)). Industrial production of NH3 by Haber-Bosch is an energy-demanding process and only operates in very high temperatures and pressure. In contrast to this energy-demanding industrial process, a single enzyme complex "nitrogenase" can reduce dinitrogen in the biological environment by the process called "nitrogen fixation." Nitrogenase is a complex oxidoreductase enzyme having two components, namely dinitrogenase (catalytic protein-reducing dinitrogen) and dinitrogenase reductase (electron transferring reductase protein). Nitrogenase enzymes have useful applications in bioelectrocatalysis for chemical synthesis and engineering biological nitrogen fixation (BNF).

11.4.1 Enzymatic Bioelectrocatalysis

Oxidoreductase is a class of enzyme that catalyzes redox reaction (i.e., catalyzes the transfer of electrons from one molecule (the oxidant) to another molecule (the reductant)). In recent decades, researchers have shown that oxidoreductase enzymes can be interfaces between an electrode on solid surfaces to facilitate the transfer of an electron to control catalysis (Chen et al., 2020). This type of catalysis by oxidoreductase using electrodes is called bioelectrocatalysis. So basically, in this process, the electrochemical potential between electrodes is used instead of ATP for electron transfer. Bioelectrocatalysis has several applications in the area of biosensors, biofuel cells, and chemical syntheses. One of the examples of the biosensor using bioelectrocatalysis is commercially available glucose sensors that are used by diabetic patients to test their blood glucose levels.

Based on the enzyme cofactors, oxidoreductase can be categorized into two categories: a) Metalloenzyme (use metal cofactors like Fe centers, Fe-S clusters, Mo-Fe Centers, etc.) and b) Non-metalloenzyme (use cofactors like FAD, FMN, or pyrroloquinoline quinone) (Milton and Minteer, 2017). Nitrogenases are metalloenzyme and can be classified into three types based on their metal cofactor (FeMo-co, FeFe-co, and VFe-co). FeMO-co nitrogenase enzyme is the most widely studied system and is also employed for NH_3 production by bioelectrocatalysis. Milton et al. (2016) successfully established a bio-electrocatalytic N2 fixation system. They immobilized the Mo-Fe component of wild-type nitrogenase purified from *Azotobactorvinelandii* on the electrode surface and derived reduction of protons to dihydrogen, azide to ammonia, and nitrite to ammonia. The reduction was independent of the Fe protein and ATP hydrolysis and utilizes electron mediator cobaltocene for electron shuttling. For electron transfer an enzyme can be wired to the electrode or a small molecule like cobaltocene can be used to shutting electrons (Mediated Electron transfer (MET)) (Jenner and Butt, 2018). They also used a mutant version of nitrogenase (β-98$^{Tyr-His}$). Evaluation of apparent enzymatic kinetics resulted in an approximate twofold increase in the apparent maximum velocity (V_{MAX}) through the application of the Michaelis–Menten model. Before the use of cobaltocene, small molecule electron mediators like low-potential Eu chelates [Eu^{2+}EGTA (ethylene glycol tetraacetic acid) were also used to artificially reduce the MoFe protein, which resulted in the reduction of some nitrogenase substrates such as H^+, C_2H_2, and N_2H_2 efficiency of which is further increased by utilizing mutant nitrogenase (b-98Tyr-His) (Danyal et al., 2010). Cai et al. (2018) investigated the use of cobaltocene/cobaltocenium (Cc/Cc$^+$) as an electron mediator for the vanadium-dependent VFe protein. VFe protein was of interest due to its ability to reduce CO and CO_2. This experiment confirmed that the VFe protein was able to form ethylene (C_2H_4) and propylene (C_3H_6) from CO_2 as the substrate. For the production of upgraded nitrogen-based products or value-added products like pharmaceuticals and agrochemicals Chen et al. (2019b) also developed an enzyme cascade utilizing nitrogenase, diaphorase, and alanine dehydrogenase. This enzyme cascade electrochemically generates transaminase, which further synthesizes chiral amines used in pharmaceuticals.

11.4.2 Engineering Biological Nitrogen Fixation

A group of microbes called diazotrophs from the bacteria and archaea domain carries the mechanism of biological nitrogen fixation. These microbes may be free-living, symbiotic (e.g., *Rhizobium*), or associative (e.g., *Azospirillum*) based on their habitat, but enzyme complex nitrogenase performs N-fixation in all the cases. The unique feature of the nitrogenase complex is its extreme sensitivity towards oxygen and to counter that many bacteria are obligate anaerobe while some are facultative anaerobe where they perform N fixation only in anaerobic conditions. In aerobic diazotrophs, different oxygen protection mechanisms exist. In *A. vinelandii* decoupling of respiration and ATP production during N2 fixation results in consuming O_2 by cytochrome oxidases and hence protecting nitrogenase from O_2 damage (Poole and Hill, 1997). FeSII protein also plays a major role in nitrogenase protection in *A. vinelandii, Gluconacetobacterdiazotrophicus*, and *Azobacter* species by binding to nitrogenase to give conformational protection (Schlesier et al., 2016; Ureta and Nordlund, 2002; Scherings et al., 1983). Oxygenic photosynthetic microbe *Cyanobacteria* protect nitrogenase from O2 by spatial separation of N fixation in thick-walled, efficient O2 barrier, specialized cells called heterocyst. In addition to spatial separation into heterocyst oxygenic photosynthesis is temporally separated through circadian control. Symbiotic diazotroph *Rhizobium*, performs N fixation in nodules where oxygen protection is provided by 1) oxygen-binding heme protein (leghemoglobin) (Downie, 2005), 2) enhanced respiration (Delgado et al., 1998), and 3) oxygen diffusion barrier in the nodule cortex (Minchin et al., 2008).

Structural genes encoding Mo-Fe-dependent component 1 of nitrogenase is *nifD* and *nifK* and homodimer component 2 is a product of *nifH*. Along with the structural *nifHDK* gene, several other *nif* genes play roles in its regulation, maturation, electron transport, co-factor maturation, and assembly (Table 11.1). All diazotrophs contain Mo-Fe type nitrogenase but some diazotrophs additionally harbor VFe and FeFe type nitrogenase. VFe and MoFe type nitrogenase share very high homology among their structural genes. Fe protein homodimer is encoded by *vnfH* and shares 91% sequence similarity with *nifH*. *vnfD* and *vnfK* encode component 1 and have sequence similarity of 33% and 32%, respectively,with nifD and nifK (Hu et al., 2012). An additional subunit (δ subunit) is present in component 1 of VFe and FeFe type nitrogenase is encoded by *vnfG* and *anfG* respectively FeFe Type nitrogenase is encoded by *anfHDK* gene *anfH* shares high similarity to both *vnfH* and *NifH* however *anfD* and *anfK*are very less similar to both its counterpart. An assembly apparatus NifEN is encoded by *nifE* and *nifN* genes, which are indispensable for nitrogenase assembly.

TABLE 11.1
Genes Involved in Nitrogenase Synthesis, Maturation, Assembly, and Electron Transport

Gene	Function
nifH	Structural gene for Fe protein (component 2)
nifD	Structural gene for α subunit of MoFe protein (Component 1)
nifH	Structural gene for β subunit of MoFe protein (Component 1)
nifB	Encode NifB-co, a complex [8Fe-9S-C] required for FeMo-co, synthesis
nifE	Together with *nifN*gene product forms NifEN assembly apparatus; assembly of MoFe-co
nifN	Together with *nifE* gene product forms NifEN assembly apparatus; assembly of MoFe-co
nifX	biosyntheses of FeMo-co
nifQ	NifQ donates molybdenum to NifEN/NifH for the biosynthesis of FeMo-co
nif V	Homocitrate synthesis involved in FeMo-co synthesis
nifY	Its product forms the third subunit in the hexamericapodinitrogenase in *Klebsiella pneumonia*
nifM	maturation of the Fe-protein component (*nifH* product)
nifU	mobilization of Fe–S cluster synthesis and repair
nifS	mobilization of S for Fe–S cluster synthesis and repair
nifW	Involved in the stability of dinitrogenase. Proposed to protect dinitrogenase from O_2 inactivation
nifA	N-responsive positive regulator of *nif* gene; In low N condition *NtrC* trigger *nifA*expression
nifL	Negative regulator of *nif* genes; In high N and High O_2 environment it deregulates *nifA*stopping*nif* gene expression
nifF	Flavodoxin, electron transport to *nifH*
nifQ	NifQ donates molybdenum to NifEN/NifH for the biosynthesis of FeMo-co
nifJ	Pyruvate flavodoxin (ferredoxin) oxidoreductase is involved in electron transport to nitrogenase
fdxN	Ferredoxin serves as an electron donor to nitrogenase

To increase BNF into crop species of economic importance mainly cereals (maize, rice, wheat) are a potentially attractive alternative to decrease nitrogen fertilizer use. Three types of strategies have been put forward:

1) Improving cereal diazotroph association
2) Development of legumes like root nodule symbiosis
3) Heterologous expression of nitrogenase from diazotrophs to crop species.

The third strategy involves transferring prokaryotic *nif* genes to the plant system to build complete nitrogen fixation machinery. Transfer of *nif* genes is a very daunting task that requires an elegant synthetic biology approach. Maintaining proper orientation of *nif* genes, maintaining the stoichiometric ratio of all the Nif proteins to get the functional final product, and grooming the prokaryotic genes to express in a eukaryotic environment are some of the challenges in this task (Buren et al., 2017). Another major challenge is the O_2 sensitivity of nitrogenase as well as the accessory proteins needed for maturation (Buren and Rubio, 2018). Chloroplast and mitochondria are proposed to be best for expressing *nif* genes because nitrogen fixation is an energy-demanding process and also due to the prokaryotic origin of these organelles, bacterial genes can be expressed with greater efficiency. To date, we have not achieved functional nitrogenase transfer to plant cells because this approach needs a rich toolbox. There have been some phenomenal successes in this area. For example, Lopez-Torrej´ et al. (2016) were able to isolate active NifH from mitochondria of yeast cultures growing under highly aerobic conditions. This study also highlights respiration provides a protective function in mitochondria. Yang et al. (2017) proved that the native Fe-s cluster Biosynthesis process already present in mitochondria eliminates the need for transferring *the nifU* and *nifS* genes in mitochondria. In a venture to generate activable MoFe-co Buren et al. (2017) attempted to express additional Nif components in yeast, which resulted in the generation of 96 yeast strains with mitochondria targeting NifH, NifDK, NifU, NifS, NifM, NifB, and NifEN. Their study proves yeast to be a good model system for screening the expression of *nif* genes. Xiang et al. (2020) engineered NifD variants that are resistant to cleavage by native peptidases of mitochondria and retain high levels of nitrogenase activity in mitochondria of yeast.

11.5 WAY FORWARD

Since nitrogenase is an important enzyme that has a prominent role in food production, unraveling the complexities involved in the biological nitrogen fixation process and decoding the molecular understanding of the enzyme must be the prime focus of nitrogenase research. There are a few research gaps that need to be addressed to enhance these understanding of the enzyme. The hydrolysis of MgATP coupled with substrate reduction must be established at a molecular level. Further studies must be conducted on the P-clusters such as the functional oxidation states. Owing to the unstable nature of FeMo co-factor clusters outside the protein environment, research in its extraction seems to be a limited area of research. Although huge advances have been made in nitrogenase research, still there are more gray areas to explore.

REFERENCES

Burén, S.; Jiang, X.; López-Torrejón, G.; Echavarri-Erasun, C.; Rubio, L. M. Purification and in Vitro Activity of Mitochondria Targeted Nitrogenase Cofactor Maturase NifB. *Frontiers in Plant Science* 2017, *8*, 1567.

Burén, S.; Jiménez-Vicente, E.; Echavarri-Erasun, C.; Rubio, L. M. Biosynthesis of Nitrogenase Cofactors. *Chemical Reviews* 2020, *120* (12), 4921–4968.

Burén, S.; Rubio, L. M. State of the Art in Eukaryotic Nitrogenase Engineering. *FEMS Microbiology Letters* 2018, *365* (2), fnx274.

Burén, S.; Young, E. M.; Sweeny, E. A.; Lopez-Torrejón, G.; Veldhuizen, M.; Voigt, C. A.; Rubio, L. M. Formation of Nitrogenase NifDK Tetramers in the Mitochondria of Saccharomyces Cerevisiae. *ACS Synthetic Biology* 2017, *6* (6), 1043–1055.

Cai, R.; Milton, R. D.; Abdellaoui, S.; Park, T.; Patel, J.; Alkotaini, B.; Minteer, S. D. Electroenzymatic C–C Bond Formation from CO2. *Journal of the American Chemical Society* 2018, *140* (15), 5041–5044.

Chen, H.; Cai, R.; Patel, J.; Dong, F.; Chen, H.; Minteer, S. D. Upgraded Bioelectrocatalytic N2 Fixation: From N2 to Chiral Amine Intermediates. *Journal of the American Chemical Society* 2019, *141* (12), 4963–4971.

Chen, H.; Dong, F.; Minteer, S. D. The Progress and Outlook of Bioelectrocatalysis for the Production of Chemicals, Fuels and Materials. *Nature Catalysis* 2020, *3* (3), 225–244.

Danyal, K.; Inglet, B. S.; Vincent, K. A.; Barney, B. M.; Hoffman, B. M.; Armstrong, F. A.; Dean, D. R.; Seefeldt, L. C. Uncoupling Nitrogenase: Catalytic Reduction of Hydrazine to Ammonia by a MoFe Protein in the Absence of Fe Protein-ATP. *Journal of the American Chemical Society* 2010, *132* (38), 13197–13199.

Delgado, M. J.; Bedmar, E. J.; Downie, J. A. Genes Involved in the Formation and Assembly of Rhizobial Cytochromes and Their Role in Symbiotic Nitrogen Fixation. *Advances in Microbial Physiology* 1998, *40*, 191–231.

Dixon, R.; Kahn, D. Genetic Regulation of Biological Nitrogen Fixation. *Nature Reviews Microbiology* 2004, *2* (8), 621–631.

Dodsworth, J. A.; Leigh, J. A. Regulation of Nitrogenase by 2-Oxoglutarate-Reversible, Direct Binding of a PII-like Nitrogen Sensor Protein to Dinitrogenase. *Proceedings of the National Academy of Sciences* 2006, *103* (26), 9779–9784.

Downie, J. A. Legume Haemoglobins: Symbiotic Nitrogen Fixation Needs Bloody Nodules. *Current Biology* 2005, *15* (6), R196–R198.

Grabbe, R.; Klopprogge, K. A. I.; Schmitz, R. A. Fnr Is Required for NifL-Dependent Oxygen Control of Nif Gene Expression in Klebsiella Pneumoniae. *Journal of Bacteriology* 2001, *183* (4), 1385–1393.

Grunwald, S. K.; Lies, D. P.; Roberts, G. P.; Ludden, P. W. Posttranslational Regulation of Nitrogenase in Rhodospirillum Rubrum Strains Overexpressing the Regulatory Enzymes Dinitrogenase Reductase ADP-Ribosyltransferase and Dinitrogenase Reductase Activating Glycohydrolase. *Journal of Bacteriology* 1995, *177* (3), 628–635.

Halbleib, C. M.; Zhang, Y.; Roberts, G. P.; Ludden, P. W. Effects of Perturbations of the Nitrogenase Electron Transfer Chain on Reversible ADP-Ribosylation of Nitrogenase Fe Protein in Klebsiella Pneumoniae Strains Bearing the Rhodospirillum Rubrum Dra Operon. *Journal of Bacteriology* 2000, *182* (13), 3681–3687.

Heiniger, E. K.; Oda, Y.; Samanta, S. K.; Harwood, C. S. How Posttranslational Modification of Nitrogenase Is Circumvented in Rhodopseudomonas Palustris Strains That Produce Hydrogen Gas Constitutively. *Applied and Environmental Microbiology* 2012, *78* (4), 1023–1032.

Hoffmann, M.-C.; Pfänder, Y.; Fehringer, M.; Narberhaus, F.; Masepohl, B. NifA-and CooA-Coordinated CowN Expression Sustains Nitrogen Fixation by Rhodobacter Capsulatus in the Presence of Carbon Monoxide. *Journal of Bacteriology* 2014, *196* (19), 3494–3502.

Hu, Y.; Lee, C. C.; Ribbe, M. W. Vanadium Nitrogenase: A Two-Hit Wonder? *Dalton Transactions* 2012, *41* (4), 1118–1127.

Huergo, L. F.; Chandra, G.; Merrick, M. PII Signal Transduction Proteins: Nitrogen Regulation and Beyond. *FEMS Microbiology Reviews* 2013, *37* (2), 251–283.

Huergo, L. F.; Pedrosa, F. O.; Muller-Santos, M.; Chubatsu, L. S.; Monteiro, R. A.; Merrick, M.; Souza, E. M. PII Signal Transduction Proteins: Pivotal Players in Post-Translational Control of Nitrogenase Activity. *Microbiology* 2012, *158* (1), 176–190.

Jenner, L. P.; Butt, J. N. Electrochemistry of Surface-Confined Enzymes: Inspiration, Insight and Opportunity for Sustainable Biotechnology. *Current Opinion in Electrochemistry* 2018, *8*, 81–88.

Jiang, Y.-L.; Wang, X.-P.; Sun, H.; Han, S.-J.; Li, W.-F.; Cui, N.; Lin, G.-M.; Zhang, J.-Y.; Cheng, W.; Cao, D.-D. Coordinating Carbon and Nitrogen Metabolic Signaling through the Cyanobacterial Global Repressor NdhR. *Proceedings of the National Academy of Sciences* 2018, *115* (2), 403–408.

Klassen, G.; Souza, E. M.; Yates, M. G.; Rigo, L. U.; Costa, R. M.; Inaba, J.; Pedrosa, F. O. Nitrogenase Switch-off by Ammonium Ions in Azospirillum Brasilense Requires the GlnB Nitrogen Signal-Transducing Protein. *Applied and Environmental Microbiology* 2005, *71* (9), 5637–5641.

Koch-Nolte, F.; Reche, P.; Haag, F.; Bazan, F. ADP-Ribosyltransferases: Plastic Tools for Inactivating Protein and Small Molecular Weight Targets. *Journal of Biotechnology* 2001, *92* (2), 81–87.

López-Torrejón, G.; Jiménez-Vicente, E.; Buesa, J. M.; Hernandez, J. A.; Verma, H. K.; Rubio, L. M. Expression of a Functional Oxygen-Labile Nitrogenase Component in the Mitochondrial Matrix of Aerobically Grown Yeast. *Nature Communications* 2016, *7* (1), 1–6.

Martinez-Argudo, I.; Little, R.; Shearer, N.; Johnson, P.; Dixon, R. The NifL-NifA System: A Multidomain Transcriptional Regulatory Complex That Integrates Environmental Signals. *Journal of Bacteriology* 2004, *186* (3), 601–610.

Masepohl, B.; Forchhammer, K. Regulatory Cascades to Express Nitrogenases. *Biology of the Nitrogen Cycle* 2007, 131–145.

Masepohl, B.; Hallenbeck, P. C. Nitrogen and Molybdenum Control of Nitrogen Fixation in the Phototrophic Bacterium Rhodobacter Capsulatus. *Recent Advances in Phototrophic Prokaryotes* 2010, 49–70.

Milton, R. D.; Minteer, S. D. Direct Enzymatic Bioelectrocatalysis: Differentiating between Myth and Reality. *Journal of the Royal Society Interface* 2017, *14* (131), 20170253.

Minchin, F. R.; James, E. K.; Becana, M. Oxygen diffusion, production of reactive oxygen and nitrogen species, and antioxidants in legume nodules. In: M.J. Dilworth, E.K. James, J.I. Sprent, and W.E. Newton (Eds.), *Nitrogen-Fixing Leguminous Symbioses*, Dordrecht, Springer, 2008, pp. 321–362.

Moure, V. R.; Costa, F. F.; Cruz, L. M.; Pedrosa, F. O.; Souza, E. M.; Li, X.-D.; Winkler, F.; Huergo, L. F. Regulation of Nitrogenase by Reversible Mono-ADP-Ribosylation. *Endogenous ADP-Ribosylation* 2014, 89–106.

Moure, V. R.; Danyal, K.; Yang, Z.-Y.; Wendroth, S.; Müller-Santos, M.; Pedrosa, F. O.; Scarduelli, M.; Gerhardt, E. C.; Huergo, L. F.; Souza, E. M. The Nitrogenase Regulatory Enzyme Dinitrogenase Reductase ADP-Ribosyltransferase (DraT) Is Activated by Direct Interaction with the Signal Transduction Protein GlnB. *Journal of Bacteriology* 2013, *195* (2), 279–286.

Moure, V. R.; Siöberg, C. L.; Valdameri, G.; Nji, E.; Oliveira, M. A. S.; Gerdhardt, E. C.; Pedrosa, F. O.; Mitchell, D. A.; Seefeldt, L. C.; Huergo, L. F. The Ammonium Transporter AmtB and the PII Signal Transduction Protein GlnZ Are Required to Inhibit DraG in Azospirillum Brasilense. *The FEBS Journal* 2019, *286* (6), 1214–1229.

Nordlund, S.; Högbom, M. ADP-ribosylation, a Mechanism Regulating Nitrogenase Activity. *The FEBS Journal* 2013, *280* (15), 3484–3490.

Oetjen, J.; Reinhold-Hurek, B. Characterization of the DraT/DraG System for Posttranslational Regulation of Nitrogenase in the Endophytic Betaproteobacterium Azoarcus Sp. Strain BH72. *Journal of Bacteriology* 2009, *191* (11), 3726–3735.

Poole, R. K.; Hill, S. Respiratory Protection of Nitrogenase Activity in Azotobacter Vinelandii—Roles of the Terminal Oxidases. *Bioscience Reports* 1997, *17* (3), 303–317.

Poza-Carrión, C.; Jiménez-Vicente, E.; Navarro-Rodríguez, M.; Echavarri-Erasun, C.; Rubio, L. M. Kinetics of Nif Gene Expression in a Nitrogen-Fixing Bacterium. *Journal of Bacteriology* 2014, *196* (3), 595–603.

Rao, D. L. N. Recent Advances in Biological Nitrogen Fixation in Agricultural Systems. In: *Proceedings of the Indian National Science Academy*, 2014, Vol. 80, pp. 359–378.

Rutledge, H. L.; Tezcan, F. A. Electron Transfer in Nitrogenase. *Chemical Reviews* 2020, *120* (12), 5158–5193.

Scherings, G.; Haaker, H.; Wassink, H.; Veeger, C. On the Formation of an Oxygen-tolerant Three-Component Nitrogenase Complex from Azotobacter Vinelandii. *European Journal of Biochemistry* 1983, *135* (3), 591–599.

Schlesier, J.; Rohde, M.; Gerhardt, S.; Einsle, O. A Conformational Switch Triggers Nitrogenase Protection from Oxygen Damage by Shethna Protein II (FeSII). *Journal of the American Chemical Society* 2016, *138* (1), 239–247.

Schnabel, T.; Sattely, E. Engineering Post-Translational Regulation of Glutamine Synthetase for Controllable Ammonia Production in the Plant-Symbiont A. Brasilense. *Applied and Environmental Microbiology* 2021, AEM. 00582-21.

Stal, L. J. The Effect of Oxygen Concentration and Temperature on Nitrogenase Activity in the Heterocystous Cyanobacterium Fischerella Sp. *Scientific Reports* 2017, *7* (1), 1–10.

Truan, D.; Huergo, L. F.; Chubatsu, L. S.; Merrick, M.; Li, X.-D.; Winkler, F. K. A New PII Protein Structure Identifies the 2-Oxoglutarate Binding Site. *Journal of Molecular Biology* 2010, *400* (3), 531–539.

Tsuge, H.; Nagahama, M.; Nishimura, H.; Hisatsune, J.; Sakaguchi, Y.; Itogawa, Y.; Katunuma, N.; Sakurai, J. Crystal Structure and Site-Directed Mutagenesis of Enzymatic Components from Clostridium Perfringens Iota-Toxin. *Journal of Molecular Biology* 2003, *325* (3), 471–483.

Ureta, A.; Nordlund, S. Evidence for Conformational Protection of Nitrogenase against Oxygen in Gluconacetobacter Diazotrophicus by a Putative FeSII Protein. *Journal of Bacteriology* 2002, *184* (20), 5805–5809.

Wang, T.; Zhao, X.; Shi, H.; Sun, L.; Li, Y.; Li, Q.; Zhang, H.; Chen, S.; Li, J. Positive and Negative Regulation of Transferred Nif Genes Mediated by Indigenous GlnR in Gram-Positive Paenibacillus Polymyxa. *PLoS Genetics* 2018, *14* (9), e1007629.

Watzer, B.; Spät, P.; Neumann, N.; Koch, M.; Sobotka, R.; Macek, B.; Hennrich, O.; Forchhammer, K. The Signal Transduction Protein PII Controls Ammonium, Nitrate and Urea Uptake in Cyanobacteria. *Frontiers in Microbiology* 2019, *10*, 1428.

Xiang, N.; Guo, C.; Liu, J.; Xu, H.; Dixon, R.; Yang, J.; Wang, Y.-P. Using Synthetic Biology to Overcome Barriers to Stable Expression of Nitrogenase in Eukaryotic Organelles. *Proceedings of the National Academy of Sciences* 2020, *117* (28), 16537–16545.

Yang, J.; Xie, X.; Yang, M.; Dixon, R.; Wang, Y.-P. Modular Electron-Transport Chains from Eukaryotic Organelles Function to Support Nitrogenase Activity. *Proceedings of the National Academy of Sciences* 2017, *114* (12), E2460–E2465.

Zhang, Y.; Cummings, A. D.; Burris, R. H.; Ludden, P. W.; Roberts, G. P. Effect of an NtrBC Mutation on the Posttranslational Regulation of Nitrogenase Activity in Rhodospirillum Rubrum. *Journal of Bacteriology* 1995, *177* (18), 5322–5326.

Zhang, Y.; Pohlmann, E. L.; Roberts, G. P. GlnD Is Essential for NifA Activation, NtrB/NtrC-Regulated Gene Expression, and Posttranslational Regulation of Nitrogenase Activity in the Photosynthetic, Nitrogen-Fixing Bacterium Rhodospirillum Rubrum. *Journal of Bacteriology* 2005, *187* (4), 1254–1265.

Zhang, Y.; Pohlmann, E. L.; Roberts, G. P. Identification of Critical Residues in GlnB for Its Activation of NifA Activity in the Photosynthetic Bacterium Rhodospirillum Rubrum. *Proceedings of the National Academy of Sciences* 2004, *101* (9), 2782–2787.

12 *nod*, *nif*, and *fix* Genes

Irfan Iqbal Sofi[1]*, Shabir Ahmad Zargar*[1], *and Shivali Verma*[2]
[1]Department of Botany, University of Kashmir, Srinagar, J&K, India
[2]Department of Botany, University of Jammu, Jammu Tawi, J&K, India
*Corresponding author: sofi.irfan98@gmail.com

CONTENTS

12.1 Introduction 155
12.2 Nitrogen Fixation Process 156
12.3 Location of SNF Genes 156
    12.3.1 nod Genes 157
    12.3.2 nif Genes 160
    12.3.3 fix Genes 161
12.4 Conclusion 162
References 163

12.1 INTRODUCTION

The symbiotic relationship of plants with microbes plays an important role in the plant's survival approach (i.e., growth and development). Biological nitrogen fixation (BNF), an important symbiotic relationship between species of microbes and plants, involves the enzymatic reduction of atmospheric dinitrogen (N_2) to ammonia (NH_3). Nitrogen fixation is one of the primary sources of nitrogen for plants and an important part of allocating this nutrient in the ecosystem (Sur et al., 2010). This symbiotic association is important for agriculture by increasing soil quality, mitigating greenhouse gases, lowering farmers' costs, and increasing soil fertility of infertile and stressed soil. A group of free-living microbes called diazotrophs including bacteria and archaea occupying many habitats (water, soil, hot vents) can fix nitrogen when certain physiological and nutritional conditions are achieved (Mus et al., 2019; Dos Santos and Addo 2020). Diazotrophs from more than 100 genera are responsible for the fixation of atmospheric nitrogen gas into a usable form like NH_3 (Young, 1992). BNF microbes can be aerobic (e.g., *Azotobacter chroococcum*), facultative anaerobic (e.g., *Rhodobacter capsulatus*) or anaerobic (*Clostridium kluyveri*) heterotrophs, oxygenic or anoxygenic phototrophs (e.g., *Rhodobacter capsulatus, Anabaena variabilis*, respectively), and chemo-lithrotrophs (e.g., *Acidithiobacillus ferrooxidans*) (Mus et al., 2019). Legumes form symbiotic connexion with nitrogen-fixing soil bacteria collectively called rhizobia, which infect the inter- or intra-cellularly tissues of legumes influenced by the release of certain flavonoids and form outgrowths called nodules particularly on the roots of legumes. The phylum proteobacteria contain diazotrophs in which 18 different genera of rhizobium with hundreds of species have been defined among α and β and proteobacteria. On the transformation of rhizobia into bacteroids, N_2 of the atmosphere is converted into NH_3 (Oldroyd and Downie, 2008, Peix et al., 2015; Mus et al., 2016). A dynamic process of BNF involves a broad energy supply (Postgate, 1982), requiring 16 molecules of ATP to reduce a mole of nitrogen (Hubbell and Kidder, 2009). Nitrogenase, an oxygen-sensitive metalloenzyme with three isoforms containing Mo-, V- and Fe- as catalytic clusters, is the main player in the process of biological nitrogen fixation (Eady, 1996; Rutledge and Tezcan, 2020). The most common form of the enzyme contains an electron supplier Fe protein and catalytic center MoFe protein (Hoffmann et al., 2013).

The advancement in DNA sequencing has helped in the detailed genetic breakthrough of gene structure and genetic regulation of BNF with a focus on model organisms of the process like *Bradyrhizobium japonicum*, *Rhizobium leguminosarium* (Barney et al., 2020). The relationship between the two associates that carry on symbiosis is started by a molecular cross-talk (De'narie' et al., 1992; Schlaman et al., 1992). From the entry of rhizobia inside the host to fixation of N2 involves the release of important signaling molecules responsible to carry out the important process of nitrogen fixation (Lindström et al., 2010). These signaling molecules are important factors that help in the activation of important genes and involve a more complex exchange of signals between symbionts and the development of a root nodule. Nodulation and nitrogen fixation genes are the primary and important bacterial genes involved in BNF. The *nod* (nodulation gene), *nif*, and *fix* (nitrogen fixation gene) genes and their role in legume-rhizobial symbiosis will be an important focus of the current chapter.

12.2 NITROGEN FIXATION PROCESS

One of the most substantial and best-known types of nitrogen fixation is that of legumes and rhizobia symbiosis (Coyne and Frye, 2005). The process of symbiotic association initiates with the perception of signaling molecules (mainly flavonoids) by bacteria released by the host into the rhizosphere, which persuade the synthesis and secretion of nodulation factors, mainly lipo-chitooligosaccharides, by the bacteria (Oldroyd, 2011; Hassan et al., 2012) that behave as mitogens prompting rapid multiplication of root cortex cells, and the development of the nodule primordium (Geurts et al., 2005; Cooper, 2007). On infection, the tips of root hair curl and entrap the bacteria, then infection thread-harboring bacteria infiltrate inwards towards the inner portion of root cortex to occupy a nodule primordium, which develops into a fully grown outgrowth called a nodule accommodating the bacteria (Dazzo, and Gardiol, 1984). Peribacteroid or symbiosome membrane encloses nodule cells and a new structure shaped by the intracellular bacteria encased within a peribacteroid membrane is referred to as a symbiosome (Coba de la Peña et al., 2018) where fast multiplication of bacteria occurs that differentiate into nitrogen-fixing bacteroids (Oldroyd et al., 2011). During the root hair infection process, the upregulation of many genes occurs that controls diverse developments including host-symbiont signaling and cell cycle and cell growth (Libault et al., 2010; Breakspear et al., 2014). The symbiosome is the main nitrogen-fixing unit where processed nitrogen in the form of ammonia and ammonium ions are carried towards the cytoplasm of macro-symbiont (Oldroyd et al., 2011).

12.3 LOCATION OF SNF GENES

The rhizobia are members of both α (alpha) and β (beta) class proteobacteria with genome size nearly double or more than the average bacterial genome size (Geddes et al., 2020). The genome of each group is divided into two or more large replicons, with one replicon like the main chromosome of bacteria (*Escherichia coli or Bacillus subtilis*) and constituting many of the core genes. The remaining replicons of one or more are called plasmids, megaplasmids, chromatids, or 2nd chromosomes. In the case of *Sinorhizobium meliloti*, a megaplasmid pSymA of 1.34 Mb contains several SNF genes including *nod*, *nif*, and *fix*. Likewise, in the case of other rhizobia genera including *Rhizobium, Sinorhizobium, Phyllobacterium,* and *Burkholderia*, plasmid or megaplasmid (pSym) possess genes of SNF. Some rhizobia genera (*Bradyrhizobium, Mesorhizobium,* and *Azorhizobium*) with a single large chromosome contain regions called symbiotic islands that are sites for the occurrence of SNF genes (Geddes et al., 2020).

12.3.1 NOD GENES

The studies on the molecular genetics of rhizobia and symbiotic associations of mutant bacterial strains have opened the way for the identification of nodulation genes (nod, nul, and noe). These nodulation genes are involved in the infection process, formation of nodules, and host specificity regulation (Göttfert, 1993; van Rhijn and Vanderleyden, 1995; Dénarié, 1996). The nodulation (*nod*) genes of bacterial partners are the important bacterial determinants responsible for the production of lipo-chitinous Nod factors that initiate *de novo* organogenesis of special root structures called nodules in photosynthetic microsymbiont. The Nod factors possess the backbone of β 1,4-linked

TABLE 12.1
Nodulation Gene Products Function

nod gene	Gene product function	References
nodD	LysR-type regulator, Transcriptional regulator of common nod genes	Fisher and Long (1992)
nodV	two-component family sensor	Sanjuan et al., 1994
nodW	two-component family regulator	Sanjuan et al., 1994
nolA	MerR-type regulator	Sadowsky et al., 1991
Synthesis of the Chito oligosaccharide backbone		
nodM	D-glucosamine synthase	Marie et al., 1992
nodC	UDP-GlcNAc transferase	Spaink et al., 1994; Inon de Iannino et al., 1995; Bloemberg et al., 1995a
nodB	De-N-acetylase	Spaink et al., 1994; Atkinson et al., 1994; John et al., 1993
N-substitutions at the nonreducing end		
nodE	Beta-ketoacyl synthase	Spaink et al., 1991; Bloemberg et al., 1995a
nodF	Acyl carrier protein	Shearman et al., 1986;
nodA	N-acyltransferase	Atkinson et al., 1994; Röhrig et al., 1994
nodS	S-adenosyl methionine methyltransferase	Mergaert et al., 1995; Jabbouri et al., 1995; Geelen et al., 1993, Geelen et al., 1995
O-substitutions at the nonreducing end		
nodL	6-0-acetyltransferase	Downie 1989; Ardourel et al., 1995; Bloemberg et al., 1995b
nodU	6-0 carbamoyltransferase	Jabbouri 1995;
O-substitutions at reducing end		
nodP	ATP sulfurylase	Schwedock et al., 1992; et al., 1990
nodQ	ATP sulfurylase, APS kinase	Schwedock et al., 1992; Schwedock et al., 1990
nodH	Sulfotransferase	Bourdineaud et al., 1995; Schultze et al., 1995
nodZ	Fucosyl transferase, Glycosyltransferase	Stacey et al., 1994; Fellay et al., 1995
nolK	Sugar epimerase	Geelen et al., 1995
nodX	Acetyltransferase	Goethals et al., 1992
Secretion of Nod factors		
nodI	ATP-binding protein	Evans and Downie 1986; Vázquez et al., 1993
nodJ	Membrane protein	Evans and Downie 1986; Vázquez et al., 1993
nodT	Outer membrane protein	Rivilla et al., 1995; Surin et al., 1990
nolFGHI	Membrane proteins	Baev et al., 1992; Saier et al., 1984

N-acetyl glucosamine residues, which require UDP-*N*-acetyl-D-glucosamine. The *nod* genes and various combinations in different species are shown in Table 12.1. For the biosynthesis of the core structure of Nod factors, three important genes (*nodA, nodB, nod C*) play an important role in which an acyltransferase linking acyl chain to the oligosaccharides non-reducing end is encoded by *nodA*, while deacetylase is involved in the removal of N-acetyl moiety from these oligosaccharides non-reducing end and is encoded by *nodB*, and N-acetyl-glucosaminyl transferase linking N-acetyl-D-glucosamine into oligosaccharides is encoded by *nodA* (Bonaldi et al., 2010; Lindström and Mousavi, 2019). The development of pseudo-nodules in alfalfa, sweet clover, or Afghanistan pea roots *sym* mutants when treated with 2,3,5-triiodobenzoic acid or NPA showed the presence of nodulin genes *ENOD2*, *ENOD8*, and *ENOD12* (Wu et al., 1996). Similarly, the application of cytokinin on the roots *Sesbania rostrata* and alfalfa activated the expression of *ENOD2* (Dehio and deBruijn, 1992) and *ENOD2* and *ENOD12A*, respectively.

The development of nodules on the root and stem of *Aeschynomene indica* and *A. sensitive* by BTAi1 and ORS278 *Bradyrhizobium* strains has been found in the absence of Nod factors and *nodABC* gene (Giraud et al., 2007). Moreover, Nod factor mutant in *B. elkanii* and *Glycine max* cv. Enrei and symbiotic relationship of *Lotus japonicus* and NF mutants of *Mesorhizobium loti* show the same phenomenon (Madsen et al., 2010; Okazaki et al., 2013). This process of nodule organogenesis in non-photosynthetic bradyrhizobia has been assigned to the type III secretion system (T3SS). Furthermore, the symbiotic studies in T3SS mutant *Bradyrhizobium* strain ORS285 showed a possible positive or negative role of T3SS in NF-dependent *Aeschynomene* species but is dispensable for the interaction with all studied NF-independent *Aeschynomene* species (Okazaki et al., 2016).

Rhizobia synthesize some Nod factors that are strain-specific and occur in number from two in the case of *Rhizobium etli* to 60 in *Rhizobium galegae* (Poupot et al., 1995; Yang et al., 1999). The Nod factors of different rhizobia vary by the presence of some modifications that play a key role in plant host-specificity as the genes carrying the synthesis of these factors diverge among the rhizobia species (Denarie et al., 1996; Gage, 2004). The presence of a different class of *nod* genes is the primary factor for the variation in the structural backbone of Nod factors, which is initiated by the different substitutions, including sulfation, methylation, glycosylation, and acetylation by the *nod* genes (Bonaldi et al., 2010; Andrews and Andrews, 2017).

Nodulation genes of various rhizobia are organized as clusters in several operons and a similar mode of regulation among many nodulation operons of rhizobia is due to the presence of *nod*-box, a highly conserved sequence of DNA found in the upstream of promoter regions (Göttfert, 1993). The expression of *nod* genes occurs in response to flavonoid compounds produced by the plant (Mulligan and Long, 1985; Liu et al., 1998). Flavonoids interact with constitutively expressed LysR-type transcriptional regulator NodD, a product of *nodD* gene, which is responsible for the regulation of other nodulation genes (e.g., *nod*, *nol*, and *noe*) in the presence of flavonoid as an inducer (Cooper, 2004; Bonaldi et al., 2012). The proper combination of this interaction allows NodD to bind at *nod*-box, a regulatory sequence for the synthesis of other Nod factors. The binding of NodD protein at *nod* genes causes DNA bending and leads to transcriptional activation after finding the inducer (Kosslak, 1987). The Nod factor signal molecules bind on the specific receptor kinases with a lysine motif in their extracellular domain called LysM-RLK (Lysin Motif Receptor-Like Kinase). Nod factor receptors containing serine/threonine receptor kinases located on the plasma membrane recognize the Nod factors in legumes (Okazaki et al., 2016). In legumes, the genome encodes a large number of LysM receptor families with a known function to only a few (e.g., *Medicago truncatula* and *Lotus japonicus* have 26 and 19 receptor kinases of such type, respectively). In Lotus and Medicago, the impaired function of Nfr5/Nfp and Nfr1/Lyk3 show complete loss of symbiotic association with the presence of mutant of Nfr5 and Nfr1 (Smit et al., 2007; Amor et al., 2003), and overexpression of the same genes leads to the spontaneous formation of nodules (Ried et al., 2014). In *Rhizobium leguminosarum*, the bending of DNA *nod* promoters also involves a histone-like protein, Px (Liu1998), and Px association with the promoter of *nod* genes leads to amplified transcription *of nod* genes by NodD. In *B. japonicum, two proteins* of the NodD family including NodD1 and NodD2 *show discrete expression and roles. In the presence of* daidzein and genistein (isoflavones) produced by the plant, an autoregulated NodD1 acts as a positive transcriptional activator (Banfalvi, 1988; Göttfert, 1992; Kosslak et al., 1987). In contrast, NodD2 acts by repressing *nod* gene expression (Garcia et al., 1996; Gillette, 1996). Moreover, *nodD1* is also induced by glycosylated isoflavones (e.g., 6-O-malonyl daidzin and 6-O-malonyl genistin), which are unable to induce the expression of *nodYABC* transcription which gives a specific edge to the plant for regulating the expression of *nodD1* (Smit et al., 1992, Loh and Stacey, 2003). All nodulation factors producing rhizobia possess the *nodABC*, *nodD*, and *nodIJ* genes; nevertheless, in the case of other *nod, noe,* and *nol* genes, a substantial variation in the presence or absence occurs between the various rhizobia and is a reason for the difference in Nod factors from different rhizobia for carrying substituents and modifications (Dénarié et al., 1996; Gage, 2004; Broughton et al., 2000; Geddes et al., 2020). The acetylation and sulfation of nonreducing and reducing terminal in *S. meliloti* nodulation factors are carried by *nod*L and *nodH*, respectively (Ardourel et al., 1994; Lerouge et al., 1990; Roche et al., 1991). Similarly, rhizobia with carbamoyl group, *nod*U, encodes an alternate N-terminal modification to it. (D'Haeze et al., 1999; Jabbouri et al., 1995), and fucosylation of nonreducing terminal residue is encoded by nodZ (Mergaert et al., 1997). The genomic studies in rhizobia have revealed that the number of *nod*D copies may vary between one and five among species. Similarly, the different NodD vary in the manner they prefer to bind their flavonoid in the same strain (Broughton et al., 2000; del Cerro et al., 2015). The Nod factors are excreted and play an important role in the symbiotic developmental process from root hair curling, through the expression of important genes unto formation of nodules (Oldroyd et al., 2013; Laranjo et al., 2014). The released flavonoids of host and Nod factors of bacterial partners are specific, and interaction of these chemicals match them to play a part in the symbiotic process. A single mutation may result in complete loss of nodulation in *Rhizobium leguminosarum* bv. trifolii as the bacteria carry only one copy of *nodD* gene contrary to *Rhizobium leguminosarum* bv. phaseoli and *Bradyrhizobium japonicum* where nodulation is affected but not suppressed completely due to the presence of multiple copies of the *nodD* gene (Broughton et al., 2000, Garcia et al., 1996, del Cerro et al., 2015). While the role of NodD in the regulation of *nod* genes is fundamental, the control of nodulation genes is now recognized to be regulated by some additional regulators (Kondorosi et al., 1991).

The finding of NodVW in *B. japonicum* suggested an alternative mechanism for the activation of *nod* genes (Sanjuan et al., 1994). NodVW plays an important role in the nodulation of mungbean, siratro, and cowpea but not soybean and gives *B. japonicum* the flexibility to nodulate a wider range of hosts and the combined role of NodDVW is important for the amplified synthesis of Nod signals (Loh and Stacey, 2003). The NodV and NodW of *B. japonicum* act as a two-component regulatory system in which NodV acts as sensor kinase and NodW as response regulator protein (Göttfert et al., 1992; Loh et al., 1997; Stock et al., 2000). The autophosphorylation of sensor kinase communicates the phosphoryl group to the response regulator and is later used to activate the target *nodYABC* operon (Loh and Stacey, 2003). Besides NodW, NwsB acts as a second response regulator in *B. japonicum* shares 65% amino acid similarity with NodW and has a conserved DNA binding motif (Grob et al., 1993). NwsB has been suggested to be involved in quorum regulation and activation of nodulation genes (Loh and Stacey, 2003). The *nod* genes are subjected to negative regulation (Sadowsky et al., 1991). In *Rhizobium* species, *B. elkani* and strain NGR234 and NodD2 act as a repressor while in *S. meliloti* NolR mediates repression of *nod* genes. The two important components of negative regulation in *B. japonicum* are NolA and NodD2; NolA in presence of ambient conditions induces NodD2 expression, which represses the nodulation genes. The *nod*IJ genes are present in rhizobial species that have been studied genetically. NodI is associated with the intermembrane and is related to an ATP-binding cassette (ABC transporters) family of proteins while NodJ is very hydrophobic and probably an integral membrane protein (Evans and Downie, 1986). Both NodI and NodJ show a robust resemblance with proteins involved in polysaccharide secretion in bacteria (Vázquez et al.,

1993). The *nod*IJ mutants of *Rhizobium leguminosurum* bv. trifolii show poor nodulation in the host, however, in *Rhizobium leguminosurum* bv. viciae with alternate transport system the absence of *nod*IJ genes shows poor effects (Dénarié et al., 1996).

12.3.2 NIF GENES

The genes involved in the encoding and assembly of an important enzyme complex nitrogenase, which is responsible for the converting of N2 gas to ammonia, are included in the family of *nif* genes (Lindström and Mousavi, 2019). According to the presence of molybdenum (Mo), vanadium (V), and iron (Fe) cofactors at their active site, three types of nitrogenase exist in nature. The most common and abundant is the molybdenum (Mo) nitrogenase, and the other two include vanadium (V) and iron-only (Fe) nitrogenases (Bishop and Joerger, 1990). The *nif* genes are expressed in response to a low concentration of oxygen, which is maintained by leghaemoglobin inside the nodules of the host. In *S. meliloti*, *nifDK* encodes the two subunits of the molybdenum nitrogenase protein and *nifH* encodes the nitrogenase reductase.

The regulation of nitrogen fixation in rhizobia is under the master control of the NifA-RpoN regulator. NifA protein, which in association with the RNA polymerase RpoN (σ^{54}), binds to an upstream activating sequence of *nif* genes and activates them. For the biosynthesis of the FeMo cofactor located inside the active site of nitrogenase protein, the *nifB*, *nifX*, *nifE*, and *nifN* genes of *the nif* family are involved (Masson-Boivin et al., 2009; Rubio and Ludden, 2008). Moreover, *nifS*, a gene located on chromosome-encoding cysteine desulfurase, has some role to play in FeMo cofactor synthesis (Capela et al., 2001). The transcriptional regulator involved in transcription of *nifB*, *nifN*, *nifHDKEX*, and *fixABCX* during *rpoN* (sigma54) transcription is encoded by *nifA* (Galardini et al., 2013; Gong et al., 2006). Diazotrophs carrying alternative nitrogenases based on V or Fe at the position of Mo carry an additional subunit encoded by *vnfG or anfG* genes (Eady, 1996). The number of genes that are important for the synthesis of efficient nitrogenase protein varies between species and usually ranges from 10 to 20 (Temme et al., 2012; Poza-Carrion et al., 2015). The rhizobia generally have a single copy of *the nifA* gene, but in the case of *Mesorhizobium loti*, two *nifA* genes including *nifA1* and *nifA2* have been found that are located on the symbiotic islands (Nukui et al., 2006). The *nifA1* gene has most of its similarity to *the nifA* gene of *Rhizobium etli*, *Rhizobium leguminosarum*, and *S. meliloti* while *nifA2* is similar to *nifA* from *Bradyrhizobium japonicum* (Sullivan et al., 2013). Similarly, through the genomic analysis of non-photosynthetic *Bradyrhizobium* sp. DOA9 has two *nifA* genes that have been found to occur in it (Okazaki et al., 2015).

In symbiotic diazotrophs (e.g., *Sinorhizobium meliloti*) FixL-FixJ, an oxygen-responsive two-component system, controls the transcriptional regulation of nifA and fix genes (Bobik et al., 2006). Similarly, RegS-RegR controls the NifA-RpoN regulatory system in *Bradyrhizobium japonicum* (Hawkins and Johnston, 1988), while as in *Azorhizobium caulinodans* it is directly controlled by FixK (Kaminski and Elmerich, 1998), but in *Rhizobium leguminosarum*, nifA is autoregulated (Martinez et al., 2004).

There is enormous variation in the organization and complexity of *nif* genes in various groups of microorganisms (Downie, 1989) (Table 12.2). In most cases, NifA (a positive activator of transcription) and NifL (a negative regulator) control the regulation of all *nif* genes. The regulation of *nif* genes is affected by the concentration of both oxygen and nitrogen (Merrick and Edwards, 1995). When the soil concentrations of ammonia (NH3 or NH4) are high, nitrogen fixation is slowed down by NifL which acts as a negative controller by preventing NifA to act as an activator. When O2 concentration is high, then the synthesis of nitrogenase decreases and leads to a decrease in nitrogen fixation. Using genetic and biochemical techniques, the *nif* genes of the first non-symbiotic diazotroph *Klebsiella pneumoniae* have been studied (Arnold et al., 1988, MacNeil et al., 1978; Roberts et al., 1978). The 20 *nif* genes (*nifJHDKTYENXUSVWZMFLABQ*) present in the species occur in a simple cluster of a single 23-kb region in the chromosome (Rubio and Ludden, 2008).

TABLE 12.2
The List of *nif* Genes and Their Function in *Enterobacter* sp. R4-368 and *K. pneumonia*

nif genes	Function
nifJ	pyruvate-flavodoxin oxidoreductase
nifH	Fe protein
nifD	MoFe protein
nifK	MoFe protein, beta subunit
nifT	Nitrogen fixation protein
nifY, nifE, nifN, nifX	Iron-molybdenum cofactor biosynthesis
nifU	Fe-S cluster assembly
nifS	Nitrogenase metalloclusters biosynthesis
nifV	Homocitrate synthase
nifW	Nitrogenase stabilizing/protection
nifZ	Fe-S cofactor synthesis
nifM	Peptidyl-prolyl cis/trans isomerase
nifF	Flavodoxin for electron transfer
nifL	Negative regulatory protein
nifA	Positive regulatory protein
nifB	FeMo cofactor biosynthesis

In the case of *Azotobacter vinelandii*, nif genes are arranged in two major and minor clusters of different chromosomal linkage groups depending on physiological conditions encountered by nitrogenase enzyme in these bacteria. The *nifHDKTYENX, iscA^{nif}, nifUSV, cysE1^{nif}, nifWZM, nifF* genes and some interspersed open reading frames between these genes are present on the major cluster (Jacobson et al., 1989). While as *rnfABCDGEH, nafY* genes in one direction (Rubio et al., 2002; Curattiet al.,2005) and the *niflAB, fdxN, nifOQ* genes in the opposite orientation of DNA (Joerger and Bishop, 1988).

12.3.3 Fix Genes

The genes that are required to support the process of N2 fixation in nodules but not involved in the nodule formation are included in the fix genes. These genes are present in both the α-and β-rhizobia (Edgren and Nordlund, 2004; Ledbetter et al., 2017) (Table 13.3).

The *fixABCX* genes are present in *R. meliloti* (Pühler et al., 1984; Ruvkun et al., 1982), *B. japonicum* (Fuhrmann et al., 1985; Gubler et al., 1986), *A. caulinodans* (Arigoni et al., 1991), *R. leguminosarum* bv. viciae (Gronger et al., 1987), *R. leguminosarum* bv. trifolii (Iismaa et al., 1989), and *R. leguminosarum* bv. phaseoli (Michiels and Vanderleyden, 1993) and are ordered in a single operon with with exception of *B. japonicum*, in which fixA and fixBCX are present in clusters II and I, respectively, and form discrete transcriptional units. The *fixBCX* operon of *B. japonicum* has a proximal open reading frame (ORF35), whose translation significantly increase mRNA of *fixBCX* (Gubler and Hennecke, 1986) besides playing a role in nitrogen fixation. The loss of function in any one of the *fixABCX* genes of *R. meliloti, B. japonicum,* and *A. caulinodans* eliminates nitrogen fixation (Earl et al., 1987; Gubler et al., 1986). The proteins including FixA and FixB are homologs of electron transfer flavoproteins of *E.Coli* while FixC is like oxidoreductase and FixX as a ferredoxin-like protein (Costas et al., 2017). In *S. meliloti*, ferredoxin encoded by *fdxN* is essential for nitrogen fixation (Klipp et al., 1989) and FdxN in *Rhodospirillum rubrum* acts as an electron donor to nitrogenase (Edgren and Nordlund, 2004).

TABLE 12.3
fix Genes and Their Function

fix gene	Function
fixABCX	Required for nitrogen fixation, fix shows similarity to ferredoxins
fixD	Transcription activator of *nif, fix*, and additional genes
fixF	Codes for a polypeptide homologous to the *nifK* gene product
fixGHIS	Required for the formation of high-affinity cbb_3 cytochrome oxidase
fixLJ	The regulatory two-component system involved in oxygen regulation of fixK and nifA transcription
fixK	The regulatory protein belongs to the Crp/Fnr family of prokaryotic transcriptional activators
fixNOQP	Microaerobically induced, membrane-bound high-affinity cbb_3 cytochrome oxidase
fixR	Sequence similarity to NAD-dependent dehydrogenases not essential for symbiotic nitrogen fixation in *B. japonicum*
fixT	Negative regulator of *fixL*

The *fixNOQP* linked to the regulatory genes *fixLJ* and *fixK* was first studied in *R. meliloti* and its homologous genes are present in *B. japonicum* (Preisig et al., 1993), *A. caulinodans* (Mandon et al., 1994), and *R. leguminosarum* bv. viciae (Hynes et al., 1992). A high-affinity cbb_3 type cytochrome c oxidase encoded by the *fixNOQP* genes serves as the terminal oxidase functioning at very low O2 concentration in legume nodules (Preisig et al., 1996; Torres et al., 2014). Three copies of fixNOQP (fixNOQP1, fixNOQP2, and fixNOQP3) genes on the pSymA plasmid are found in *S. meliloti*. The fixNOQP1 and fixNOQP2 show 95% sequence homology while fixNOQP3 shares 61% homology with no role in symbiotic nitrogen fixation (Torres et al., 2014). The expression of nitrogen fixation genes is under the coordination of *fixLJ* and *fixK* genes (Dixon and Kahn, 2004). In response to specific environmental signals (i.e., oxygen-limiting conditions), the autophosphorylation of the histidine kinase FixLsensor, transfer phosphate to FixJ, which acts as response regulator and activates transcription of the *nifA* and *fixK* genes, which are involved in the expression of other *nif* and *fix* genes, respectively (Reyrat et al., 1993). The *fixT* acts as a negative regulator of FixLJ-dependent genes by preventing FixL autophosphorylation (Garnerone et al., 1999) while *fixM* encodes a flavoprotein that controls the inhibition of gene expression involved in respiratory and nitrogen fixation and subsequent decrease of fixK transcription in *Ensifer meliloti* (Cosseau et al., 2002). In *S. meliloti*, a membrane-anchored hemoprotein FixL acts as an oxygen sensor in which oxygen binds a heme group on histidine located within PAS structural motif, which senses diverse stimuli including light, oxygen, and redox potential (Vanderheyden et al., 2001). FixJ phosphate stimulates the expression of *the fixk* gene, which acts as an activator of *fixNOQP* genes that code for *fixGHIS* and *fixT* genes. These genes (*fixGHIS* and *fixT*) then form respiratory oxidase complex (Vanderheyden et al., 2001)

12.4 CONCLUSION

The important role of microorganisms in the ecosystem for growth and survival of plants is still largely vague. The recent advances in biological science have helped in identifying major and important genetic components that are required during symbiotic N2 fixation. The discovery of the role of new players including small RNAs (Ren et al., 2019) and receptors like Epr3 (Kawaharada et al., 2015) has opened a new way to uncover many novel facets of the symbiotic relationship. However, detailed studies on the role of new receptors and their respective ligands are still needed. Genomic and proteomic tools will help to add knowledge on various stages of symbiosis and will help to elucidate the expression of different genes and proteins at different stages of the symbiotic relationship.

REFERENCES

Amor, B. B., Shaw, S. L., Oldroyd, G. E., Maillet, F., Penmetsa, R. V., Cook, D.,... and Gough, C. (2003). The NFP locus of Medicago truncatula controls an early step of Nod factor signal transduction upstream of rapid calcium flux and root hair deformation. The Plant Journal 34(4), 495–506.

Andrews, M., and Andrews, M. E. (2017). Specificity in legume-rhizobia symbioses. International Journal of Molecular Sciences 18(4), 705.

Ardourel, M., Demont, N., Debellé, F., Maillet, F., de Billy, F., Promé, J. C.,... and Truchet, G. (1994). Rhizobium meliloti lipooligosaccharide nodulation factors: different structural requirements for bacterial entry into target root hair cells and induction of plant symbiotic developmental responses. The Plant Cell 6(10), 1357–1374.

Ardourel, M., Lortet, G., Maillet, F., Roche, P., Truchet, G., Promé, J. C. and Rosenberg, C. (1995). In *Rhizobium meliloti*, the operon associated with the nod box n5 comprises *nodL, noeA* and *noeB*, three host-range genes specifically required for the nodulation of particular Medicago species. Molecular Microbiology 17(4), 687–699.

Arigoni, F., Kaminski, P. A., Hennecke, H., and Elmerich, C. (1991). Nucleotide sequence of the fixABC region of *Azorhizobium caulinodans* ORS571: similarity of the fixB product with eukaryotic flavoproteins, characterization of fixX, and identification of nifW. Molecular and General Genetics MGG 225(3), 514–520.

Arnold, W., Rump, A., Klipp, W., Priefer, U. B. and Pühler, A. (1988). Nucleotide-sequence of a 24,206-base-pair DNA fragment carrying the entire nitrogen-fixation gene-cluster of Klebsiella-Pneumoniae. Journal of Molecular Biology 203(3).

Atkinson, E. M., Palcic, M. M., Hindsgaul, O. and Long, S. R. (1994). Biosynthesis of *Rhizobium meliloti* lipooligosaccharide Nod factors: NodA is required for an N-acyltransferase activity. Proceedings of the National Academy of Sciences 91(18), 8418–8422.

Baev, N., Schultze, M., Barlier, I., Ha, D. C., Virelizier, H., Kondorosi, E. and Kondorosi, A. (1992). Rhizobium nodM and nodN genes are common nod genes: nodM encodes functions for efficiency of nod signal production and bacteroid maturation. Journal of Bacteriology 174(23), 7555–7565.

Banfalvi, Z., Nieuwkoop, A., Schell, M., Besl, L., and Stacey, G. (1988). Regulation of nod gene expression in *Bradyrhizobium japonicum*. Molecular and General Genetics 214(3), 420–424.

Barney A.G., Jason K., Richard M., George C.D. and Turlough M.F (2020) The genomes of rhizobia. Advances in Botanical Research, 94.

Bishop, P. E. and Joerger, R. D. (1990). Genetics and molecular biology of alternative nitrogen fixation systems. Annual Review of Plant Biology 41(1), 109–125.

Bloemberg, G. V., Kamst, E., Harteveld, M., van der Drift, K. M., Haverkamp, J., Thomas-Oates, J. E., Lugtenburg, B.J. and Spaink, H. P. (1995a). A central domain of Rhizobium NodE protein mediates host specificity by determining the hydrophobicity of fatty acyl moieties of nodulation factors. Molecular Microbiology 16(6), 1123–1136.

Bloemberg, G. V., Lagas, R. M., van Leeuwen, S., Van der Marel, G. A., Van Boom, J. H., Lugtenberg, B. J. and Spaink, H. P. (1995b). Substrate specificity and kinetic studies of nodulation protein NodL of *Rhizobium leguminosarum*. Biochemistry 34(39), 12712–12720.

Bobik, C., Meilhoc, E. and Batut, J. (2006). FixJ: A major regulator of the oxygen limitation response and late symbiotic functions of *Sinorhizobium meliloti*. Journal of Bacteriology 188(13), 4890–4902.

Bonaldi, K., Gourion, B., Fardoux, J., Hannibal, L., Cartieaux, F., Boursot, M.,... and Giraud, E. (2010). Large-scale transposon mutagenesis of photosynthetic Bradyrhizobium sp. strain ORS278 reveals new genetic loci putatively important for nod-independent symbiosis with Aeschynomene indica. Molecular Plant-Microbe Interactions 23(6), 760–770.

Bourdineaud, J. P., Bono, J. J., Ranjeva, R. and Cullimore, J. V. (1995). Enzymatic radiolabelling to a high specific activity of legume lipo-oligosaccharidic nodulation factors from *Rhizobium meliloti*. Biochemical Journal 306(1), 259–264.

Breakspear, A., Liu, C., Roy, S., Stacey, N., Rogers, C., Trick, M., Morieri, G., Mysore, K.S., Wen,J., Oldroyd, G.E and Downie, J. A. (2014). The root hair "infectome" of *Medicago truncatula* uncovers changes in cell cycle genes and reveals a requirement for auxin signaling in rhizobial infection. The Plant Cell 26(12), 4680–4701.

Broughton, W. J., Jabbouri, S., and Perret, X. (2000). Keys to symbiotic harmony. Journal of Bacteriology 182(20), 5641–5652.

Capela, D., Barloy-Hubler, F., Gouzy, J., Bothe, G., Ampe, F., Batut, J., Boistard, P., Becker, A., Boutry, M, Cadieu, E., and Dréano, S. (2001). Analysis of the chromosome sequence of the legume symbiont *Sinorhizobium meliloti* strain 1021. Proceedings of the National Academy of Sciences 98(17), 9877–9882.

Coba de la Peña, T., Fedorova, E., Pueyo, J. J. and Lucas, M. M. (2018). The symbiosome: Legume and rhizobia co-evolution toward a nitrogen-fixing organelle? Frontiers in Plant Science 8, 2229.

Cooper, J. E. (2004). Multiple responses of rhizobia to flavonoids during legume root infection. Advances in Botanical Research 41, 1–62.

Cooper, J. E. (2007). Early interactions between legumes and rhizobia: Disclosing complexity in a molecular dialogue. Journal of Applied Microbiology 103(5), 1355–1365.

Cosseau, C., Garnerone, A. M. and Batut, J. (2002). The fixM flavoprotein modulates inhibition by AICAR or 5′ AMP of respiratory and nitrogen fixation gene expression in Sinorhizobium meliloti. Molecular Plant–Microbe Interactions 15(6), 598–607.

Costas, A. M. G., Poudel, S., Miller, A. F., Schut, G. J., Ledbetter, R. N., Fixen, K. R., Seefeldt, L.C., Adams, M.W., Harwood, C.S., Boyd, E.S. and Peters, J. W. (2017). Defining electron bifurcation in the electron-transferring flavoprotein family. Journal of Bacteriology 199(21).

Coyne, M. S. and Frye, W.W. (2005). *Nitrogen in soil*. In D. Hillel (Ed.), *Cycle. Encyclopedia of Soil in the Environment*, Elsevier Ltd., 13–21.

Curatti, L., Brown, C. S., Ludden, P. W. and Rubio, L. M. (2005). Genes required for rapid expression of nitrogenase activity in *Azotobacter vinelandii*. Proceedings of the National Academy of Sciences 102(18), 6291–6296.

D'Haeze, W., Van Montagu, M., Promé, J. C. and Holsters, M. (1999). Carbamoylation of azorhizobial Nod factors is mediated by NodU. Molecular Plant–Microbe Interactions 12(1), 68–73.

Dazzo, F. B. and Gardiol, A. E. (1984). Host specificity in Rhizobium-legume interactions. In: D. P. S. Verma & T. Hohn (eds.), *Genes Involved in Microbe–Plant Interactions* (pp. 3–31). Springer-Verlag, Vienna.

Dehio, C. and de Bruijn, F. J. (1992). The early nodulin gene SrEnod2 from *Sesbania rostrata* is inducible by cytokinin. The Plant Journal 2(1), 117–128.

Del Cerro, P., Rolla-Santos, A. A. P., Gomes, D. F., Marks, B. B., Pérez-Montaño, F., Rodríguez-Carvajal, M. Á., Nakatani, A.S., Gil-Serrano, A., Megías, M., Ollero, F.J. and Hungria, M. (2015). Regulatory nodD1 and nodD2 genes of Rhizobium tropici strain CIAT 899 and their roles in the early stages of molecular signaling and host-legume nodulation. BMC Genomics 16(1), 251.

Dénarié, J., Debellé, F., and Promé, J. C. (1996). Rhizobium lipo-chitooligosaccharide nodulation factors: signaling molecules mediating recognition and morphogenesis. Annual Review of Biochemistry 65(1), 503–535.

Dixon, R., and Kahn, D. (2004). Genetic regulation of biological nitrogen fixation. Nature Reviews Microbiology 2(8), 621–631.

Dos Santos, P. C. and Addo, M. A. (2020). Distribution of nitrogen fixation genes in prokaryotes containing alternative nitrogenases. ChemBioChem 21, 1749–1759.

Downie, J. A. (1989). The nodL gene from Rhizobium leguminosarum is homologous to the acetyl transferases encoded by lacA and cysE. Molecular Microbiology 3(11), 1649–1651.

Eady, R. R. (1996). Structure–function relationships of alternative nitrogenases. Chemical Reviews 96(7), 3013–3030.

Earl, C. D., Ronson, C. W., and Ausubel, F. M. (1987). Genetic and structural analysis of the Rhizobium meliloti fixA, fixB, fixC, and fixX genes. Journal of Bacteriology 169(3), 1127–1136.

Edgren, T. and Nordlund, S. (2004). The *fixABCX* genes in *Rhodospirillum rubrum* encode a putative membrane complex participating in electron transfer to nitrogenase. Journal of Bacteriology 186(7), 2052–2060.

Evans, I. J., and Downie, J. A. (1986). The nodI gene product of Rhizobium leguminosarum is closely related to ATP-binding bacterial transport proteins; nucleotide sequence analysis of the nodI and nodJ genes. Gene 43(1–2), 95–101.

Fellay, R., Perret, X., Viprey, V., Broughton, W. J. and Brenner, S. (1995). Organization of host-inducible transcripts on the symbiotic plasmid of Rhizobium sp. NGR234. Molecular microbiology 16(4), 657–667.

Fisher, R. F., and Long, S. R. (1992). Rhizobium–plant signal exchange. Nature 357(6380), 655–660.

Fuhrmann, M., Fischer, H. M. and Hennecke, H. (1985). Mapping of *Rhizobium japonicum* nifB-, fixBC-, and fixA-like genes and identification of the fixA promoter. Molecular and General Genetics MGG 199(2), 315–322.

Gage, D. J. (2004). Infection and invasion of roots by symbiotic, nitrogen-fixing rhizobia during nodulation of temperate legumes. Microbiology and Molecular Biology Review 68(2), 280–300.

Galardini, M., Pini, F., Bazzicalupo, M., Biondi, E. G. and Mengoni, A. (2013). Replicon-dependent bacterial genome evolution: The case of *Sinorhizobium meliloti*. Genome Biology and Evolution 5(3), 542–558.

Garcia, M., Dunlap, J., Loh, J., and Stacey, G. (1996). Phenotypic characterization and regulation of the nolA gene of *Bradyrhizobium japonicum*. MPMI-Molecular Plant Microbe Interactions 9(7), 625–636.

Garcia, M., Dunlap, J., Loh, J., and Stacey, G. (1996). Phenotypic characterization and regulation of the *nolA* gene of *Bradyrhizobium japonicum*. MPMI-Molecular Plant Microbe Interactions 9(7), 625–636.

Garnerone, A. M., Cabanes, D., Foussard, M., Boistard, P. and Batut, J. (1999). Inhibition of the FixL Sensor Kinase by the FixT Protein in *Sinorhizobium meliloti*. Journal of Biological Chemistry 274(45), 32500–32506.

Geddes, B. A., Kearsley, J., Morton, R. and Finan, T. M. (2020). The genomes of rhizobia. In: P. Frendo, F. Frugier, & C. Masson-Boivin (eds.), *Advances in Botanical Research. Regulation of Nitrogen-Fixing Symbioses in Legumes*, Vol. 94, Academic Press, pp. 213–249.

Geelen, D., Leyman, B., Mergaert, P., Klarskov, K., Van Montagu, M., Geremia, R. and Holsters, M. (1995). NodS is an S-adenosyl-l-methionine-dependent methyltransferase that methylates chitooligosaccharides deacetylated at the non-reducing end. Molecular Microbiology 17(2), 387–397.

Geelen, D., Mergaert, P., Geremia, R. A., Goormachtig, S., Van Montagu, M. and Holsters, M. (1993). Identification of *nodSUIJ* genes in Nod locus 1 of *Azorhizobium caulinodans*: Evidence that nodS encodes a methyltransferase involved in Nod factor modification. Molecular Microbiology 9(1), 145–154.

Geurts, R., Fedorova, E. and Bisseling, T. (2005). Nod factor signaling genes and their function in the early stages of Rhizobium infection. Current Opinion in Plant Biology 8(4), 346–352.

Gillette, W. K. and Elkan, G. H. (1996). Bradyrhizobium (Arachis) sp. strain NC92 contains two *nodD* genes involved in the repression of *nodA* and a *nolA* gene required for the efficient nodulation of host plants. Journal of Bacteriology 178(10), 2757–2766.

Giraud, E., Moulin, L., Vallenet, D., Barbe, V., Cytryn, E., Avarre, J.C., Jaubert, M., Simon, D., Cartieaux, F., Prin, Y. and Bena, G. (2007) Legumes symbioses: absence of nod genes in photosynthetic bradyrhizobia. Science 316, 1307–1312.

Goethals, K., Mergaert, P., Gao, M., Geelen, D., Van Montagu, M. and Holsters, M. (1992). Identification of a new inducible nodulation gene in *Azorhizobium caulinodans*. Molecular Plant Microbe Interactions 5, 405–405.

Gong, Z. Y., He, Z. S., Zhu, J. B., Yu, G. Q. and Zou, H. S. (2006). *Sinorhizobium meliloti* nifA mutant induces different gene expression profile from wild type in Alfalfa nodules. Cell Research 16(10), 818–829.

Göttfert, M. (1993). Regulation and function of rhizobial nodulation genes. FEMS Microbiology Reviews 10(1–2), 39–63.

Göttfert, M., Holzhäuser, D., Bäni, D. and Hennecke, H. (1992). Structural and functional analysis of two different nodD genes in *Bradyrhizobium japonicum* USDA110. Mol. Plant–Microbe Interact 5(3), 257–265.

Grob, P., Michel, P., Hennecke, H. and Göttfert, M. (1993). A novel response-regulator is able to suppress the nodulation defect of a *Bradyrhizobium japonicum* nodW mutant. Molecular and General Genetics MGG 241(5–6), 531–541.

Grönger, P., Manian, S. S., Reiländer, H., O'Connell, M., Priefer, U. B. and Pühler, A. (1987). Organization and partial sequence of a DNA region of the Rhizobium leguminosarum symbiotic plasmid pRL6JI containing the genes fix ABC, nif A, nif B and a novel open reading frame. Nucleic Acids Research 15(1), 31–49.

Gubler, M. and Hennecke, H. (1986). *fixA*, *B* and *C* genes are essential for symbiotic and free-living, microaerobic nitrogen fixation. FEBS Letters 200(1), 186–192.

Hassan, S. and Mathesius, U. (2012). The role of flavonoids in root–rhizosphere signalling: Opportunities and challenges for improving plant–microbe interactions. Journal of Experimental Botany 63(9), 3429–3444.

Hawkins, F. K. L. and Johnston, A. W. B. (1988). Transcription of a *Rhizobium Ieguminosarum* biovar phaseoli gene needed for melanin synthesis is activated by nifA of Rhizobium and *Klebsiella pneumoniae*. Molecular microbiology 2(3), 331–337.

Hoffmann, A., Citek, C., Binder, S., Goos, A., Rübhausen, M., Troeppner, Oliver T., Ivana I. B., Erik C. W., Stack, T. D. and Sonja H. P. (2013). Catalytic phenol hydroxylation with dioxygen: extension of the tyrosinase mechanism beyond the protein matrix. AngewandteChemie International Edition 52(20), 5398–5401.

Hubbell, D.H. and Kidder, G. (2009). *Biological Nitrogen Fixation*. University of Florida IFAS Extension Publication.

Hynes, M. F., J. Quandt, A. Schiuter, T. Paschkowski, S. Weidner, and U. B. Priefer. 1992. Abstract Book Ninth International Congress on Nitrogen Fixation, abstract 464.

Iismaa, S. E., Ealing, P. M., Scott, K. F. and Watson, J. M. (1989). Molecular linkage of the nifl fix and nod gene regions in *Rhizobium leguminosarum* biovear trifolii. Molecular Microbiology 3(12), 1753–1764.

Inon de Iannino, N., Pueppke, S. G. and Ugalde, R. A. (1995). Biosynthesis of the Nod factor chito-oligosaccharide backbone in Rhizobium fredii is controlled by the concentration of UDP-N-acetyl-D-glucosamine. Molecular plant microbe interactions: MPMI.

Jabbouri, S., Fellay, R., Talmont, F., Kamalaprija, P., Burger, U., Relic, B., Promé, J.C. and Broughton, W. J. (1995). Involvement of nodS in N-methylation and nodU in 6-O-carbamoylation of Rhizobium sp. NGR234 Nod factors. Journal of Biological Chemistry 270(39), 22968–22973.

Jacobson, M. R., Brigle, K. E., Bennett, L. T., Setterquist, R. A., Wilson, M. S., Cash, V. L. and Dean, D. R. (1989). Physical and genetic map of the major nif gene cluster from *Azotobacter vinelandii*. Journal of Bacteriology 171(2), 1017–1027.

Joerger, R. D. and Bishop, P. E. (1988). Nucleotide sequence and genetic analysis of the nifB-nifQ region from *Azotobacter vinelandii*. Journal of Bacteriology 170(4), 1475–1487.

John, M., Röhrig, H., Schmidt, J., Wieneke, U. and Schell, J. (1993). Rhizobium NodB protein involved in nodulation signal synthesis is a chitooligosaccharide deacetylase. Proceedings of the National Academy of Sciences 90(2), 625–629.

Kaminski, P. A. and Elmerich, C. (1998). The control of *Azorhizobium caulinodans* nifA expression by oxygen, ammonia and by the HF-I-like protein, NrfA. Molecular Microbiology 28(3), 603–613.

Kawaharada, Y., Kelly, S., Nielsen, M. W., Hjuler, C. T., Gysel, K., Muszynski, A., Carlson, R.W., Thygesen, M.B., Sandal, N., Asmussen, M.H. and Vinther, M. (2015). Receptor-mediated exopolysaccharide perception controls bacterial infection. Nature, 523(7560), 308.

Klipp, W., Reiländer, H., Schlüter, A., Krey, R. and Pühler, A. (1989). The *Rhizobium meliloti fdxN* gene encoding a ferredoxin-like protein is necessary for nitrogen fixation and is cotranscribed with nifA and nifB. Molecular and General Genetics MGG 216(2–3), 293–302.

Kondorosi, E., Buire, M., Cren, M., Iyer, N., Hoffmann, B. and Kondorosi, A. (1991). Involvement of the *syrM* and *nodD3* genes of *Rhizobium meliloti* in nod gene activation and in optimal nodulation of the plant host. Molecular Microbiology 5(12), 3035–3048.

Kosslak, R. M., Bookland, R., Barkei, J., Paaren, H. E. and Appelbaum, E. R. (1987). Induction of *Bradyrhizobium japonicum* common nod genes by isoflavones isolated from *Glycine max*. Proceedings of the National Academy of Sciences 84(21), 7428–7432.

Laranjo, M., Alexandre, A., & Oliveira, S. (2014). Legume growth-promoting rhizobia: An overview on the Mesorhizobium genus. Microbiological Research, 169(1), 2–17.

Ledbetter, R. N., Garcia Costas, A. M., Lubner, C. E., Mulder, D. W., Tokmina-Lukaszewska, M., Artz, J.H., Patterson, A., Magnuson, T.S., Jay, Z.J., Duan, H.D. and Miller, J. (2017). The electron bifurcating FixABCX protein complex from Azotobacter vinelandii: Generation of low-potential reducing equivalents for nitrogenase catalysis. Biochemistry 56(32), 4177–4190.

Lerouge, P., Roche, P., Faucher, C., Maillet, F., Truchet, G., Promé, J. C. and Dénarié, J. (1990). Symbiotic host-specificity of Rhizobium meliloti is determined by a sulphated and acylated glucosamine oligosaccharide signal. Nature 344(6268), 781–784.

Libault, M., Farmer, A., Brechenmacher, L., Drnevich, J., Langley, R. J., Bilgin, D. D., Radwan, O., Neece, D.J., Clough, S.J., May, G.D. and Stacey, G. (2010). Complete transcriptome of the soybean root hair cell, a single-cell model, and its alteration in response to Bradyrhizobium japonicum infection. Plant Physiology 152(2), 541–552.

Lindström, K. and Mousavi, S. A. (2019). *Effectiveness of Nitrogen Fixation in Rhizobia*. Microbial Biotechnology.

Lindström, K., Murwira, M., Willems, A., and Altier, N. (2010). The biodiversity of beneficial microbe-host mutualism: The case of rhizobia. Research in Microbiology 161(6), 453–463.

Liu, S. T., Chang, W. Z., Cao, H. M., Hu, H. L., Chen, Z. H., Ni, F. D., Liu, S.T., Chang, W.Z., Cao, H.M., Hu, H.L., Chen, Z.H., Ni, F.D., Lu, H.F. and Hong, G.F. (1998). A HU-like Protein Binds to Specific Sites within nodPromoters of *Rhizobium leguminosarum*. Journal of Biological Chemistry 273(32), 20568–20574.

Loh, J. and Stacey, G. (2003). Nodulation gene regulation in *Bradyrhizobium japonicum*: A unique integration of global regulatory circuits. Applied and Environmental Microbiology 69(1), 10–17.

Loh, J., Garcia, M. and Stacey, G. (1997). NodV and NodW, a second flavonoid recognition system regulating nod gene expression in *Bradyrhizobium japonicum*. Journal of bacteriology 179(9), 3013–3020.

MacNeil, T., MacNeil, D., Roberts, G. P., Supiano, M. A. and Brill, W. J. (1978). Fine-structure mapping and complementation analysis of *nif* (nitrogen fixation) genes in *Klebsiella pneumoniae*. Journal of Bacteriology 136(1), 253–266.

Madsen, L. H., Tirichine, L., Jurkiewicz, A., Sullivan, J. T., Heckmann, A. B., Bek, A. S., Ronson, C.W., James, E.K. and Stougaard, J. (2010). The molecular network governing nodule organogenesis and infection in the model legume *Lotus japonicus*. Nature Communications 1(1), 1–12.

Mandon, K., Kaminski, P. A. and Elmerich, C. (1994). Functional analysis of the fixNOQP region of *Azorhizobium caulinodans*. Journal of Bacteriology 176(9), 2560–2568.

Marie, C., Barny, M. A. and Downie, J. A. (1992). *Rhizobium leguminosarum* has two glucosamine synthases, GImS and NodM, required for nodulation and development of nitrogen-fixing nodules. Molecular Microbiology 6(7), 843–851.

Martínez, M., Palacios, J. M., Imperial, J. and Ruiz-Argüeso, T. (2004). Symbiotic autoregulation of nifA expression in *Rhizobium leguminosarum* bv. viciae. Journal of Bacteriology 186(19), 6586–6594.

Masson-Boivin, C., Giraud, E., Perret, X. and Batut, J. (2009). Establishing nitrogen-fixing symbiosis with legumes: how many rhizobium recipes? Trends in Microbiology 17(10), 458–466.

Mergaert, P., D'Haeze, W., Geelen, D., Promé, D., Van Montagu, M., Geremia, R., Promé, J.C and Holsters, M. (1995). Biosynthesis of *Azorhizobium caulinodans* Nod factors study of the activity of the Nodabcs proteins by expression of the genes in *Escherichia coli*. Journal of Biological Chemistry 270(49), 29217–29223.

Mergaert, P., Van Montagu, M. and Holsters, M. (1997). Molecular mechanisms of Nod factor diversity. Molecular Microbiology 25(5), 811–817.

Merrick, M. J., and Edwards, R. (1995). Nitrogen control in bacteria. Microbiological Reviews 59(4), 604–622.

Michiels, J. and Vanderleyden, J. (1993). Cloning and sequence of the *Rhizobium leguminosarum* biovar phaseoli fixA gene. Biochimica et Biophysica Acta (Bba)-Bioenergetics 1144(2), 232–233.

Mulligan, J. T. and Long, S. R. (1985). Induction of *Rhizobium meliloti* nodC expression by plant exudate requires nodD. Proceedings of the National Academy of Sciences 82(19), 6609–6613.

Mus, F., Colman D.R., Peters J.W. and Boyd ES. (2019). Geobiological feedbacks, oxygen, and the evolution of nitrogenase. Free Radical Biology and Medicine 140, 250–259.

Nukui, N., Minamisawa, K., Ayabe, S. I., and Aoki, T. (2006). Expression of the 1-aminocyclopropane-1-carboxylic acid deaminase gene requires symbiotic nitrogen-fixing regulator gene nifA2 in *Mesorhizobium loti* MAFF303099. Applied and Environmental Microbiology 72(7), 4964–4969.

Okazaki, S., Noisangiam, R., Okubo, T., Kaneko, T., Oshima, K., Hattori, M., Teamtisong, K., Songwattana, P., Tittabutr, P., Boonkerd, N. and Saeki, K. (2015). Genome analysis of a novel Bradyrhizobium sp. DOA9 carrying a symbiotic plasmid. PloS One 10(2), e0117392.

Okazaki, S., Tittabutr, P., Teulet, A., Thouin, J., Fardoux, J., Chaintreuil, C Gully, D., Arrighi, J.F., Furuta, N., Miwa, H. and Yasuda, M. (2016). Rhizobium–legume symbiosis in the absence of Nod factors: Two possible scenarios with or without the T3SS. The ISME Journal 10(1), 64–74.

Oldroyd, G. E. (2013). Speak, friend, and enter: Signalling systems that promote beneficial symbiotic associations in plants. Nature Reviews Microbiology 11(4), 252–263.

Oldroyd, G. E. and Downie, J. A. (2008). Coordinating nodule morphogenesis with rhizobial infection in legumes. Annual Review Plant Biology 59, 519–546.

Oldroyd, G. E., Murray, J. D., Poole, P. S. and Downie, J. A. (2011). The rules of engagement in the legume-rhizobial symbiosis. Annual Review of Genetics 45, 119–144.

Peix, A., Ramírez-Bahena, M. H., Velázquez, E., and Bedmar, E. J. (2015). Bacterial associations with legumes. Critical Reviews in Plant Sciences 34(1–3), 17–42.

Postgate, J. R. (1982). *The Fundamentals of Nitrogen Fixation*. Cambridge, Cambridge University Press.

Poupot, R., Martinez-Romero, E., Maillet, F. and Prome, J. C. (1995). Rhizobium tropici nodulation factor sulfation is limited by the quantity of activated form of sulfate. FEBS Letters, 368(3), 536–540.

Poza-Carrión, C., Echavarri-Erasun, C. and Rubio, L. M. (2015). Regulation of *nif* gene expression in *Azotobacter vinelandii*. Biological Nitrogen Fixation 1, 101–107.

Preisig, O., Anthamatten, D. and Hennecke, H. (1993). Genes for a microaerobically induced oxidase complex in *Bradyrhizobium japonicum* are essential for a nitrogen-fixing endosymbiosis. Proceedings of the National Academy of Sciences 90(8), 3309–3313.

Preisig, O., Zufferey, R., Thöny-Meyer, L., Appleby, C. A. and Hennecke, H. (1996). A high-affinity cbb3-type cytochrome oxidase terminates the symbiosis-specific respiratory chain of *Bradyrhizobium japonicum*. Journal of Bacteriology 178(6), 1532–1538.

Pühler, Alfred, M. O. Aguilar, M. Hynes, P. Müller, W. Klipp, U. Priefer, R. Simon, and G. Weber (1984). Advances in the genetics of free-living and symbiotic nitrogen fixing bacteria. In: C. Veeger & W.E. Newton (eds.), *Advances in Nitrogen Fixation Research. Advances in Agricultural Biotechnology*, Vol. 4, Springer, Dordrecht.

Ren, B., Wang, X., Duan, J. and Ma, J. (2019). Rhizobial tRNA-derived small RNAs are signal molecules regulating plant nodulation. Science 365(6456), 919–922.

Reyrat, J. M., David, M., Blonski, C., Boistard, P. and Batut, J. (1993). Oxygen-regulated in vitro transcription of *Rhizobium meliloti nifA* and *fixK* genes. Journal of Bacteriology 175(21), 6867–6872.

Ried, M. K., Antolín-Llovera, M. and Parniske, M. (2014). Spontaneous symbiotic reprogramming of plant roots triggered by receptor-like kinases. Elife 3, e03891.

Rivilla, R., Sutton, J. M. and Downie, J. A. (1995). Rhizobium leguminosarum NodT is related to a family of outer-membrane transport proteins that includes TolC, PrtF, CyaE and AprF. Gene 161(1), 27–31.

Roberts, G. P., MAcNEIL, T. A. N. Y. A., MAcNEIL, D. O. U. G. L. A. S. and Brill, W. J. (1978). Regulation and characterization of protein products coded by the *nif* (nitrogen fixation) genes of *Klebsiella pneumoniae*. Journal of Bacteriology 136(1), 267–279.

Roche, P., Debellé, F., Maillet, F., Lerouge, P., Faucher, C., Truchet, G., Dénarié, J. and Promé, J. C. (1991). Molecular basis of symbiotic host specificity in Rhizobium meliloti: nodH and nodPQ genes encode the sulfation of lipo-oligosaccharide signals. Cell 67(6), 1131–1143.

Röhrig, H., Schmidt, J., Wieneke, U., Kondorosi, E., Barlier, I., Schell, J. and John, M. (1994). Biosynthesis of lipooligosaccharide nodulation factors: Rhizobium NodA protein is involved in N-acylation of the chitooligosaccharide backbone. Proceedings of the National Academy of Sciences 91(8), 3122–3126.

Rubio, L. M. and Ludden, P. W. (2008). Biosynthesis of the iron-molybdenum cofactor of nitrogenase. Annual Review of Microbiology, 62, 93–111.

Rubio, L. M., Rangaraj, P., Homer, M. J., Roberts, G. P. and Ludden, P. W. (2002). Cloning and mutational analysis of the γ Gene from *Azotobacter vinelandii* defines a new family of proteins capable of metallocluster binding and protein stabilization. Journal of Biological Chemistry 277(16), 14299–14305.

Rutledge, H. L. and Tezcan, F. A. (2020). *Electron Transfer in Nitrogenase*. Chemical Reviews 120, 5158–5193.

Ruvkun, G. B., Sundaresan, V. and Ausubel, F. M. (1982). Directed transposon Tn5 mutagenesis and complementation analysis of *Rhizobium meliloti* symbiotic nitrogen fixation genes. Cell 29(2), 551–559.

Sadowsky, M. J., Cregan, P. B., Gottfert, M., Sharma, A., Gerhold, D., Rodriguez-Quinones, F., Keyser, H.H., Hennecke, H. and Stacey, G. (1991). The *Bradyrhizobium japonicum* nolA gene and its involvement in the genotype-specific nodulation of soybeans. Proceedings of the National Academy of Sciences 88(2), 637–641.

Saier Jr, M. H., Tam, R., Reizer, A. and Reizer, J. (1994). Two novel families of bacterial membrane proteins concerned with nodulation, cell division and transport. Molecular microbiology 11(5), 841–847.

Sanjuan, J., Grob, P., Goettfert, M., Hennecke, H. and Stacey, G. (1994). NodW is essential for full expression of the common nodulation genes in *Bradyrhizobium japonicum*. MPMI-Molecular Plant Microbe Interactions 7(3), 364–369.

Schlaman, H. R., Okker, R. J. and Lugtenberg, B. J. (1992). Regulation of nodulation gene expression by NodD in rhizobia. Journal of Bacteriology 174(16), 5177.

Schultze, M., Staehelin, C., Röhrig, H., John, M., Schmidt, J., Kondorosi, E., Schell, J. and Kondorosi, A. (1995). In vitro sulfotransferase activity of Rhizobium meliloti NodH protein: Lipochitooligosaccharide nodulation signals are sulfated after synthesis of the core structure. Proceedings of the National Academy of Sciences 92(7), 2706–2709.

Schwedock, J. and Long, S. R. (1990). ATP sulphurylase activity of the nodP and nodQ gene products of *Rhizobium meliloti*. Nature 348(6302), 644–647.

Schwedock, J. S. and Long, S. R. (1992). *Rhizobium meliloti* genes involved in sulfate activation: the two copies of nodPQ and a new locus, saa. Genetics 132(4), 899–909.

Shearman, C. A., Rossen, L., Johnston, A. W. B. and Downie, J. A. (1986). The *Rhizobium leguminosarum* nodulation gene *nodF* encodes a polypeptide similar to acyl-carrier protein and is regulated by nodD plus a factor in pea root exudate. The EMBO Journal 5(4), 647–652.

Smit, G., Puvanesarajah, V., Carlson, R. W., Barbour, W. M., and Stacey, G. (1992). *Bradyrhizobium japonicum* nodD1 can be specifically induced by soybean flavonoids that do not induce the nodYABCSUIJ operon. Journal of Biological Chemistry 267(1), 310–318.

Smit, P., Limpens, E., Geurts, R., Fedorova, E., Dolgikh, E., Gough, C. and Bisseling, T. (2007). Medicago LYK3, an entry receptor in rhizobial nodulation factor signaling. Plant Physiology 145(1), 183–191.

Spaink, H. P., Sheeley, D. M., van Brussel, A. A., Glushka, J., York, W. S., Tak, T Geiger, O., Kennedy, E.P., Reinhold, V.N. and Lugtenberg, B.J. (1991). A novel highly unsaturated fatty acid moiety of lipo-oligosaccharide signals determines host specificity of Rhizobium. Nature 354(6349), 125–130.

Spaink, H. P., Wijfjes, A. H., der van Drift, K. M., Haverkamp, J., Thomas-Oates, J. E. and Lugtenberg, B. J. (1994). Structural identification of metabolites produced by the NodB and NodC proteins of *Rhizobium leguminosarum*. Molecular Microbiology 13(5), 821–831.

Stacey, G., Luka, S., Sanjuan, J., Banfalvi, Z., Nieuwkoop, A. J., Chun, J. Y., Forsberg, L.S and Carlson, R. (1994). nodZ, a unique host-specific nodulation gene, is involved in the fucosylation of the lipooligosaccharide nodulation signal of *Bradyrhizobium japonicum*. Journal of Bacteriology 176(3), 620–633.

Stock, A. M., Robinson, V. L. and Goudreau, P. N. (2000). Two-component signal transduction. Annual Review of Biochemistry 69(1), 183–215.

Sullivan, J. T., Brown, S. D. and Ronson, C. W. (2013). The NifA-RpoN regulon of *Mesorhizobium loti* strain R7A and its symbiotic activation by a novel LacI/GalR-family regulator. PLoS One 8(1), e53762.

Sur, S., Bothra, A. K. and Sen, A. (2010). Symbiotic nitrogen fixation-a bioinformatics perspective. Biotechnology 9(3), 257–273.

Surin, B. P., Watson, J. M., Hamilton, W. D. O., Economou, A. and Downie, J. A. (1990). Molecular characterization of the nodulation gene, nodT, from two biovars of *Rhizobium leguminosarum*. Molecular Microbiology 4(2), 245–252.

Temme, K., Zhao, D. and Voigt, C. A. (2012). Refactoring the nitrogen fixation gene cluster from *Klebsiella oxytoca*. Proceedings of the National Academy of Sciences 109(18), 7085–7090.

Torres Júnior, C. V., Leite, J., Santos, C. D. R., Fernandes Junior, P. I., Zilli, J. E., Rumjanek, N. G. and Xavier, G. R. (2014). Diversity and symbiotic performance of peanut rhizobia from Southeast region of Brazil. Academic Journals 8(6), 566–577.

van Rhijn, P. and Vanderleyden, J. (1995). The Rhizobium-plant symbiosis. Microbiological Reviews 59(1), 124–142.

Vanderleyden, J., Dommelen, A.V. and Michiels, J. (2001). *fix* Genes. Academic Press.

Vázquez, M., Santana, O. and Quinto, C. (1993). The NodI and NodJ proteins from Rhizobium and Bradyrhizobium strains are similar to capsular polysaccharide secretion proteins from Gram-negative bacteria. Molecular Microbiology 8(2), 369–377.

Wu, C., Dickstein, R., Cary, A. J. and Norris, J. H. (1996). The auxin transport inhibitor N-(1-naphthyl) phthalamic acid elicits pseudonodules on nonnodulating mutants of white sweetclover. Plant Physiology 110(2), 501–510.

Yang, G.P., Debellé, F., Savagnac, A., Ferro, M., Schiltz, O., Maillet, F., Promé, D., Treilhou, M., Vialas, C., Lindstrom, K. and Dénarié, J. (1999). Structure of the Mesorhizobiumhuakuii and Rhizobium galegae Nod factors: A cluster of phylogenetically related legumes are nodulated by rhizobia producing Nod factors with α, β-unsaturated N-acyl substitutions. Molecular Microbiology, 34(2), 227–237.

Young, J. P. W. (1992). Phylogenetic classification of nitrogen-fixing organisms. Biological Nitrogen Fixation 1544, 43–86.

13 Nitrogen Biofertilizers
Role in Sustainable Agriculture

Ishfaq Ul Rehmaan*, Bushra Jan, Nafeesa Farooq Khan, Tajamul Islam, Summia Rehman, Ishfaq Ahmad Sheergojri, Irfan Rashid, Aadil Farooq War, Subzar Ahmad Nanda, and Abid Hussain Wani

Department of Botany, University of Kashmir, Srinagar, Jammu, and Kashmir, India
*Correspondence: ishfaqbotany56@gmail.com

CONTENTS

- 13.1 Introduction 171
- 13.2 Nitrogen Biofertilizers: Why Their Need is Inevitable 172
- 13.3 Nitrogen-Fixing Biofertilizers 173
  - 13.3.1 Rhizobium 173
  - 13.3.2 Azospirillum 173
  - 13.3.3 Azotobacter 174
  - 13.3.4 Blue-Green Algae (Cyanobacteria) and Azolla 174
  - 13.3.5 Nitrogen-Fixing Endophytes 175
- 13.4 Different Methods of Application of Biofertilizers to Crops 175
  - 13.4.1 Seedling Root Dip 175
  - 13.4.2 Seed Treatment 175
  - 13.4.3 Soil Treatment 175
  - 13.4.4 Use of Vam Biofertilizer 175
  - 13.4.5 Use of Blue-Green Algae (BGA) 176
- 13.5 The Nitrogen Fixation Process 176
- 13.6 Molecular Analysis Of Nitrogen Fixation 177
- 13.7 Symbiotic Nitrogen Fixation 177
- 13.8 Potential Role of Biofertilizers in Agriculture 179
- 13.9 Nitrogen Biofertilizers: Environmental Stresses 180
  - 13.9.1 Constraints in Biofertilizer Use 180
- 13.10 Conclusion 181
- References 181

13.1 INTRODUCTION

Approximately 7 billion people are living in the world and this number is undoubtedly expected to rise to approximately 8 billion around 2020 (Conway, 2012). With this continued rise in worldwide population, there is increasing environmental damage as a consequence of rapid growth in industrialization and urbanization (Glick, 2012). Moreover, it is a significant challenge to feed the large population at present, which inevitably will increase with time. Subsequently, to meet the demand, countries become dependent to enormously use the chemical fertilizers in agriculture,

TABLE 13.1
Microorganisms Used as Nitrogen Biofertilizers

Groups	Examples
Free-living	*Beijerinkia, Azotobacter, Anabaena, Nostoc*
Symbiotic	*Rhizobium, Frankia, Anabaena azollae*
Associative Symbiotic	*Azospirillum*

but at the same time significantly damages the environment, and adverse effects on living beings (Sujanya and Chandra, 2011). The chemical fertilizers alter the buffering capacity of soil and subsequently become extremely toxic for humans (Savci, 2012). Since the plants cannot take up these toxic chemicals, they start collecting in groundwater and some of these chemicals are also responsible for causing water bodies to eutrophize (Savci, 2012). These chemicals have a negative water effect on soil, depletion of energy, soil fertility, increased salinity, and a deficit of soil nutrients (Savci, 2012). Taking into account all the adverse effects of extended use of chemical fertilizers, organic farming has emerged as a powerful alternative in terms of the demand for safe food supply, long-term sustainability, and environmental pollution issues (Reddy, 2013). Although the use of chemical fertilizers is unavoidable to meet the rising demand for food in the world, there are opportunities where some selected crops and niche areas can be well flourished through organic farming (Macilwain, 2004; Chen, 2006).

A biofertilizer is a material that contains living microorganisms that colonize the rhizosphere or the interior of the plants when applied to seeds, plants, or soil, and promote plant growth by the nutrient supply to the host plant (Vessey, 2003; Bardi and Malusà, 2012; Malusa and Vassilev, 2014). Biofertilizers are commonly used to speed up microbial processes that increase nutrient supply that plants can quickly assimilate. By fixing the atmospheric nitrogen and solubilizing insoluble phosphates, they increase soil fertility and produce plant growth-promoting substances in the soil (Mazid and Khan, 2015). Such biofertilizers have been promoted to harvest the naturally available nutrient mobilization biological system that greatly increases soil productivity and ultimately crop yield (Pandey and Singh, 2012).

Nitrogen biofertilizers work by fixing atmospheric nitrogen and converting it to organic (plant usable) forms in the soil and root nodules of legumes, thereby making them available to plants. Nitrogen-fixing biofertilizers are crop-specific biofertilizers (Choudhary and Kennedy, 2004). This form of biofertilizer assists farmers in determining the amount of nitrogen in the soil. Nitrogen is a necessary component that is needed for plant production. Examples of biofertilizers include *Rhizobium* Sp., *Azospirillum* sp., and blue-green algae; Azotobacteria is used for non-legume crops; Rhizobium is needed for legume crops. Similarly, rice growth needs blue-green algae while Acetobacter is used to grow sugarcane. Different crops require different types of biofertilizers. Some of the types of biofertilizers are given in Table 13.1.

13.2 NITROGEN BIOFERTILIZERS: WHY THEIR NEED IS INEVITABLE

Indiscriminate use of chemical fertilizers to meet the growing demand for food supply has undoubtedly led to contamination and severely damaged microbial habitats and friendly insects. The excess use of chemicals has also made crops more susceptible to disease and reduced soil fertility (Tilman et al., 2002; Aktar et al., 2009). It is estimated that by 2020, to achieve the target production of 321 million tons of food grain for feeding 8 billion populations worldwide, the nutrient requirement will be 28.8 million tons while the availability will be only 21.6 million tons, creating a deficit of around 7.2 million tons of nutrients required (Arun, 2007). To feed the growing population with the deficit of nutrients available, agricultural productivity needs to increase in a sustainable and environmentally friendly manner. Thus, many of the existing agricultural approaches that comprise the use of chemical fertilizers, pesticides, herbicides, fungicides, and insecticides need to be re-evaluated (Pretty and Bharucha, 2015). Considering the harmful effects of chemical fertilizers, biofertilizers are expected to be a healthy alternative to chemical inputs and to a large degree to reduce ecological disturbance. Biofertilizers are cost-effective, environmentally friendly, and their prolonged use substantially enhances soil fertility (Mahdi et al., 2010; Singh et al., 2011). It was reported that the use of biofertilizers elevates crop yield around 10–40% by increasing the contents of proteins, essential amino acids, vitamins, and nitrogen fixation (Bhardwaj et al., 2014). The advantages of using biofertilizers include inexpensive nutrient sources, excellent suppliers of microchemicals and micronutrients, organic matter suppliers, growth hormone secretion, and counteracting the harmful effects of chemical fertilizers (Gaur, 2010). Different microbes are vital components of soil and play a crucial role in different soil ecosystem biotic activities that make soil dynamic for mobilization of nutrients and sustainable crop production (Ahemad and Kibret, 2014).

13.3 NITROGEN-FIXING BIOFERTILIZERS

One means of transforming elemental nitrogen into a plant-usable form is biological nitrogen fixation (Gothwal et al., 2008). Nitrogen-fixing bacteria (NFB) function by transforming inert atmospheric N_2 to organic compounds (Bakulin, 2007). As a living fertilizer composed of microbial inoculants or groups of microorganisms that can fix atmospheric nitrogen, nitrogen fixer organisms are used in biofertilizers. N2 fixers are categorized as free-living bacteria (*Azospirillium* and *Azotobacter*), blue-green algae, and symbionts (*Rhizobium, Frankia*, and *Azolla*) (Gupta, 2004). Some N_2-fixing bacteria are associated with non-legumes (e.g., *Achromobacter, Arthrobacter, Alcaligenes, Azomonas, Beijerinckia, Bacillus, Acetobacter, Erwinia, Derxia, Clostridium, Enterobacter, Desulfovibrio, Corynebacterium, Herbaspirillum, Klebsiella, Campylobacter, Mycobacterium, Rhodospirillum, Lignobacter, Xanthobacter, Mycobacterium, Rhodo-pseudomonas*, and *Methylosinus*) (Wani, 1990). Although many N_2-fixing genera and species are isolated from the different cereals' rhizosphere, under field conditions, mainly members of *Azospirillum* and *Azotobacter* genera have been widely tested to increase the yield of legumes and cereals.

13.3.1 Rhizobium

Rhizobium belongs to the *Rhizobiaceae* family, and is symbiotic in nature, with 50–100 kg/ha nitrogen fixation ability with legumes only. In specific legumes, it colonizes the roots to form growths that are known as root nodules, which act as ammonia production factories. In a symbiotic association with legumes and certain non-legumes like Paraponia, Rhizobium can fix atmospheric nitrogen. Six species of Rhizobium are defined according to the legume host(s) that they nodulate, based on cross-inoculation studies (Table 13.2). A collection of leguminous species that develop nodules on any member of that particular group of plants is referred to as a cross-inoculation group. Therefore, almost all host species infected by an individual bacterial strain are ideally included in a single cross-inoculation group (Franche et al., 2009).

13.3.2 Azospirillum

Azospirillum belongs to the family *Spirilaceae* and is heterotrophic and associative. They also produce growth-regulating substances, in addition to their nitrogen-fixing ability of about 20–40 kg/ha. Although there are many species under this genus like *A. amazonense, A. halopraeferens, A. amazonense*, and *A. brasilense*, the benefits of inoculation and worldwide distribution have been proved mainly with *A. brasilense* and *A. lipoferum*. With many plants, the *Azospirillum* forms

TABLE 13.2
Species of Rhizobium and Cross Inoculation Groups of Host Plants

Rhizobium sp.	Host genera	Cross inoculation
R. japonicum	*Glycine*	Soybean groups
R. trifolii	*Trifolium*	Clover group
R. lupini	*Lupinus, Ornithopus*	Lupin group
R. meliloti	*Melilotus, Medicago, Trigonella*	Alfalfa group
R. leguminosarum	*Pisum, Lathyrus, Vicia, Lens*	Pea group
R. phaseolus	*Phaseolus*	Bean group
Other species of Rhizobium	*A rack is, Crotalaria, Vigna, Pueraria*	Cowpea group

Source Franche et al., 2009.

associative symbiosis, especially with those with the C_4-dicarboxyliac photosynthesis pathway (Hatch and Slack pathway), because they grow and fix nitrogen on salts of organic acids such as aspartic acid and malic acid. It is therefore recommended primarily for maize, sugarcane, sorghum, pearl millet, etc. Not only does the *Azotobacter* colonize the root surface, but a substantial proportion of them also penetrate the root tissues and live in harmony with the plants. However, they do not produce any visible nodules or root tissue growth.

13.3.3 AZOTOBACTER

Azotobacter belongs to the *Azotobacteriaceae* family, which is aerobic, free, and heterotrophic in nature. *Azotobacters* are found in soils that are neutral or alkaline, and *A. Chroococcum* is the species most commonly found in arable soils. *A. beijerinckii, A. vinelandii, A. macrocytogenes,* and *A. insignis* are other reported species. Due to the lack of organic matter and the presence of antagonistic soil microorganisms, the number of Azotobacter rarely exceeds 104 to 105 g-1 in soil. The bacterium generates anti-fungal antibiotics that inhibit the growth in the root region of several pathogenic fungi, thereby preventing seedling mortality to some extent. In the rhizosphere of several crops such as sugarcane, rice, maize, bajra vegetables, and plantation crops, the occurrence of this organism has been reported.

13.3.4 BLUE-GREEN ALGAE (CYANOBACTERIA) AND AZOLLA

These belong to eight separate families, producing auxin, gibberellic acid, and indole acetic acid, are phototrophic in nature and in submerged rice fields they fix 20–30 kg N/ha as they are abundant in paddy, and are also known as "paddy species." For low-land rice production, N is the main input needed in large quantities. Soil N and biological nitrogen fixers (BNF) are major sources of N for low-land rice. The 50–60% of N requirement is fulfilled by the combination of soil organic N mineralization and BNF by free-living bacteria associated with rice plants (Mutch et al., 2004). The demand for fixed nitrogen must be progressively fulfilled by BNF rather than by industrial nitrogen fixation, to achieve food security through sustainable agriculture. The majority of N-fixing BGA are filamentous, consisting of a vegetative cell chain containing specialized heterocyst cells that act as micronodules for synthesis and N-fixing machinery. BGA forms a symbiotic relationship with mushrooms, ferns, liverworts, and flowering plants capable of fixing nitrogen, but the most common symbiotic relationship has been found between *Azolla* and *Anabaena azollae* (BGA), a free-floating aquatic fern. The rapid decomposition in soil and the effective supply of its nitrogen to rice plants makes *Azolla* a good biofertilizer for rice crops. In addition to N fixation, large quantities of P, S,

Zn, K, Fe, Mb, and other micronutrients are contributed by these biofertilizers or biomanures. *Azolla* can be used as green manure before rice planting by incorporating it into the fields. *A. pinnata* is the most common species occurring in India and can be propagated on a commercial scale by vegetative means.

13.3.5 NITROGEN-FIXING ENDOPHYTES

A various number of studies identify the presence of nitrogen-fixing bacteria inside the host plant tissues that do not exhibit signs of diseases, with *Gluconacetobacter, Azoarcus sp.*, and *Herbaspirillum* being the most studied genera (Lery et al., 2008). Without causing damage, endophytes multiply and disperse within plant tissues. Early infection steps may be comparable to those documented in rhizospheric bacteria, initially involving surface colonization at the site of root hair emergence (Tan et al., 2003). Type IV pili have been identified as important for this process in the case of *Azoarcus* and hydrolytic enzymes or endoglucanases are involved in penetration of tissue (Krause et al., 2006). The concentration of bacteria recovered after sterilization of the root system can reach up to 108 colony forming unit (CFU) per g of dry weight. The systemic spread found in plant xylem vessels and shoots, as defined in the case of sugarcane infection with *G. diazotrophicus* species, is another characteristic feature of bacteria (James et al., 1998).

13.4 DIFFERENT METHODS OF APPLICATION OF BIOFERTILIZERS TO CROPS

13.4.1 SEEDLING ROOT DIP

The use of this method for transplanted crops is common. Five packets (1.0 kg) of inoculants are needed for one hectare combined with 40 liters of water. The seedling's root is soaked in the solution for 5–10 minutes and then transplanted. In rice, *Azospirillum* is commonly used as seedling root dip. This method is used mostly for rice crop cultivation. In the field on which the crop has to grow a bed of water is spread. The rice seedlings are planted and held for 8–10 hours in the water.

13.4.2 SEED TREATMENT

In the seed treatment method, the phosphorus and nitrogen fertilizers are combined in the water. The seeds are then dipped in this mixture. The seeds are then dried after applying this paste to the seeds. After they dry out, they must be sown as early as possible before dangerous microorganisms affect them. To create a slurry, each packet (200 g) of inoculant is mixed with 200 ml of rice gruel or jaggery solution. The required quantity of seeds for one hectare is mixed in the slurry to coat them uniformly with inoculants, followed by shade dry for 30 minutes. The seeds should be sown within 24 hours as per the recommendations. One packet (200 g) of inoculant is adequate to treat 10 kg of seeds. The most common inoculants used for seed treatment are *Rhizobium, Azotobacter, Azospirillum*, and *Phosphobacteria*.

13.4.3 SOIL TREATMENT

In this method, both the compost fertilizers and biofertilizers are mixed followed by spread on the soil where seeds are to be sown. Recommended biofertilizers (4 kg each) are mixed with compost (200 kg) and kept overnight. At the time of sowing or planting, this combination is mixed into the soil.

13.4.4 USE OF VAM BIOFERTILIZER

The inoculum should be supplied at the time of sowing, 2–3 cm below the surface. The cuttings or seeds are sown just above the VAM inoculums so that the roots can come into contact with the inoculums and induce infection. One hundred gm bulk inoculum is appropriate for an area of one square meter. Seedlings grown in polythene bags need 5–10 g of bulk inoculum. VAM inoculums should be added at a rate of 20 g/seedling in each spot at the time of planting of saplings. For existing trees, 200 g of inoculum is required.

13.4.5 USE OF BLUE-GREEN ALGAE (BGA)

In field rice, algal culture is dispersed over the standing water as dried flakes at 10 kg/ha. This is achieved in loamy soils two days after transplantation and in clay soils 6 days after planting. Immediately after algal application, the field is kept waterlogged for a few days. The biofertilizer must be used in the same area for 3–4 consecutive seasons.

Azolla is applied @ 0.6–1.0 kg/m2 (6.25–10.0 t/ha) and incorporated before transplanting of rice.

Dual crop: *Azolla* is applied @ of 100 g/m2 (1.25t/ha), 1 to 3 days after transplanting of rice and allowed to multiply for 25–30 days. *Azolla* fronds can be incorporated into the soil at the time of the first weeding.

13.5 THE NITROGEN FIXATION PROCESS

Nitrogenase, a complex of oxygen-labile enzymes strongly conserved in free-living and symbiotic diazotrophs, catalyzes the enzymatic conversion of molecular nitrogen to ammonia. The most common form of nitrogenase, called Mo-nitrogenase or conventional nitrogenase, contains a molybdenum-prothetic group, FeMoCo (Iron-Molybdenum-Cobalt). Some bacteria, like *Azotobacter* and several photosynthetic nitrogen fixers (including some cyanobacteria), carry additional nitrogenase forms whose cofactor contains vanadium (V-nitrogenase) or just iron (Newton, 2007). The nitrogenase enzyme, isolated from different sources, is composed of two metalloproteins. Component 1, also referred to as the Mo-Fe protein, is a 220,000-Da tetramer composed of two non-identical subunits α and β, while component 2, also referred to as the Fe protein, is a 68,000-Da dimer composed of the same subunits. Two FeMoCo are bound to the MoFe protein by α subunits. There are also two other prosthetic groups comprising clusters 4Fe–4S. "P-clusters" are covalently bound to cysteine residues of MoFe protein bridging α and β subunits. The third member of the group Fe-S is linked to the protein (Fe Hu, Yilin, et al., 2007). The reduction of nitrogen is a very complex mechanism that is not yet fully elucidated. The result of a net reduction in ammonia from molecular nitrogen is generally accounted for by the following equation:

$$N_2 + 16 \text{ Mg-ATP} + 8e^- + 8H^+ = 2NH_3 + H_2 + 16 \text{ Mg-ADP} + 16 \text{ Pi}$$

Two metalloproteins (i.e., larger Mo-Fe-protein and smaller Fe-protein components) are involved in N_2 fixation. Fe-protein interacts with ATP and Mg^{++}, and when oxidized, receives an electron from ferredoxin or flavodoxin. The nitrogenase complex Mo-Fe-protein interacts with the reducible substrates (i.e., N_2) and yields two NH_3 molecules. It appears that N_2 is reduced step-wise without breaking the N-N bond before the final reduction and ammonia production is achieved. Two molecules of NH_3 are finally released from the enzyme. Finally, the electron is transferred to oxidize Mo-Fe-protein, which becomes reduced and Fe-protein is oxidized. The reduced form of Mo-Fe-protein combines with N_2 in addition to other substrates and results in the formation of NH_3 and other various products (Figure 13.1). In this reaction, the produced H_2 is used by some microorganisms having hydrogenase activity (Arnold et al., 1988). Reutilization of H_2 enhances nitrogenase activity by protecting the enzyme from inhibition of H_2. In microbial cells, ammonia is further synthesized into a variety of metabolic products. Within the cell ammonia is not accumulated, although it may be created by a few species; rather the ammonia combines with organic acids (a-keto-glutaric acid) to form amino acid (e.g., glutamic acid). Ammonia may also form glutamine or alanine by combining with organic molecules.

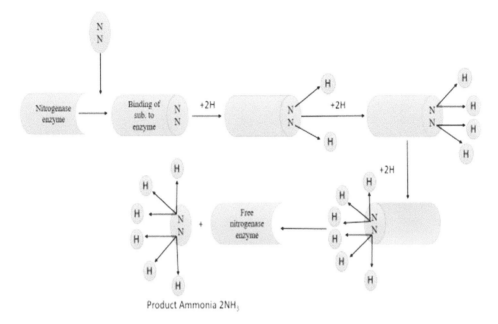

FIGURE 13.1 Model to show the reduction of atmospheric nitrogen to ammonia by nitrogenase enzyme. N, Nitrogen; H, Hydrogen; and NH_3, Ammonia.

13.6 MOLECULAR ANALYSIS OF NITROGEN FIXATION

There is significant phenotypic and genetic diversity in bacterial communities (Ovreas and Torsvik, 1998). Because most microorganisms cannot be cultivated in the environment, it is very difficult to quantify and evaluate natural diazotrophic bacterial populations (Borneman et al., 1996). This problem can, however, be solved by using separate cultivation techniques using universal gene amplification primers encoding the primary enzyme nitrogenase (Kirk et al., 2004).

Based on the presence of core metal [(iron (Fe), molybdenum (Mo), and vanadium (V)], which link two units of this enzyme, there are three types of nitrogenases (Zehr et al., 2003; Raymond et al., 2004). Mo-nitrogenase is the most predominant of these three forms. Three *nifHDK* genes encode the structural components of the nitrogenase complex. Nitrogenase expression and function depend on many other genes, apart from *nifHDK* (20 in the case of *K. pneumoniae*). In the larger segment of the nitrogenase complex called dinitrogenase, the *nifHDK* genes encode for α and β fragments, while smaller segment Fe protein is encoded by *nifH* (dinitrogenase reductase).

13.7 SYMBIOTIC NITROGEN FIXATION

Rhizobia is the best-known group of nitrogen-fixing symbiotic bacteria. However, nitrogen can also be fixed in symbiosis with plants by two other classes of bacteria, including *Frankia* and *Cyanobacteria*. The symbiotic fixation of nitrogen in rhizobia is mainly regulated by genes such as *nod, nif*, and *fix* (Table 13.3). In bacteria, the accessory genes that are placed in transmissible genetic elements such as plasmids, symbiotic islands, and chromids encode Nod and Nif proteins. Thus, these sets of genes can be transferred within the species of a bacteria genus in high frequencies and infrequently between genera (Remigi et al., 2016). *Rhizobial* lipo-chitooligosaccharide signal molecules or Nod factors are encoded by a special group of rhizobial genes called nod genes. In the early stages of nodulation, these are important genes for bacterial invasion and induction of nodules by inducing host responses such as root hair deformation and cortical cell division (Moulin

TABLE 13.3
A List of the Most Common Rhizobial *nod*, *nif*, and *fix* Genes

Genes	Function of gene products
Nodulation genes	
nodA	Acyltransferase
nodB	Chitooligosaccharide deacetylase
nodC	N-acetylglucosaminyltransferase
Nod	Transcriptional regulator of common *nod* genes
nodIJ	Nod factor transport
nodPQ	Synthesis of Nod factor substituents
nodX	Synthesis of Nod factor substituents
nodEF	Synthesis of Nod factor substituents
Other *nod* genes	Several functions in the synthesis of Nod factors
nol genes	Several functions in the synthesis of Nod factor substituents and secretion
noe genes	Synthesis of Nod factor substituents
Nitrogen fixation genes	
nifH	Dinitrogenase reductase (Fe protein)
nifD	α subunits of dinitrogenase (MoFe protein)
nifK	β subunits of dinitrogenase (MoFe protein)
nifA	Transcriptional regulator of the other *nif* genes
nifBEN	Biosynthesis of the Fe-Mo cofactor
fixABCX	Electron transport chain to nitrogenase
FixNOPQ	Cytochrome oxidase
fixLJ	Transcriptional regulators
fixK	Transcriptional regulator
fixGHIS	Copper uptake and metabolism
fdxN	Ferredoxin

Source: Laranjo et al., 2014.

et al., 2004) (Figure 13.2). The role of rhizobial NFs has recently been revisited, as their association with NF receptors and chitinases suggests a putative role in the equilibrium between symbiosis and defense (Kelly et al., 2017). Legumes exudate the flavonoids that are recognized by rhizobial NodD, initiate the interaction between rhizobia and legume roots, the rhizobial NodD, and in turn, regulates the transcription of several other nodulation genes in rhizobia (*nod, nol,* and *noe*) (Bonaldi et al., 2010; Rogel et al., 2011). In most rhizobia, the three *nod* genes, *nodA, nodB, and nodC*, occur as single-copy genes. They are involved in the Nod factor backbone synthesis (Moulin et al., 2004). For the synthesis of the NF core structure, three nod genes are important, according to Bonaldi et al. (2010): i) *NodA* gene encodes acyltransferase that links an acyl chain to the oligosaccharide non-reducing end; ii) *NodB* encodes a deacetylase that eliminates the N-acetyl moiety from these oligosaccharides non-reducing terminus; and iii) N-acetyl-glucosaminyltransferase is encoded by *nodC*, which polymerizes UDP-N-acetyl-D-glucosamine into oligosaccharide chains. In distinct rhizobial species, other nodulation genes (*nod, nol,* or *noe*) can be found that are more or less essential for the host specificity, by decorating the backbone of the essential Nod factor, resulting in multiple substitutions, such as methylation, acetylation, sulfation, and glycosylation (Bonaldi et al., 2010; Andrews and Andrews, 2017).

Rhizobia colonization of legume roots results in the curling of root hairs and deformation, and the expression of several genes (e.g., early nodulation (*ENOD*) genes and *Msrip1*) in the epidermis of legumes (Franssen et al., 1995; McAdam et al., 2018). Several recent studies discuss the dynamic

Nitrogen Biofertilizers

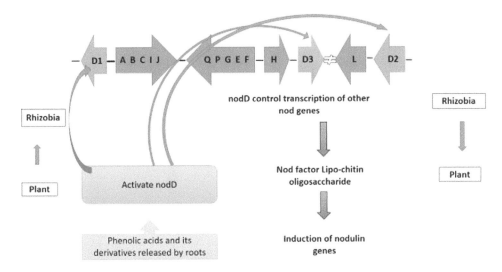

FIGURE 13.2 Diagrammatic representation of Plant–Bacteria interaction.

biological processes leading to rhizobia infection and nodulation, such as how legumes suppress the major rhizobia infection (Berrabah et al., 2019; Buhian and Bensmihen, 2018; Poole et al., 2018). By encoding the nitrogenase complex as well as a variety of regulatory proteins involved in nitrogen fixation, the Nif family of rhizobia plays a key role in symbiotic nitrogen fixation (SNF). *Nif* genes can also be present in other bacterial communities other than rhizobia. In addition to *nif* genes, nitrogen fixation regulation is dependent on the so-called fix genes. In rhizobia, the two important regulatory cascades are the oxygen-responsive two-component FixL-FixJ system and RpoN-NifA, together with FixK. NifA is the nitrogen fixation master regulator and belongs to the transcriptional regulator family–enhancer-binding proteins that interact with RNA polymerase sigma factor, σ54 (RpoN). The Hauke Henneke group made an important contribution to the early discovery of nif and fix genes in *Bradyrhizobium japonicum* (e.g., Fischer and Hennecke, 1984; Fischer et al., 1986; Fischer, 1994). In the case of Rhizobium etli CFN42, the two copies of rpoN are represented: RpoN1 and RpoN2 (Salazar et al., 2010). *Rpon1* is encoded in free-living conditions in the chromosome, while *rpoN2* is found in the symbiotic plasmid and is expressed in bacteroids independently of rpoN1 (Salazar et al., 2010). In a study it was demonstrated that in *Neorhizobium galegae* sv. Orientalis, *rpoN2* is required for nitrogen fixation (Österman et al., 2015). Interestingly, *rpoN2* could be found only in symbiovar orientalis but not in the other symbiovars, *N. galegae* sv. officinalis. The other regulator cascade (FixL-FixJ-FixK) is one of the two regulators of transcription of *nifA* and *fix* genes in SNF (Dixon and Kahn, 2004).

13.8 POTENTIAL ROLE OF BIOFERTILIZERS IN AGRICULTURE

Biofertilizers play a major role in increasing soil fertility (Kachroo and Razdan, 2006; Son et al., 2007). In addition, their soil application enhances soil structure and minimizes the single use of chemical fertilizers. The application of BGA + *Azospirillum* proved to be substantially advantageous in improving leaf area index (LAI) under low land conditions. Through the use of biofertilizers, the crop yield and harvest index will also improve. The highest increase in grain yield and straw of wheat plants with rock phosphate as a P fertilizer was achieved with inoculation with *Rhizobium* + *Azotobacter* + VAM. *Azolla* is economical, inexpensive, and friendly, and can provide benefit in terms of nitrogen and carbon enrichment (Kaushik and Prassana, 1989). Some commercially available biofertilizers are also used for crops. Raj (2007) claimed that microorganisms (*Thiobacillus thioxidans B. subtilis* and *Saccharomyces sp.*) can be used to solubilize fixed micronutrients such as zinc as bio-fertilizers. Like many other legumes, soybean plants can symbiotically fix atmospheric nitrogen and soybeans might provide about 80 to 90% of nitrogen demand by symbiosis (Bieranvand et al., 2003). Biocontrol, a new approach to disease prevention, may play an important role in farming (Senthilkumar and Rajendran, 2004; Li-Bin et al., 2005; Hossain et al., 2009). In French beans, the *Trichoderma*-based BAU-bio fungicide has been found to be promising to control root-knot diseases (Rahman, 2005). The use of antagonist bacteria such as *Bradyrhizobium* and *Rhizobium* has important in controlling mungbean root knots (Khan et al., 2006). Biofertilizers containing bacterial nitrogen fixators, phosphate- and potassium-solubilizing bacteria, and microbial strains of certain bacteria have greatly improved the growth, yield, and quality parameters of some plants (Youssef and Eissa, 2014).

13.9 NITROGEN BIOFERTILIZERS: ENVIRONMENTAL STRESSES

The main constraints that impact the production of crops are biotic and abiotic stresses. Several modern scientific tools have been extensively used for the improvement of crops under stress conditions, in which the role of plant growth-promoting rhizobacteria (PGPRs) as bioprotectants has become indispensable in this regard (Yang et al., 2009). There are two mechanisms through which PGPRs support the growth of the plant: (1) direct and (2) indirect. In a direct mechanism, plant growth increases when there is no plant pathogen and other rhizosphere microorganisms. In indirect mechanisms, plant growth has been reinforced by the elimination of the deleterious effects of plant pathogens. Diazotrophic *Rhizobium* bacteria present in the coastal regions have been shown to increase productivity, including in saline soil (Zahran, 1999). Flavonoids and Nod factors are produced by *Phaseolus vulgaris* under salt stress in combination with *Azospirillum brasilense* co-inoculated with *Rhizobium* (Dardanelli et al., 2008). Inoculated with *Trifolium alexandrinum*, *Rhizobium trifolii* displayed higher biomass and increased nodulation under the salinity stress situation (Hussain et al., 2002; Antoun, et al., 2005). *Phosphobacteria spp.*, *Glucanaceto-bacter spp.*, and *Azospirillium spp.* isolated from the rhizosphere of mangroves and rice fields were found to be more resistant to heavy metal particularly iron (Gill et al., 2012; Bhardwaj et al., 2014). Distinct types of adaptation mechanisms are used by plants to resist drought stress. In this process, phytohormones are very important (Potters et al., 2007). PGPRs produce osmolytes, siderophores, sugars, antibiotics, organic acids, nitric oxide, etc., to combat stress. Ethylene (a negative regulator) is regulated by legume plant nodulation. 1-aminocyclopropane- 1-carboxylic acid (ACC) is a precursor of ethylene. Thus, the ACC level is necessary to be maintained in plants. PGPR has an enzyme called ACC deaminase that degrades ACC and promotes root growth (Long et al., 2008; Belimov et al., 2009; Belimov et al., 2007). In this way, PGPRs decrease plant ethylene levels and support plant growth.

The analysis of PGPR-elicited responses should now aim at numerous other crops to assess their spectrum of action against numerous biotic and abiotic stresses. To use this instrument as a powerful tool against conventional and chemical approaches to combat environmental pressures, the mechanism involved in the triggering of defensive signals and different proteins must also be considered.

13.9.1 Constraints in Biofertilizer Use

- Production-level constraints: Lack of suitable and successful strains, unavailability of the suitable carrier, mutation during fermentation.
- Market-level constraints: Lack of awareness among farmers, inadequate and inexperienced personnel, quality assurance is not up to the mark, seasonal and unassured demand.
- Resource constraints: Limited resource generation for biofertilizer production.
- Field-level constraints: Soil and climatic factors, the native population of microbials, etc.

13.10 CONCLUSION

An integrated approach to evaluating the most desirable plant–microorganism relationship is vital to increasing and sustaining agricultural land productivity. Biofertilizers have an important role to play in increasing nutrient supplies and the quality of crops in the years ahead They are non-bulky, environmentally friendly, and low-cost agricultural inputs. A biofertilizer is an agricultural commodity in a concentrated form containing a particular microorganism extracted either from plant roots or from root zone soil (Rhizosphere). Among the biofertilizers the important ones for nitrogen fixation are *Azotobacter, Azospirillum,* and *Acetobacter.* The current trend of low-input chemicals in sustainable agricultural systems will contribute to the goal. It is estimated that the application of inorganic chemical fertilizers can be considerably reduced by 30–50%. This assists in achieving environmentally friendly and sustainable agriculture.

ACKNOWLEDGMENTS

We acknowledge the facilities provided by the Department of Botany, University of Kashmir, Srinagar, J & K, India, to carry out this work. Also thankful to University Grants Kashmir (UGC) for providing fellowship to Ishfaq Ul Rehmaan through the NET–JRF Fellowship scheme and also thankful to all those who made available their literature, which helped us to compile the catalog.

REFERENCES

Ahemad, M., & Kibret, M. (2014). Mechanisms and applications of plant growth promoting rhizobacteria: Current perspective. *Journal of King Saud University-Science*, 26(1), 1–20.

Aktar, W., Sengupta, D., & Chowdhury, A. (2009). Impact of pesticides use in agriculture: Their benefits and hazards. *Interdisciplinary Toxicology*, 2(1), 1–12.

Andrews, M., & Andrews, M. E. (2017). Specificity in legume-rhizobia symbioses. *International Journal of Molecular Sciences*, 18(4), 705.

Antoun, H., & Prévost, D. (2005). Ecology of plant growth promoting rhizobacteria. In: Z.A. Siddiqui (Ed.), *PGPR: Biocontrol and Biofertilization*. Dordrecht, Springer, pp. 1–38.

Arnold, W., Rump, A., Klipp, W., Priefer, U. B., & Pühler, A. (1988). Nucleotide-sequence of a 24,206-base-pair DNA fragment carrying the entire nitrogen-fixation gene-cluster of Klebsiella-Pneumoniae. *Journal of Molecular Biology*, 203(3).

Arun, K.S. (2007). Bio-fertilizers for sustainable agriculture. In K.S. Arun, *Bio-Fertilizers for Sustainable Agriculture Jodhpur, India*. Jodhpur, Agribios Publishers, pp. 196–197.

Bakulin, M. K., Grudtsyna, A. S., & Pletneva, A. Y. (2007). Biological fixation of nitrogen and growth of bacteria of the genus Azotobacter in liquid media in the presence of perfluorocarbons. *Applied Biochemistry and Microbiology*, 43(4), 399–402.

Bardi, L., & Malusà, E. (2012). Drought and nutritional stresses in plant: alleviating role of rhizospheric microorganisms. In: N. Haryana and S. Punj (Eds.), *Abiotic Stress: New Research*. Hauppauge, Nova Science Publishers Inc., pp. 1–57.

Belimov, A. A., Dodd, I. C., Hontzeas, N., Theobald, J. C., Safronova, V. I., & Davies, W. J. (2009). Rhizosphere bacteria containing 1-aminocyclopropane-1-carboxylate deaminase increase yield of plants grown in drying soil via both local and systemic hormone signalling. *New Phytologist*, 181(2), 413–423.

Belimov, A. A., Dodd, I. C., Safronova, V. I., Hontzeas, N., & Davies, W. J. (2007). Pseudomonas brassicacearum strain Am3 containing 1-aminocyclopropane-1-carboxylate deaminase can show both pathogenic and growth-promoting properties in its interaction with tomato. *Journal of Experimental Botany*, 58(6), 1485–1495.

Berrabah, F., Ratet, P., & Gourion, B. (2019). Legume nodules: Massive infection in the absence of defense induction. *Molecular Plant-Microbe Interactions*, 32(1), 35–44.

Bhardwaj, D., Ansari, M. W., Sahoo, R. K., & Tuteja, N. (2014). Biofertilizers function as key player in sustainable agriculture by improving soil fertility, plant tolerance and crop productivity. *Microbial Cell Factories*, 13(1), 1–10.

Bieranvand, N. P., Rastin, N. S., Afarideh, H., & Sagheb, N. (2003). *An evaluation of the N-fixation capacity of some Bradyrhizobium japonicum strains for soybean cultivars* (No. RESEARCH).

Bonaldi, K., Gourion, B., Fardoux, J., Hannibal, L., Cartieaux, F., Boursot, M.,... & Giraud, E. (2010). Large-scale transposon mutagenesis of photosynthetic Bradyrhizobium sp. strain ORS278 reveals new genetic loci putatively important for nod-independent symbiosis with *Aeschynomene indica*. Molecular *Plant-Microbe Interactions*, 23(6), 760–770.

Borneman, J., Skroch, P. W., O'Sullivan, K. M., Palus, J. A., Rumjanek, N. G., Jansen, J. L.,... & Triplett, E. W. (1996). Molecular microbial diversity of an agricultural soil in Wisconsin. *Applied and Environmental Microbiology*, 62(6), 1935–1943.

Buhian, W. P., & Bensmihen, S. (2018). Mini-review: nod factor regulation of phytohormone signaling and homeostasis during rhizobia-legume symbiosis. *Frontiers in Plant Science*, 9, 1247.

Chen, J. H. (2006, October). The combined use of chemical and organic fertilizers and/or biofertilizer for crop growth and soil fertility. In *International workshop on sustained management of the soil-rhizosphere system for efficient crop production and fertilizer use* (Vol. 16, No. 20, pp. 1–11). Land Development Department Bangkok Thailand.

Choudhury, A. T. M. A., & Kennedy, I. R. (2004). Prospects and potentials for systems of biological nitrogen fixation in sustainable rice production. *Biology and Fertility of Soils*, 39(4), 219–227.

Conway, G. (2012). *One Billion Hungry: Can We Feed the World?* Cornell University Press, Ithaca, New York.

Dardanelli, M. S., de Cordoba, F. J. F., Espuny, M. R., Carvajal, M. A. R., Díaz, M. E. S., Serrano, A. M. G.,... & Megías, M. (2008). Effect of Azospirillum brasilense coinoculated with Rhizobium on *Phaseolus vulgaris* flavonoids and Nod factor production under salt stress. *Soil Biology and Biochemistry*, 40(11), 2713–2721.

Dixon, R., & Kahn, D. (2004). Genetic regulation of biological nitrogen fixation. *Nature Reviews Microbiology*, 2(8), 621–631.

Fischer, H. M. (1994). Genetic regulation of nitrogen fixation in rhizobia. *Microbiological Reviews*, 58(3), 352–386.

Fischer, H. M., & Hennecke, H. (1984). Linkage map of the *Rhizobium japonicum* nifH and nifDK operons encoding the polypeptides of the nitrogenase enzyme complex. *Molecular and General Genetics MGG*, 196(3), 537–540.

Fischer, H. M., Alvarez-Morales, A., & Hennecke, H. (1986). The pleiotropic nature of symbiotic regulatory mutants: *Bradyrhizobium japonicum nifA* gene is involved in control of nif gene expression and formation of determinate symbiosis. *The EMBO Journal*, 5(6), 1165–1173.

Franche, C., Lindström, K., & Elmerich, 0. (2009). Nitrogen-fixing bacteria associated with leguminous and non-leguminous plants. *Plant and Soil*, 321(1–2), 35–59.

Franssen, H., Mylona, P., Pawlowski, K., Van De Sande, K., Heidstra, R., Geurts, R., & Martinez-Abarca, F. (1995). Plant genes involved in root-nodule development on legumes. *Philosophical Transactions of the Royal Society of London. Series B: Biological Sciences*, 350(1331), 101–107.

Gaur, V. (2010). Biofertilizer–necessity for sustainability. International Journal of Advanced Research and Development, 1, 7–8.

Gill, S. S., Khan, N. A., & Tuteja, N. (2012). Cadmium at high dose perturbs growth, photosynthesis and nitrogen metabolism while at low dose it up regulates sulfur assimilation and antioxidant machinery in garden cress (*Lepidium sativum* L.). *Plant Science*, 182, 112–120.

Glick, B. R. (2012). Plant growth-promoting bacteria: Mechanisms and applications. *Scientifica*, 2012, 1–15.

Gothwal, R. K., Nigam, V. K., Mohan, M. K., Sasmal, D., & Ghosh, P. (2008). Screening of nitrogen fixers from rhizospheric bacterial isolates associated with important desert plants. *Applied Ecology and Environmental Research*, 6(2), 101–109.

Gupta, A. K. (2004). The complete technology book on biofertilizers and organic farming. *National Institute of Industrial Research Press. India*, 168.

Hossain, M. A., Mahbub, M., Khanam, N., Hossain, M. S., & Islam, M. M. (2008). Effect of Bio-agents on growth and root-knot (*Meloidogyne javanica*) disease of soybean. *Journal of Agroforestry and Environment*, 3(1), 77–80.

Hu, Y., Fay, A. W., Lee, C. C., & Ribbe, M. W. (2007). P-cluster maturation on nitrogenase MoFe protein. *Proceedings of the National Academy of Sciences*, 104(25), 10424–10429.

Hussain, N., Mujeeb, F., Tahir, M., Khan, G. D., Hassan, N. M., & Bari, A. (2002). Effectiveness of Rhizobium under salinity stress. *Asian Journal of Plant Sciences*, 1(1), 12–14.

James, E. K., & Olivares, F. L. (1998). Infection and colonization of sugar cane and other graminaceous plants by endophytic diazotrophs. *Critical Reviews in Plant Sciences*, *17*(1), 77–119.

Kachroo, D., & Razdan, R. (2006). Growth, nutrient uptake and yield of wheat (Triticum aestivum) as influenced by biofertilizers and nitrogen. *Indian Journal of Agronomy*, *51*(1), 37–39.

Kaushik, B. D., & Prasanna, R. (1989). Status of biological nitrogen fixation by cyanobacteria and Azolla in India. *Biological Nitrogen Fixation Research Status in India*, *1989*, 141–208.

Kelly, S., Radutoiu, S., & Stougaard, J. (2017). Legume LysM receptors mediate symbiotic and pathogenic signalling. *Current Opinion in Plant Biology*, *39*, 152–158.

Khan, A., Zaki, M. J., & Tariq, M. (2006). Seed treatment with nematicidal Rhizobium species for the suppression of Meloidogyne javanica root infection on mungbean. *International Journal of Biology and Biotechnology*, *3*(3), 575–578.

Kirk, J. L., & Beaudette, L. A. M. Har t, P. Moutoglis, JN Klironomos, H. Lee, and JT Trevors. 2004. *Methods of studying soil microbial diversity. Journal of Microbiological Methods*, *58*, 169–188.

Krause, A., Ramakumar, A., Bartels, D., Battistoni, F., Bekel, T., Boch, J.,... & Linke, B. (2006). Complete genome of the mutualistic, N 2-fixing grass endophyte Azoarcus sp. strain BH72. *Nature Biotechnology*, *24*(11), 1384–1390.

Laranjo, M., Alexandre, A., & Oliveira, S. (2014). Legume growth-promoting rhizobia: an overview on the Mesorhizobium genus. *Microbiological Research*, *169*(1), 2–17.

Lery, L. M. S., von Krüger, W. M. A., Viana, F. C., Teixeira, K. R. S., & Bisch, P. M. (2008). A comparative proteomic analysis of Gluconacetobacter diazotrophicus PAL5 at exponential and stationary phases of cultures in the presence of high and low levels of inorganic nitrogen compound. *Biochimica et Biophysica Acta (BBA)-Proteins and Proteomics*, *1784*(11), 1578–1589.

Li, B., Xie, G. L., Soad, A., & Coosemans, J. (2005). Suppression of Meloidogyne javanica by antagonistic and plant growth-promoting rhizobacteria. *Journal of Zhejiang University. Science. B*, *6*(6), 496.

Long, H. H., Schmidt, D. D., & Baldwin, I. T. (2008). Native bacterial endophytes promote host growth in a species-specific manner; phytohormone manipulations do not result in common growth responses. *PLoS One*, *3*(7), e2702.

Macilwain, C. (2004). Organic: is it the future of farming? *Nature*, *428*(6985), 792–793.

Mahdi, S. S., Hassan, G. I., Samoon, S. A., Rather, H. A., Dar, S. A., & Zehra, B. (2010). Bio-fertilizers in organic agriculture. *Journal of Phytology,* *2*(10), 42-54.

Malusá, E., & Vassilev, N. (2014). A contribution to set a legal framework for biofertilisers. *Applied Microbiology and Biotechnology*, *98*(15), 6599–6607.

Mazid, M., & Khan, T. A. (2015). Future of bio-fertilizers in Indian agriculture: An overview. *International Journal of Agricultural and Food Research*, *3*(3).

McAdam, E. L., Reid, J. B., & Foo, E. (2018). Gibberellins promote nodule organogenesis but inhibit the infection stages of nodulation. *Journal of Experimental Botany*, *69*(8), 2117–2130.

Moulin, L., Béna, G., Boivin-Masson, C., & Stępkowski, T. (2004). Phylogenetic analyses of symbiotic nodulation genes support vertical and lateral gene co-transfer within the Bradyrhizobium genus. *Molecular Phylogenetics and Evolution*, *30*(3), 720–732.

Mutch, L. A., & Young, J. P. W. (2004). Diversity and specificity of Rhizobium leguminosarum biovar viciae on wild and cultivated legumes. *Molecular Ecology*, *13*(8), 2435–2444.

Newton, W. E. (2007). Physiology, biochemistry, and molecular biology of nitrogen fixation. In: H. Bothe, S. J. Ferguson and W. E. Newton (Eds.), *Biology of the Nitrogen Cycle* (pp. 109–129). Amsterdam, Elsevier.

Øvreås, L., & Torsvik, V. (1998). Microbial diversity and community structure in two different agricultural soil communities. *Microbial Ecology*, *36*(3–4), 303–315.

Pandey, J., & Singh, A. (2012). Opportunities and constraints in organic farming: an Indian perspective. *Journal of Scientific Research*, *56*, 47–72.

Poole, P., Ramachandran, V., & Terpolilli, J. (2018). Rhizobia: From saprophytes to endosymbionts. *Nature Reviews Microbiology*, *16*(5), 291.

Potters, G., Pasternak, T. P., Guisez, Y., Palme, K. J., & Jansen, M. A. (2007). Stress-induced morphogenic responses: Growing out of trouble?. *Trends In Plant Science*, *12*(3), 98–105.

Pretty, J., & Bharucha, Z. P. (2015). Integrated pest management for sustainable intensification of agriculture in Asia and Africa. *Insects*, *6*(1), 152–182.

Rahman, M. (2005). *Effect of BAU-Biofungicide and nematicide Curaterr against Root-Knot of French Bean.* Doctoral dissertation, MS Thesis, Department of Plant Pathology, Bangladesh Agricultural University, Mymensingh.

Raj SA (2007). Bio-fertilizers for micronutrients. Biofertilizer Newsletter (July), 8–10.

Raymond, J., Siefert, J. L., & Staples, C. R. (2004). Blankenship RE. The natural history of nitrogen fixation. Molecular Biology and Evolution, *21*(3), 541–54.

Reddy, B. S. (2013). *Soil Health: Issues and Concerns-A Review* (No. 131). Working Paper.

Remigi, P., Zhu, J., Young, J. P. W., & Masson-Boivin, C. (2016). Symbiosis within symbiosis: Evolving nitrogen-fixing legume symbionts. *Trends in Microbiology*, *24*(1), 63–75.

Rogel, M. A., Ormeno-Orrillo, E., & Romero, E. M. (2011). Symbiovars in rhizobia reflect bacterial adaptation to legumes. *Systematic and Applied Microbiology*, *34*(2), 96–104.

Salazar, E., Díaz-Mejía, J. J., Moreno-Hagelsieb, G., Martínez-Batallar, G., Mora, Y., Mora, J., & Encarnación, S. (2010). Characterization of the NifA-RpoN regulon in Rhizobium etli in free life and in symbiosis with *Phaseolus vulgaris*. *Applied and Environmental Microbiology*, *76*(13), 4510–4520.

Savci, S. (2012). An agricultural pollutant: Chemical fertilizer. *International Journal of Environmental Science and Development*, *3*(1), 73.

Senthilkumar, T., & Rajendran, G. (2004). Biocontrol agents for the management of disease complex involving root-knot nematode, Meloidogyne incognita and Fusarium moniliforme on grapevine (Vitis vinifera). *Indian Journal of Nematology*, *34*(1), 49–51.

Singh, J. S., Pandey, V. C., & Singh, D. P. (2011). Efficient soil microorganisms: a new dimension for sustainable agriculture and environmental development. *Agriculture, Ecosystems & Environment*, *140*(3-4), 339–353.

Son, T. N., Thu, V. V., Duong, V. C., & Hiraoka, H. (2007). *Effect of Organic and Bio-Fertilizers on Soybean and Rice Cropping System.* Japan International Research Center for Agricultural Sciences, Tsukuba, Ibaraki, Japan.

Sujanya, S., & Chandra, S. (2011). Effect of part replacement of chemical fertilizers with organic and bio-organic agents in ground nut, Arachis hypogea. *Journal of Algal Biomass Utilization*, *2*(4), 38–41.

Tan, Z., Hurek, T., & Reinhold-Hurek, B. (2003). Effect of N-fertilization, plant genotype and environmental conditions on nifH gene pools in roots of rice. *Environmental Microbiology*, *5*(10), 1009–1015.

Tilman, D., Cassman, K. G., Matson, P. A., Naylor, R., & Polasky, S. (2002). Agricultural sustainability and intensive production practices. *Nature*, *418*(6898), 671–677.

Vessey, J. K. (2003). Plant growth promoting rhizobacteria as biofertilizers. *Plant and Soil*, *255*(2), 571–586.

Wani, S. P. (1990). Inoculation with associative nitrogen-fixing bacteria: Role in cereal grain production improvement. *Indian Journal of Microbiology*, *30*(4), 363–393.

Yang, J., Kloepper, J.W. & Ryu, C.M. (2009). Rhizosphere bacteria help plants tolerate abiotic stress. *Trends in Plant Science*, *14*(1), 1–4.

Youssef, M. M. A., & Eissa, M. F. M. (2014). Biofertilizers and their role in management of plant parasitic nematodes. A review. *Journal of Biotechnology and Pharmaceutical Research*, *5*(1), 1–6.

Zahran, H. H. (1999). Rhizobium-legume symbiosis and nitrogen fixation under severe conditions and in an arid climate. *Microbiology and Molecular Biology Reviews*, *63*(4), 968–989.

Zehr, J. P., Jenkins, B. D., Short, S. M., & Steward, G. F. (2003). Nitrogenase gene diversity and microbial community structure: A cross-system comparison. *Environmental Microbiology*, *5*(7), 539–554.

14 Effect of Biotic Stresses on Plant Nitrogen Metabolism

Subzar Ahmad Nanda*, Ishfaq Ahmad Sheergojri, Aadil Farooq War, Mohd Asgar Khan, and Ishfaq ul Rehmaan
Department of Botany, University of Kashmir, Srinagar, J&K, India
Corresponding author: subzarnanda4271@gmail.com

CONTENTS

14.1 Introduction 185
14.2 Action of Different Biotic Agents 186
14.3 Defense Responses 187
14.4 Physiological and Molecular Changes 187
References 189

14.1 INTRODUCTION

Plant nitrogen metabolism is considered as the principal metabolic pathway after carbon metabolism (Chikov and Bakirova, 2000) with nitrogen playing a crucial role in all metabolic processes (Tucker, 1999), augmenting growth and reproduction in plants (Singh et al., 2016) and responsible for grain yield and crop biomass production (Bergamo Fenilli et al., 2007). However, nitrogen metabolism does not operate in isolation in plants and many other physiological processes such as amino acid synthesis, tricarboxylic (TCA) cycle, photosynthesis, respiration, and photorespiration (Makino, 2011; Foyer et al., 2011) are also connected with it. It is well known that plants growing under natural conditions frequently encounter biotic stress such as attack from viral, bacterial, and fungal pathogens, parasitic plants, and insect herbivores, resulting in huge biological consequences. Since stress causes damage and decreases the performance of plants (Osmond et al., 1987), there must be some mechanisms that enable them to cope with this particular stress (Berger et al., 2007). It has been shown that many transcription factors are used up by plants to defend against various biotic and abiotic stresses (Yanagisawa et al., 2004; Kurai et al., 2011), with around 85 transcription factors alone being involved in nitrogen deficiency response and growth in the case of rice plants (Yang et al., 2015). Above and beyond this plants have developed powerful tactics to counter biotic stress through hormonal defense mechanisms, especially jasmonic acid and ethylene (ET), which specifically act against necrotrophic pathogens and herbivores, and salicylic acid (SA) and cytokinin (CK) acting against biotrophic and hemibiotrophic pathogens (Gao et al., 2015; Conrath et al., 2015). In these defense responses, several molecules are produced by the plants such as reactive oxygen species (ROS) (Inze and Van Montagu, 2002) and GABA (Shelp et al., 1999; Kinnersley and Turano, 2000), and some defenses are also induced to neighboring plants, which are known as systemic acquired resistance (SAR) and induced systemic resistance (ISR) (Pieterse et al., 2014).

Under normal circumstances, plants assimilate inorganic nitrogen (nitrate and ammonium) and fix it into organic form (amino acids, proteins, and nucleic acids) with the help of two key enzymes: glutamine synthetase (GS) and glutamate synthase (GOGAT) (Kant et al., 2011). Conversely, when plants encounter biotic stress they respond by remobilizing their nitrogen into signaling molecules such as polyamines (PAs), proline, GABA, and glycine betaine (Majumdar et al., 2016), triggering a moderate increase in ammonia toxicity inside the cells. Most studies have suggested an increase in the levels of PAs upon viral infection and their balance after the attack (Pál and Tibor, 2017). This exchange of PAs and N is emerging as a major contributor to stress responses in plants and glutamate (Glu) serves as a precursor in the remobilization of N into other molecular nitrogen-containing molecules such as proline (Pro), ornithine (Orn), arginine (Arg), and PAs, thereby linking carbon (C) and nitrogen (N) assimilation in plants (Majumdar et al., 2016). Whereas Pro is involved in stress moderation, Arg acts as an initiator molecule for the synthesis of other nitrogen compounds, including PAs in plants. At the molecular level, the increase in ammonium toxicity inside the cells leads to activation of a gene encoding α-subunit of glutamate dehydrogenase (gdh-NAD; A1) by ROS signal, which participates in ammonia assimilation with GS/GOGAT cycle (Skopelitis et al., 2007).

Much work has been done to explore the influence of abiotic stress on plant nitrogen metabolism (Farhangi-Abriz and Torabian, 2018) while only a few studies have examined the effect of biotic stress on it. This chapter highlights some important issues concerning nitrogen metabolism including the role of different kinds of biotic agents, defense responses, and physiological and molecular changes involved.

14.2 ACTION OF DIFFERENT BIOTIC AGENTS

The process of nitrogen metabolism has been intensely studied and known in the case of plants and algae (Hachiya and Sakakibara, 2017; Kant, 2018) with *Chlamydomonas* serving as model algae in understanding the N metabolism process (Fernandez and Galvan, 2007). Biotic stress refers to a form of disturbance that arises due to the damage caused to a plant by biotic agents, such as viruses, parasites, bacteria, fungi, insect herbivores, unwanted plants as well as cultivated or native plants. Due to their immobility, several different kinds of biotic agents such as plant herbivores, viral, bacterial, and fungal pathogens, parasitic plants, and insect herbivores constantly attack these plants, leaving varying effects on nitrogen status and intense damages before and after harvesting of agricultural plants (Romero et al., 2018). Various events pertaining to nitrogen metabolism directly or indirectly on nitrogen-containing compounds have been reported in different plants upon interaction with a particular biotic agent. For example, exposure of polyamine oxidase (PAO) overexpressing tobacco plants to *Pseudomonas syringaepvtabaci* resulted in the upregulation of PAO gene

TABLE 14.1
List of Biotic Agents Affecting the Expression of Corresponding Genes Involved in Nitrogen Metabolism during Plant Biotic Interactions

Biotic agent	Affected gene	Function	References
Pseudomonas syringae	NRT2.1	High-affinity nitrate transceptors, repression of responses to biotrophic pathogens	Camañes et al., 2012
Cladosporium fulvum	NRF1	Indicator of nitrogen presence/absence	Pérez-Garcia et al., 2001
Colletotrichum lindemuthianum	GS1 α	Promotion of N mobilization	Tavernier et al., 2007
Pseudomonas syringae	PAO	Polyamine catabolism	Moschou et al., 2009
Pseudomonas syringae	NATA1	Acetylation of putrescine (Put)	Lou et al., 2016
Potato Virus Y, Potyvirus	GS1	Remobilization of protein-derived nitrogen	Pageau et al. (2006)
Potato Virus Y, Potyvirus	GDH	Deamination of glutamate in higher plants	Pageau et al. (2006)

(Moschou et al., 2009); extensive changes in the production/degradation of Pasin response to bacteria, fungi, and viruses (Moschou et al., 2009); accumulation of gamma-aminobutyric acid (GABA) at infection site upon exposure to *Cladosporium fulvum* (Solomon and Oliver, 2002) *Stagonospora nodorum*, and *Fusarium graminearum*, which is necessary for their development (Bönnighausen et al., 2015: prominent expression of glutamine synthetase (GS) upon infection with *Colletotrichum gloeosporioides* (Stephenson et al., 1997); and increased expression of genes coding for PAL (phenylalanine ammonia-lyase), CCR (cinnamoyl coenzyme A reductase), and CAD (cinnamyl alcohol dehydrogenase) upon exposure of *Linum usitatissimum* with mycelium extracts of *Phoma exigua, Fusarium oxysporum,* and *Botrytis cinerea* (Hano et al., 2006). During these plant, biotic interactions, the expression of different genes involved in nitrogen metabolism are affected by biotic stress (Table 14.1). Further, it has been shown that the resistance of a plant against the biotic agent varies with nitrogen status. For example, in rice plant it has been shown that out of three genes tested that are involved in resistance against *Magnaporthe oryzae*, the Co39/Pia and Pi2 genes are not affected but the working of the Pi1 gene is affected under high nitrogen and plants develop more necrotic lesions characteristic of the hypersensitive response (Ballini et al., 2013).

14.3 DEFENSE RESPONSES

When plants face biotic stress, their metabolic pathways are directed towards the defense system as well as to sustain growth, reproduction, and cellular maintenance (Berger et al., 2007). At first, there occurs a mutual interaction and recognition between the biotic agent and the host plant leading to activation of receptors and induction of chief defense hormones such as salicylic acid, jasmonic acid, and ethylene (Howe and Jander, 2008; Lopez et al., 2008). While salicylic acid (SA) mediates defense in response to biotrophic and hemibiotrophic pathogens, jasmonic acid and ethylene (ET) are effective against necrotrophic pathogens and herbivores (Conrath et al., 2015; Gao et al., 2015). Some other hormones that play a small role in the plant defense system include cytokinins, abscisic acid (ABA), gibberellins, auxins, and brassinosteroids (Robert-Seilaniantz et al., 2011; Pieterse et al., 2012). Biotic stress also leads to the generation of ROS in plants as a defense response (Inze and Van Montagu, 2002). These ROS molecules elicit H_2O_2 production, which functions as diffusible signaling molecule (Foyer and Harbinson, 1997), phytoalexin production, accumulation of cytosolic calcium concentration, and stimulating defense-related gene expression (Mittler et al., 2004; Tada et al., 2008). In the context of nitrogen metabolism, the downstream regulation of ROS-AzA-G3P by nitric oxide confers system acquired resistance (SAR), acting analogous to salicylic acid-mediated events (Wang et al., 2014). In Arabidopsis, it has been found that two nitrate transport (NRT2) genes, nrt2.1 and nrt2.6, are involved in defense responses against *Pseudomonas syringae* and *Erwinia amylovora* (Dechorgnat et al., 2012), with nrt2.1 curbing responses to biotrophic pathogens and nrt2.6 curbing responses to the necrotrophic pathogens. Furthermore, nitric oxide (NO), a signaling molecule that is generated from nitrate reductase (NR), is likewise involved in plant defenses against the pathogen (Bellin et al., 2013), and its formation and degradation are interconnected to N metabolism. Oliveira et al. (2009) have shown that plants that lack NR show increased susceptibility towards *Pseudomonas syringae* due to their incapability to produce NO. Thus, the accumulation of NO in response to necrotrophic fungal pathogens (Asai et al., 2010; Perchepied et al., 2010), bacteria (Modolo et al., 2006; Oliveira et al., 2009), or some pathogen-associated molecular patterns (Rasul et al., 2012; Shi and Li, 2008) depends upon the existence of nitrate reductase.

14.4 PHYSIOLOGICAL AND MOLECULAR CHANGES

Regulation of plant nitrogen metabolism is vital to coping with a particular stress and it encompasses nearly all the physiological processes of a plant (Lawlor, 2002) especially amino acid synthesis, tricarboxylic acid (TCA) cycle, photosynthesis, respiration, and photorespiration (Makino, 2011;

Foyer et al., 2011). When stress is a pathogen attack, then the plant metabolism is partly regulated by transcription of the N metabolism genes (Ward et al., 2010), which is considered vital for plant-microbial interactions. These N metabolism genes and especially those stimulated during N remobilization are well known to us (Masclaux et al., 2001; Guo et al., 2004). Normally plants absorb nitrate as their main source of nitrogen (Scheible et al., 1997) and transport it to the cytosol through NTR1 and NTR2 transporters (Xu et al., 2012) where it is first reduced to nitrite by nitrate reductase (Meyer and Stitt, 2001), which is then transported to chloroplast through NAR1 transporter (Rexach et al., 2000) and converted to ammonia by nitrite reductase; thereafter GS and GOGAT incorporate ammonia into amino acids (Lea and Miflin, 1974; Fernandez and Galvan, 2007; Liu et al., 2013). Moreover, nuclear magnetic resonance (NMR) and double labeling (13C/15N) confirm that the 2-oxoglutarate, which comes from the tricarboxylic acid cycle, is used for ammonium re-assimilation by the GS/GOGAT cycle (Yang et al., 2016; Lancien et al., 1999), thereby making carbon metabolism an important pathway for nitrogen assimilation (Gauthier et al., 2010). It has also been shown that ammonia may result from secondary metabolic processes (e.g., from the photorespiration process) (Keys, 2006; Forde and Lea, 2007) (Figure 14.1).

These two enzymes, GS and GOGAT, are highly important for the assimilation of inorganic nitrogen into organic form (amides and amino acids) in the cell, linking carbon and nitrogen metabolism together (Gauthier et al., 2010), while the glutamate dehydrogenase (GDH) is mostly involved in the nitrogen remobilization process (Lea and Miflin, 2003). Among these, GS is a fundamental enzyme that can promote nitrogen assimilation and translocation in crop plants upon edifying its activity (Wang et al., 2003). The GS enzyme is present in two isoforms: GS1, which has been found in the cytosol of *Arabidopsis thaliana*, is encoded by the GLN1 gene and appears to participate in primary ammonium assimilation from nitrate reduction and in ammonium assimilation from secondary sources (e.g., from photorespiration), while GS2, which has been found in the chloroplast of *A. thaliana*, is encoded by the GLN2 gene and appears to participate in the assimilation of ammonia from amino acid recycling (Yun et al., 2008; Cánovas et al., 2007). Accumulation of cytosolic GS enzyme in the mesophyll cells of tomato following biotic stress confirms its role in nitrogen reassimilation released during the disintegration of photosynthetic apparatus (Pérez-García et al., 1998). Likewise, two versions of GOGAT engage in reassimilation of ammonia, one of which during photorespiration is Ferredoxin-GOGAT and the other is NADH GOGAT during nitrogen

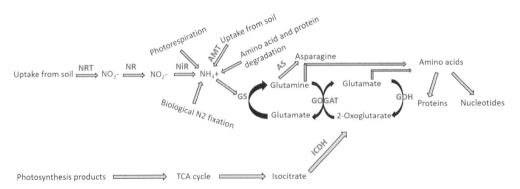

FIGURE 14.1 Pathway of Nitrogen-assimilation in higher plants. Inorganic nitrogen (nitrate or ammonia) becomes assimilated into amino acids and other organic molecules as illustrated. The particular steps revealed comprise nitrate transporters (NRT), nitrate reductase (NR), nitrite reductase (NiR), ammonium transporters (AMT), glutamine synthetase (GS), glutamate synthase (GOGAT), asparagine synthetase (AS), glutamate dehydrogenase (GDH), and isocitrate dehydrogenase (ICDH).

Source: Modified from Lu et al., 2016.

remobilization (Lea and Miflin, 2003). Apart from GS/GOGAT cycle and GDH, another enzyme involved in ammonia assimilation is asparagine synthetase (AS), which transfers the amido group from glutamine to aspartate molecule and generates glutamate and asparagine (Lam et al., 2003) (Figure 14.1).

However, during scary abiotic and biotic stress circumstances, plants react by remobilizing their N and C into signaling molecules such as PAs, Pro, GABA, glycine betaine, and β-Ala (Majumdar et al., 2016) resulting in a moderate increase in ammonium toxicity inside the cells. This N remobilization process is improved through the stimulation of GLN1, GDH, and ASN (asparagine synthetase) genes by these biotic and abiotic stresses (Pérez-Garcia et al., 1998; Olea et al., 2004). Not only during stress conditions but nitrogen is also removed from the leaf to the bolls through the hydrolysis of proteins into amino acids during the reproductive stage of a plant (Schrader and Thomas, 1981) with ammonium as a central intermediate in that metabolism (Roosta et al., 2007). Ammonia being toxic inside the cells (Britto and Kronzucker, 2002) needs immediate removal and its assimilation, thus the moderate increase in ammonium ions leads to the activation of genes responsible for its re-assimilation. At the molecular level, the stimulation of a gene encoding α-subunit of glutamate dehydrogenase (gdh-NAD; A1) by ROS stress signal acts in combination with the GS/GOGAT cycle for ammonia assimilation and glutamate is the first amino acid in which ammonium is incorporated. This stress-induced glutamate (Glu) formation by glutamate dehydrogenase is diverted towards proline formation and rise in endocellular Pas, which are later emitted out and apoplastically oxidized by PAOs resulting in the production of excess H_2O_2, which helps the plants to endure the pathogen attack (Moschou et al., 2009). In this manner, transgenic plants having increased PAOs were developed to exhibit pre-induced resistance against biotrophic and hemibiotrophic pathogens (Moschou et al., 2009). Gene expression concerning nitrogen metabolism is intensely blown up by the pathogen infection (e.g., the induction of cytosolic GS1 in tobacco (Pageau et al., 2006), GLN1.1 induction by *E. amylovora* (Fagard et al., 2014), upregulation of Gln-α in *Phaseolus vulgaris* against pathogenic and nonpathogenic fungus, induction of NRT2 family of high-affinity nitrate transporters in response to *P. syringae* and *E. amylovora* (Dechorgnat et al., 2012), and stimulation of cytosolic GS genes in leaf cells of infected plants during pathogen attack (Oléa et al., 2004; Pageau et al., 2006; Tavernier et al., 2007)). In *Phaseolus vulgaris*, it has been shown that the upregulation of Gln-α expression in response to fungal biotic stress is accompanied by phenylalanine ammonia-lyase (PAL 3) suggesting ammonia reassimilation by cytosolic GS from PAL 3 activity (Tavernier et al., 2007).

REFERENCES

Asai, S., Mase, K., & Yoshioka, H. (2010). Role of nitric oxide and reactive oxygen species in disease resistance to necrotrophic pathogens. Plant signaling & behavior, 5(7), 872–874.

Ballini, E., Nguyen, T. T., & Morel, J. B. (2013). Diversity and genetics of nitrogen-induced susceptibility to the blast fungus in rice and wheat. Rice, 6(1), 1–13.

Bellin, D., Asai, S., Delledonne, M., & Yoshioka, H. (2013). Nitric oxide as a mediator for defense responses. Molecular plant-microbe interactions, 26(3), 271–277.

Bergamo Fenilli, T. A., Reichardt, K., Ocheuze Trivelin, P. C., & Favarin, J. L. (2007). Volatilization of ammonia derived from fertilizer and its reabsorption by coffee plants. Communications in soil science and plant analysis, 38(13–14), 1741–1751.

Berger, S., Sinha, A. K., & Roitsch, T. (2007). Plant physiology meets phytopathology: plant primary metabolism and plant–pathogen interactions. Journal of experimental botany, 58(15–16), 4019–4026.

Bönnighausen, J., Gebhard, D., Kröger, C., Hadeler, B., Tumforde, T., Lieberei, R.,.... & Bormann, J. (2015). Disruption of the GABA shunt affects mitochondrial respiration and virulence in the cereal pathogen Fusarium graminearum. Molecular microbiology, 98(6), 1115–1132.

Britto, D. T., & Kronzucker, H. J. (2002). NH4+ toxicity in higher plants: a critical review. Journal of plant physiology, 159(6), 567–584.

Camanes, G., Pastor, V., Cerezo, M., Garcia-Andrade, J., Vicedo, B., García-Agustín, P., & Flors, V. (2012). A deletion in NRT2. 1 attenuates Pseudomonas syringae-induced hormonal perturbation, resulting in primed plant defenses. Plant physiology, 158(2), 1054–1066.

Cánovas, F. M., Avila, C., Canton, F. R., Canas, R. A., & de la Torre, F. (2007). Ammonium assimilation and amino acid metabolism in conifers. Journal of Experimental Botany, 58(9), 2307–2318.

Chikov, V., & Bakirova, G. (2000). Relationship between carbon and nitrogen metabolisms in photosynthesis. The role of photooxidation processes. Photosynthetica, 37(4), 519–527.

Conrath, U., Beckers, G. J., Langenbach, C. J., & Jaskiewicz, M. R. (2015). Priming for enhanced defense. Annual review of phytopathology, 53, 97–119.

Dechorgnat, J., Patrit, O., Krapp, A., Fagard, M., & Daniel-Vedele, F. (2012). Characterization of the Nrt2.6 gene in Arabidopsis thaliana: a link with plant response to biotic and abiotic stress. PloS one, 7(8), e42491.

Fagard, M., Launay, A., Clément, G., Courtial, J., Dellagi, A., Farjad, M., & Masclaux-Daubresse, C. (2014). Nitrogen metabolism meets phytopathology. Journal of experimental botany, 65(19), 5643–5656.

Farhangi-Abriz, S., & Torabian, S. (2018). Biochar improved nodulation and nitrogen metabolism of soybean under salt stress. Symbiosis, 74(3), 215–223.

Fernandez, E., & Galvan, A. (2007). Inorganic nitrogen assimilation in Chlamydomonas. Journal of experimental botany, 58(9), 2279–2287.

Forde, B. G., & Lea, P. J. (2007). Glutamate in plants: metabolism, regulation, and signalling. Journal of experimental botany, 58(9), 2339–2358.

Foyer, C. H., & Harbinson, J. E. R. E. M. Y. (1997). The photosynthetic electron transport system: efficiency and control. A molecular approach to primary metabolism in higher plants, 3, 40.

Foyer, C. H., Noctor, G., & Hodges, M. (2011). Respiration and nitrogen assimilation: targeting mitochondria-associated metabolism as a means to enhance nitrogen use efficiency. Journal of experimental botany, 62(4), 1467–1482.

Gao, Q. M., Zhu, S., Kachroo, P., & Kachroo, A. (2015). Signal regulators of systemic acquired resistance. Frontiers in plant science, 6, 228.

Gauthier, P. P., Bligny, R., Gout, E., Mahé, A., Nogués, S., Hodges, M., & Tcherkez, G. G. (2010). In folio isotopic tracing demonstrates that nitrogen assimilation into glutamate is mostly independent from current CO_2 assimilation in illuminated leaves of Brassica napus. New phytologist, 185(4), 988–999.

Guo, Y., Cai, Z., & Gan, S. (2004). Transcriptome of Arabidopsis leaf senescence. Plant, cell & environment, 27(5), 521–549.

Hachiya, T., & Sakakibara, H. (2017). Understanding plant nitrogen metabolism through metabolomics and computational approaches. Journal of experimental botany, 68(10), 2501–2512.

Hano, C., Addi, M., Bensaddek, L., Crônier, D., Baltora-Rosset, S., Doussot, J. & Lamblin, F. (2006). Differential accumulation of monolignol-derived compounds in elicited flax (Linum usitatissimum) cell suspension cultures. Planta, 223(5), 975–989.

Howe, G. A., & Jander, G. (2008). Plant immunity to insect herbivores. Annual Review of Plant Biology, 59, 41–66.

Inze, D., & Montagu, M. V. (2002). *Oxidative stress in plants*. Taylor and Froncis Press.

Kant, S. (2018, February). Understanding nitrate uptake, signaling and remobilisation for improving plant nitrogen use efficiency. In Seminars in Cell & Developmental Biology (Vol. 74, pp. 89–96). Academic Press.

Kant, S., Bi, Y. M., & Rothstein, S. J. (2011). Understanding plant response to nitrogen limitation for the improvement of crop nitrogen use efficiency. Journal of experimental botany, 62(4), 1499–1509.

Keys, A. J. (2006). The re-assimilation of ammonia produced by photorespiration and the nitrogen economy of C 3 higher plants. Photosynthesis research, 87(2), 165.

Kinnersley, A. M., & Turano, F. J. (2000). Gamma aminobutyric acid (GABA) and plant responses to stress. Critical reviews in plant sciences, 19(6), 479–509.

Kurai, T., Wakayama, M., Abiko, T., Yanagisawa, S., Aoki, N., & Ohsugi, R. (2011). Introduction of the ZmDof1 gene into rice enhances carbon and nitrogen assimilation under low-nitrogen conditions. Plant biotechnology journal, 9(8), 826–837.

Lam, H. M., Wong, P., Chan, H. K., Yam, K. M., Chen, L., Chow, C. M., & Coruzzi, G. M. (2003). Overexpression of the ASN1 gene enhances nitrogen status in seeds of Arabidopsis. Plant physiology, 132(2), 926–935.

Lancien, M., Ferrario-Méry, S., Roux, Y., Bismuth, E., Masclaux, C., Hirel, B.,... & Hodges, M. (1999). Simultaneous expression of NAD-dependent isocitrate dehydrogenase and other Krebs cycle genes after nitrate resupply to short-term nitrogen-starved tobacco. Plant physiology, 120(3), 717–726.

Lawlor, D. W. (2002). Carbon and nitrogen assimilation in relation to yield: mechanisms are the key to understanding production systems. Journal of experimental botany, 53(370), 773–787.

Lea, P. J., & Miflin, B. J. (1974). Alternative route for nitrogen assimilation in higher plants. Nature, 251(5476), 614–616.

Lea, P. J., & Miflin, B. J. (2003). Glutamate synthase and the synthesis of glutamate in plants. Plant physiology and biochemistry, 41(6–7), 555–564.

Liu, Z., Bie, Z., Huang, Y., Zhen, A., Niu, M., & Lei, B. (2013). Rootstocks improve cucumber photosynthesis through nitrogen metabolism regulation under salt stress. Acta physiologiae plantarum, 35(7), 2259–2267.

López, M. A., Bannenberg, G., & Castresana, C. (2008). Controlling hormone signaling is a plant and pathogen challenge for growth and survival. Current opinion in plant biology, 11(4), 420–427.

Lou, Y. R., Bor, M., Yan, J., Preuss, A. S., & Jander, G. (2016). Arabidopsis NATA1 acetylates putrescine and decreases defense-related hydrogen peroxide accumulation. Plant physiology, 171(2), 1443–1455.

Lu, J., Zhang, L., Lewis, R. S., Bovet, L., Goepfert, S., Jack, A. M.,... & Dewey, R. E. (2016). Expression of a constitutively active nitrate reductase variant in tobacco reduces tobacco-specific nitrosamine accumulation in cured leaves and cigarette smoke. Plant biotechnology journal, 14(7), 1500–1510.

Majumdar, R., Barchi, B., Turlapati, S. A., Gagne, M., Minocha, R., Long, S., & Minocha, S. C. (2016). Glutamate, ornithine, arginine, proline, and polyamine metabolic interactions: the pathway is regulated at the post-transcriptional level. Frontiers in plant science, 7, 78.

Makino, A. (2011). Photosynthesis, grain yield, and nitrogen utilization in rice and wheat. Plant physiology, 155(1), 125–129.

Masclaux, C., Quillere, I., Gallais, A., & Hirel, B. (2001). The challenge of remobilisation in plant nitrogen economy. A survey of physio-agronomic and molecular approaches. Annals of applied biology, 138(1), 69–81.

Meyer, C., & Stitt, M. (2001). Nitrate reduction and signalling. In *Plant nitrogen* (pp. 37–59). Springer, Berlin, Heidelberg.

Mittler, R., Vanderauwera, S., Gollery, M., & Van Breusegem, F. (2004). Reactive oxygen gene network of plants. Trends in plant science, 9(10), 490–498.

Modolo, L. V., Augusto, O., Almeida, I. M., Pinto-Maglio, C. A., Oliveira, H. C., Seligman, K., & Salgado, I. (2006). Decreased arginine and nitrite levels in nitrate reductase-deficient Arabidopsis thaliana plants impair nitric oxide synthesis and the hypersensitive response to *Pseudomonas syringae*. Plant science, 171(1), 34–40.

Moschou, P. N., Sarris, P. F., Skandalis, N., Andriopoulou, A. H., Paschalidis, K. A., Panopoulos, N. J., & Roubelakis-Angelakis, K. A. (2009). Engineered polyamine catabolism preinduces tolerance of tobacco to bacteria and oomycetes. Plant physiology, 149(4), 1970–1981.

Olea, F., Pérez-García, A., Cantón, F. R., Rivera, M. E., Cañas, R., Ávila, C.,... & de Vicente, A. (2004). Up-regulation and localization of asparagine synthetase in tomato leaves infected by the bacterial pathogen Pseudomonas syringae. Plant and cell physiology, 45(6), 770–780.

Oliveira, H. C., Justino, G. C., Sodek, L., & Salgado, I. (2009). Amino acid recovery does not prevent susceptibility to Pseudomonas syringae in nitrate reductase double-deficient Arabidopsis thaliana plants. Plant science, 176(1), 105–111.

Osmond, C. B., Austin, M. P., Berry, J. A., Billings, W. D., Boyer, J. S., Dacey, J. W. H.,... & Winner, W. E. (1987). Stress physiology and the distribution of plants. BioScience, 37(1), 38–48.

Pageau, K., Reisdorf-Cren, M., Morot-Gaudry, J. F., & Masclaux-Daubresse, C. (2006). The two senescence-related markers, GS1 (cytosolic glutamine synthetase) and GDH (glutamate dehydrogenase), involved in nitrogen mobilization, are differentially regulated during pathogen attack and by stress hormones and reactive oxygen species in Nicotiana tabacum L. leaves. Journal of experimental botany, 57(3), 547–557.

Pal, M., & Janda, T. (2017). Role of polyamine metabolism in plant pathogen interactions. Journal of Plant Science and Phytopathology, 1, 095–100.

Perchepied, L., Balagué, C., Riou, C., Claudel-Renard, C., Rivière, N., Grezes-Besset, B., & Roby, D. (2010). Nitric oxide participates in the complex interplay of defense-related signaling pathways controlling disease resistance to Sclerotinia sclerotiorum in Arabidopsis thaliana. Molecular plant-microbe interactions, 23(7), 846–860.

Pérez-García, A., Pereira, S., Pissarra, J., Gutiérrez, A. G., Cazorla, F. M., Salema, R.,... & Canovas, F. M. (1998). Cytosolic localization in tomato mesophyll cells of a novel glutamine synthetase induced in response to bacterial infection or phosphinothricin treatment. Planta, 206(3), 426–434..

Pérez-García, A., Snoeijers, S. S., Joosten, M. H., Goosen, T., & De Wit, P. J. (2001). Expression of the avirulence gene Avr9 of the fungal tomato pathogen Cladosporium fulvum is regulated by the global nitrogen response factor NRF1. Molecular plant–microbe interactions, 14(3), 316–325.

Pieterse, C. M., Van der Does, D., Zamioudis, C., Leon-Reyes, A., & Van Wees, S. C. (2012). Hormonal modulation of plant immunity. Annual review of cell and developmental biology, 28, 489–521.

Pieterse, C. M., Zamioudis, C., Berendsen, R. L., Weller, D. M., Van Wees, S. C., & Bakker, P. A. (2014). Induced systemic resistance by beneficial microbes. Annual review of phytopathology, 52, 347–375.

Rasul, S., Dubreuil-Maurizi, C., Lamotte, O., Koen, E., Poinssot, B., Alcaraz, G.,... & Jeandroz, S. (2012). Nitric oxide production mediates oligogalacturonide-triggered immunity and resistance to Botrytis cinerea in Arabidopsis thaliana. Plant, cell & environment, 35(8), 1483–1499.

Rexach, J., Fernández, E., & Galván, A. (2000). The Chlamydomonas reinhardtii Nar1 gene encodes a chloroplast membrane protein involved in nitrite transport. The plant cell, 12(8), 1441–1453.

Robert-Seilaniantz, A., Grant, M., & Jones, J. D. (2011). Hormone crosstalk in plant disease and defense: more than just jasmonate-salicylate antagonism. Annual review of phytopathology, 49, 317–343.

Romero, F. M., Maiale, S. J., Rossi, F. R., Marina, M., Ruíz, O. A., & Gárriz, A. (2018). Polyamine metabolism responses to biotic and abiotic stress. Polyamines, 37–49.

Roosta, H. R., & Schjoerring, J. K. (2007). Effects of ammonium toxicity on nitrogen metabolism and elemental profile of cucumber plants. Journal of plant nutrition, 30(11), 1933–1951.

Scheible, W. R., Gonzalez-Fontes, A., Lauerer, M., Muller-Rober, B., Caboche, M., & Stitt, M. (1997). Nitrate acts as a signal to induce organic acid metabolism and repress starch metabolism in tobacco. The plant cell, 9(5), 783–798.

Schrader, L. E., & Thomas, R. J. (1981). Nitrate uptake, reduction and transport in the whole plant. In: Nitrogen and carbon metabolism (pp. 49–93). Springer.

Shelp, B. J., Bown, A. W., & McLean, M. D. (1999). Metabolism and functions of gamma-aminobutyric acid. Trends in plant science, 4(11), 446–452.

Shi, F. M., & Li, Y. Z. (2008). Verticillium dahliae toxins-induced nitric oxide production in Arabidopsis is major dependent on nitrate reductase. BMB reports, 41(1), 79–85.

Singh, M., Singh, V. P., & Prasad, S. M. (2016). Responses of photosynthesis, nitrogen and proline metabolism to salinity stress in Solanum lycopersicum under different levels of nitrogen supplementation. Plant physiology and biochemistry, 109, 72–83.

Skopelitis, D. S., Paranychianakis, N. V., Kouvarakis, A., Spyros, A., Stephanou, E. G., & Roubelakis-Angelakis, K. A. (2007). The isoenzyme 7 of tobacco NAD (H)-dependent glutamate dehydrogenase exhibits high deaminating and low aminating activities in vivo. Plant physiology, 145(4), 1726–1734.

Solomon, P. S., & Oliver, R. P. (2002). Evidence that γ-aminobutyric acid is a major nitrogen source during Cladosporium fulvum infection of tomato. Planta, 214(3), 414–420.

Stephenson, S. A., Green, J. R., Manners, J. M., & Maclean, D. J. (1997). Cloning and characterisation of glutamine synthetase from Colletotrichum gloeosporioides and demonstration of elevated expression during pathogenesis on Stylosanthes guianensis. Current genetics, 31(5), 447–454.

Tada, Y., Spoel, S. H., Pajerowska-Mukhtar, K., Mou, Z., Song, J., Wang, C.,... & Dong, X. (2008). Plant immunity requires conformational charges of NPR1 via S-nitrosylation and thioredoxins. Science, 321(5891), 952–956.

Tavernier, V., Cadiou, S., Pageau, K., Laugé, R., Reisdorf-Cren, M., Langin, T., & Masclaux-Daubresse, C. (2007). The plant nitrogen mobilization promoted by Colletotrichum lindemuthianum in Phaseolus leaves depends on fungus pathogenicity. Journal of experimental botany, 58(12), 3351–3360.

Tucker, M. R. (1999). *Essential plant nutrients: their presence in North Carolina soils and role in plant nutrition*. Department of Agriculture and Consumer Services, Agronomic Division.

Wang, C., El-Shetehy, M., Shine, M. B., Yu, K., Navarre, D., Wendehenne, D.,... & Kachroo, P. (2014). Free radicals mediate systemic acquired resistance. Cell Reports, 7(2), 348–355.

Wang, Y. F., Jiang, D., Yu, Z. W., & Cao, W. X. (2003). Effects of nitrogen rates on grain yield and protein content of wheat and its physiological basis. Scientia Agricultura Sinica, 36(5), 513–520.

Ward, J. L., Forcat, S., Beckmann, M., Bennett, M., Miller, S. J., Baker, J. M.,... & Grant, M. (2010). The metabolic transition during disease following infection of Arabidopsis thaliana by Pseudomonas syringae pv. tomato. The plant journal, 63(3), 443–457.

Xu, G., Fan, X., & Miller, A. J. (2012). Plant nitrogen assimilation and use efficiency. Annual review of plant biology, 63, 153–182.

Yanagisawa, S., Akiyama, A., Kisaka, H., Uchimiya, H., & Miwa, T. (2004). Metabolic engineering with Dof1 transcription factor in plants: improved nitrogen assimilation and growth under low-nitrogen conditions. Proceedings of the national academy of sciences, 101(20), 7833–7838.

Yang, W., Yoon, J., Choi, H., Fan, Y., Chen, R., & An, G. (2015). Transcriptome analysis of nitrogen-starvation-responsive genes in rice. BMC plant biology, 15(1), 31.

Yang, X., Nian, J., Xie, Q., Feng, J., Zhang, F., Jing, H.,... & Wang, G. (2016). Rice ferredoxin-dependent glutamate synthase regulates nitrogen–carbon metabolomes and is genetically differentiated between japonica and indica subspecies. Molecular plant, 9(11), 1520–1534.

Yun, C. A. O., Xiao-Rong, F. A. N., Shu-Bin, S. U. N., Guo-Hua, X. U., Jiang, H. U., & Qi-Rong, S. H. E. N. (2008). Effect of nitrate on activities and transcript levels of nitrate reductase and glutamine synthetase in rice. Pedosphere, 18(5), 664–673.

15 Salt Stress and Nitrogen Metabolism in Plants

Peerzada Yasir Yousuf[1] and Arjumand Frukh[2]
[1]Department of Botany, Government Degree College Pulwama, Pulwama, Jammu and Kashmir, India
[2]Department of Botany, School of Chemical and Life Sciences, Jamia Hamdard, New Delhi, India
*Corresponding author: E-mail: syedyasar55@gmail.com

CONTENTS

15.1	Introduction	195
15.2	Salt Stress Impedes Plant Growth and Development	196
15.3	Cellular Effects of Salinity	196
	15.3.1 Osmotically Induced Water Stress	196
	15.3.2 Specific Ion Stress	197
	15.3.3 Oxidative Stress	197
	15.3.4 Nutritional Imbalance Stress	197
15.4	Effect of Salt Stress on Nitrogen Metabolism	198
	15.4.1 Effect of Salt Stress on Nitrogen Uptake	198
	15.4.2 Effect of Salt Stress on NO_3^- Assimilation	198
	15.4.3 Effect of Salt Stress on NH_4^+ Assimilation	199
15.5	Conclusion	199
References		200

15.1 INTRODUCTION

The burgeoning human population of the world has created a considerable upsurge in food demand. It is estimated that the demand for food production on a global level will increase by 35% to 56% between 2010 and the middle of the current century (Dijk et al., 2021). Food production is associated with soil, an indispensable medium for the growth and development of plants. Most of the land suitable for agricultural purposes is cultivated and any expansion of this land may have distressing effects on environment and biodiversity. Salinization, the foremost factor responsible for shrinkage of this land, is the build-up of water-soluble salts in the soil profile to a level that induces adverse influence on agricultural productivity and environmental and economic wellbeing. Soil salinization is brought about in part by the natural causes (primary salinization) such as hydrological, pedological, and geological processes including weathering of rocks, rise of saline groundwater, overflow of seawater over the coasts, obstructed drainage, and blowing of salt containing sand by sea winds. However, the leading contribution towards salinization is made by anthropogenic activities (secondary salinization). One of the human-induced salinization factors is deforestation, which accelerated the migration of salts in upper and lower layers of soil due to increasing sensitivity of these areas towards erosion and flooding. The other factor is accumulation of water- and air-borne salts in the soil through industrial emissions and municipal wastewater.

Overgrazing, improper irrigation practices, increased use of fertilizers, rise of water table, and chemical contamination via the use of contemporary agricultural systems are some of the other anthropogenic factors responsible for the rapidly increasing salinization.

15.2 SALT STRESS IMPEDES PLANT GROWTH AND DEVELOPMENT

Salt stress induces negative impact on the growth and development of plants, upsetting their productivity and survival predominantly in the arid and semi-arid regions of the world (Yousuf et al., 2016). Salt stress causes a substantial change in cellular metabolism, leading to stern crop damage. It induces various morphological, physiological, and cellular changes, which enhance toxicity and hinder the overall growth and development of plants. The main symptoms are inhibition of growth, acceleration of senescence, and even death of the plant during long-standing exposure to salt. Salt stress impacts physiological processes in plants including seed germination, biomass production, reproductive growth, photosynthesis, nitrogen metabolism, regulation of stomatal functioning, and carbohydrate metabolism. It induces a considerable decrease in seed germination (Uçarlı, 2020), root length (Kumar et al., 2021), and shoot length (Akram et al., 2022). Salt stress also induces considerable alteration in various leaf morphological parameters including specific leaf area and leaf area ratio (El-Taher et al., 2022). Biomass content of plants is also affected. A sizeable decrease in fresh weight has been noted in many plants like cucumber (Brengi et al., 2022), potato (Selem et al., 2022), spinach (Nigam et al., 2022; Kim et al., 2021), soybean (Nigam et al., 2022), lettuce (Bres et al., 2022) maize (Kubi et al., 2021), mungbean (Ali et al., 2021), and maize (Ahmad et al., 2021) among others. Although the reproductive stage of plants is less sensitive than the early vegetative growth stage, still various reproductive processes like flowering, pollination, fruit development, yield, and quality are affected by the salinity stress (Park et al., 2016; Cai et al., 2021). Plants encounter a huge decrease in their photosynthetic rate during salinity stress. Chlorophyll content, which is linked directly to the health of the plant, decreases under saline conditions (ElSayed et al., 2021; Acosta-Motos et al., 2017). Salinity stress also induces substantial changes in chloroplast ultrastructure (Hameed et al., 2021). Photosystem II is relatively more sensitive to salt stress and is repressed in higher plants (Bashir et al., 2021). The electron-transport chain and carbon-dioxide-assimilation rate is also influenced (Seleiman et al., 2022). Salinity stress hinders stomatal processes including stomatal conductance, internal carbon dioxide concentrations, and transpiration (Liao et al., 2022). Increased salt concentrations have a profound effect on the nitrogen metabolism of plants. The main nitrogen-assimilating enzyme, nitrate reductase, is sensitive to higher salt levels. Among the amino acids, proline accumulates more than others in salt-stressed plants (Ghosh et al., 2021).

15.3 CELLULAR EFFECTS OF SALINITY

The physiological changes observed at the whole plant level have been attributed to four main stresses induced by salinity: (i) osmotically induced water stress; (ii) specific ion stress; (iii) nutritional imbalance stress; and (iv) oxidative stress (Yousuf et al., 2016).

15.3.1 Osmotically Induced Water Stress

The accumulation of salts in soil decreases the water potential of soil solution. The lowering of water potential in root zone in turn hampers both the water uptake by plants and regulation of turgor. This leads to decline in the water potential of leaves. The loss of turgor affects many plant growth aspects like cell extension. Osmotic stress also brings a sizeable reduction in leaf growth, stomatal conductance, and eventually photosynthesis. The first sign in the reduction of plant growth is attributed to the salt-induced water stress.

Salt Stress and Nitrogen Metabolism in Plants

15.3.2 Specific Ion Stress

When sodium and chloride ions accumulate in the plant body, they induce their specific negative effects. The excessive accumulation of these ions upsets both structural and functional aspects of cell membranes. One of these affected systems includes transporters that are present on the cellular membranes and regulate the distribution of cellular ions. The sodium and chloride ions disrupt these important transporters causing an imbalance in the ionic equilibrium maintained within the cells (Epstein and Bloom, 2005).

15.3.3 Oxidative Stress

The balance between energy generation and consumption in plants is concurrent with three processes: photosynthesis, photorespiration and dark respiration via electron transfer, and level of substrates, reductants, and energy (Foyer et al., 2009; Pfannschmidt et al., 2009). The electron transfer is confined to three main cell organelles: chloroplasts, mitochondria, and peroxisomes. One of the main limitations of electron transport chains in aerobic organisms is that not all the electrons during their flow reduce their ultimate targets. Some of these electrons deviate from their main paths and reduce oxygen. This gives rise to new oxygen species, which are reactive and are referred to as reactive oxygen species (ROS). The ROS are also formed at other cellular sites including plasmalemma, apoplast, ER, and cell wall. There are three main types of reactive oxygen species produced by the sequential reduction of oxygen molecule by one, two, and three electrons, respectively. These include superoxide ion (O^{2-}), hydrogen peroxide (H_2O_2), and hydroxyl radicals (OH^-) (Figure 15.1).

These ROS are toxic, but their levels are maintained inside the cells by a well-demarcated antioxidant defense system. The ROS are steadily produced during normal cellular metabolism. During salinity stress, there is divergence of some electrons from main transport chains to oxygen-reduction pathways, which ultimately leads to the overproduction of ROS and impairs the balance maintained between ROS production and scavenging systems (Yousuf et al., 2017). The ROS react with indispensable biomolecules like proteins, lipids, nucleic acids, and carbohydrates, impairing their structural and functional aspects (Yousuf et al., 2012). Moreover, ROS also cause metabolic imbalances and loss of organelle integrity. The unhealthy condition caused by the overproduction of ROS is referred to as oxidative stress. The reaction of the ROS with the biomolecules can cause loss of enzyme activity, cross-linking of proteins, alteration in membrane fluidity, disruption of ion transport, inhibition of protein synthesis, carbohydrate oxidation, and nucleic acid damage, ultimately leading to cell death.

15.3.4 Nutritional Imbalance Stress

The accumulation of salts in the soil impedes mineral nutrition by affecting the nutrient availability and uptake of nutrients. Sodium chloride interferes with withholding and transformation of nutrients in the soil. Accumulation of NaCl triggers ion antagonism and decreases root growth, restricting the nutrient uptake by the roots. Many plants face decreased concentrations of potassium, calcium,

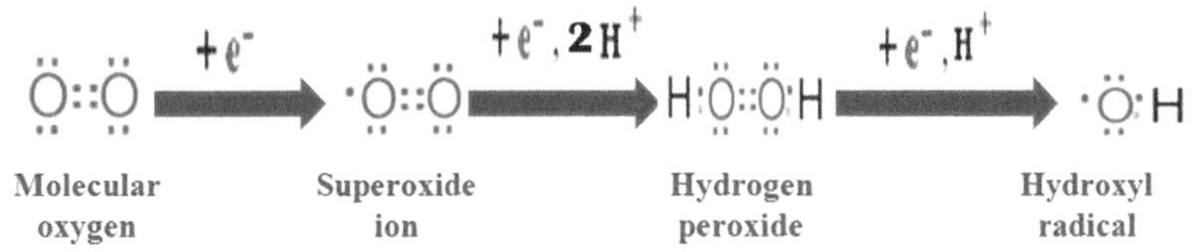

FIGURE 15.1 Reactive oxygen species: formation and structure.

manganese, nitrogen, and magnesium during the high concentrations of sodium chloride (Abbasi et al., 2016).

15.4 EFFECT OF SALT STRESS ON NITROGEN METABOLISM

Nitrogen metabolism is one of the key metabolic activities that controls the growth and development of plants (Yousuf et al., 2021). As a major constituent of cell components like chlorophyll, proteins, and nucleic acids, it regulates the main cellular activities in plants (Liao et al., 2019). Nitrogen metabolism involves three major steps including nitrogen uptake in the form of nitrate or ammonium ions, nitrate (NO_3^-) reduction, and ammonium (NH_4^+) assimilation (Ashraf et al., 2018). The absorption of the inorganic nitrogen is governed by some specific ammonium (AMTs) and nitrate transporters (NRTs) and the subsequent conversion of NO_3^- to NH_4^+ occurs via two enzymes, nitrate reductase (NR), and nitrite reductase (NiR). Finally, the assimilation of NH_4^+ is performed by glutamine synthetase (GS) and glutamate synthase (GOGAT) (Zhou et al., 2022). The assimilated nitrogen is vital component of many compounds that not only influence the plant growth but also mitigate the effect of many abiotic stresses (Sikder et al., 2020). However, salt stress has been observed to limit the plant growth and development by reducing the nitrogen uptake and metabolism in many tree and crop species.

15.4.1 EFFECT OF SALT STRESS ON NITROGEN UPTAKE

Excessive salinity results in the increase of Cl^- ions, which compete with the NO_3^- for uptake and translocation to the aerial parts of plant, and in this way decreases the rate of uptake of NO_3^- (Geilfus, 2018). It has been reported that the KCl and NaCl led accumulation of Cl^- ions inhibit the uptake of NO_3^- at higher concentrations (Zhao et al., 2019). Also, reduction in the uptake of NH_4^+ with the increase in the concentration of Na^+ has been observed in many plant species including *Saphora japonica* (Tian et al.,2021), cucumber (Li et al., 2019), rice (Sathee et al., 2021), etc.

In addition to ion imbalance, salt stress imparts changes in the soil water potential, which greatly affect nitrogen metabolism by inducing reductions in water availability and absorption (Ma et al., 2020). Also, the disruption in the root membrane integrity as a result of salt stress declines the process of nitrogen uptake in many plants (Guo et al., 2019). Salt stress also affects the process of photosynthesis, which induces s relative decrease in the growth rate of plants (Nadeem et al., 2019). The decrease in growth rate reduces the uptake of nitrogen from roots by inducing the reduction in internal nitrogen demand of plants (Farhangi-Abriz et al., 2018).

15.4.2 EFFECT OF SALT STRESS ON NO_3^- ASSIMILATION

In higher plants about 99% of the organic nitrogen is derived from NO_3^- assimilation (Ashraf et al., 2018). The first metabolic step of NO_3^- assimilation is the reduction of NO_3^- to NH_4^+ via NO_3^- assimilatory pathway. The process of NO_3^- reduction is catalyzed by two main enzymes, NR and NiR, and it occurs either in roots or leaves, or in both the organs (Cao et al., 2018). NR uses pyridine nucleotide as a source of reductant to reduce NO_3^- to nitrite (NO_2^-) in the cell cytoplasm. NO_2^- being a highly reactive molecule is immediately transported from the cytoplasm into the leaf chloroplasts and root plastids where NiR reduces it to NH_4^+. NH_4^+ is then assimilated via GS/GOGAT pathway (Feng et al., 2020).

NO_3^- assimilation is considered as a sensitive process in response to salt stress (de la Torre-González et al., 2020). It has been observed that the moderate levels of salinity lower NO_3^- loading into root xylem, which leads to higher concentration of NO_3^- in roots and reduced levels of NO_3^- in shoots (Alvarez-Aragon and Rodriguez, 2017). Since NO_3^- is a major signal affecting NR expression this process has a deleterious effect on the NO_3^- assimilation in plants (Raddatz et al., 2020). This whole process in turn leads to the inhibition of amino acid and protein metabolism. In most of the research it has been observed that high salinity levels influence NO_3^- transporters rather than NO_3^- reduction (Wang et al., 2018). Thus, the critical factor that determines the rate of plant survival under salt stress is the susceptibility of NO_3^- transporters to salinity injury. In conclusion, the decline in the NO_3^- assimilation under salt stress is mainly due to the outcome of many processes like inhibition of NO_3^- uptake by roots, reduction of NO_3^- loading into root xylem, and decreased NR leaf activity.

15.4.3 Effect of Salt Stress on NH_4^+ Assimilation

In plants, NH_4^+ is the only reduced form of nitrogen available for the assimilation in amino acids and other nitrogen containing compounds (Liu et al., 2021). NH_4^+ is combined with glutamic acid (Glu) and assimilated into glutamine (Gln) by means of glutamine synthetase (GS) enzyme. The Gln amide group is then transferred to an organic acid, 2-oxoglutarate (2-OG), by glutamate synthase (GOGAT) (Liu et al., 2021). The amino group of Glu is transferred to an organic acid and produces many amino acids (AA) via transaminases (Kishorekumar et al., 2020). Thus, the first step in the NH_4^+ assimilation involves the formation of glutamine catalyzed by GS, and the subsequent conversion of glutamine to glutamate is catalyzed by GOGAT. Therefore, GS and GOGAT are considered as the major two enzymes responsible for the NH_4^+ assimilation. The two additional enzymes also participate in ammonium assimilation including glutamate dehydrogenase (GDH) and asparagine synthetase (ASN) (Tang et al., 2020; Kishorekumar et al., 2020). However, the principal route of ammonium assimilation in plants is considered as GS/GOGAT cycle (Patel et al., 2021). A number of studies observed the reduction in the NH_4^+ assimilation under salt stress conditions (Ullah et al., 2019). The decline of NH_4^+ assimilation under salt stress is attributed to the fact that salt stress-induced inhibition of NO_3^- uptake and its subsequent effect on the nitrogen assimilation downregulates some of the genes (*OSGS1, OSGS2, OSFd-GOGAT*) responsible for the expression of GS and GOGAT enzymes, which in turn leads to the decline of the NH_4^+ assimilation in plants (Wang et al., 2012). However, the effect of salt stress at higher concentrations is more prevalent on GOGAT as compared to GS suggesting that the former is a more limiting factor for NH_4^+ assimilation under salt stress (Meng et al., 2016).

In the case of normal growth conditions in plants NH_4^+ assimilation occurs via GS/GOGAT pathway. However, under salt stress conditions the enzymatic activity of GS and GOGAT is reduced, which limits the production of amino acids but increases the NH_4^+ accumulation (Huang et al., 2020). The accumulation of NH_4^+ in saline environment leads to the activation of an alternative pathway known as GDH pathway for its assimilation (Huang et al., 2020). In various studies it was observed that the accumulation of NH_4^+ is paralleled by a sharp increase in the GDH activity. This alternate pathway may be considered as one of the means of salt tolerance (Zhang et al., 2014).

15.5 CONCLUSION

Salt stress has a profound effect on the metabolism, growth, and productivity of plants. It disrupts the redox homeostasis, osmotic balance, and mineral nutrition in plants. The accumulation of salts in the soil greatly influences the nitrogen metabolism of plants. Salt stress hampers the uptake, transport, and the assimilatory processes of nitrogen within plants.

Conflict of interest: The authors declare that there is no conflict of interest.
Contribution: Both the authors have equally contributed in this work.

REFERENCES

Abbasi, H., Jamil, M., Haq, A., Ali, S., Ahmad, R., Malik, Z., Parveen. 2016. Potassium regulation of ionic relations and redox status of salt affected plants: a review. *Zemdirbyste-Agriculture*, 103(2), 229–238.

Acosta-Motos, J.R., Ortuño, M.F., Bernal-Vicente, A., Diaz-Vivancos, P., Sanchez-Blanco, M.J., Hernandez, J.A. 2017. Plant responses to salt stress: Adaptive mechanisms. *Agronomy*, 7, 18.

Ahmad, S., Cui, W., Kamran, M. et al. 2021. Exogenous application of melatonin induces tolerance to salt stress by improving the photosynthetic efficiency and antioxidant defense system of maize seedling. *Journal of Plant Growth Regulation*, 40, 1270–1283.

Akram, W., Yasin, N.A., Shah, A.A., Khan, W.U., Li, G., Ahmad, A., Ahmed, S., Hussaan, M., Rizwan, M., Ali, S. 2022. Exogenous application of liquiritin alleviated salt stress and improved growth of Chinese kale plants. *Scientia Horticulturae*, 294, 110762.

Ali, R., Gul, H., Hamayun, M. et al. 2021. Aspergillus awamori ameliorates the physicochemical characteristics and mineral profile of mung bean under salt stress. *Chemical and Biological Technologies in Agriculture*, 8, 9.

Alvarez-Aragon, R., Rodriguez-Navarro, A. 2017. Nitrate-dependent shoot sodium accumulation and osmotic functions of sodium in *Arabidopsis* under saline conditions. *Plant Journal*, 91, 208–219.

Ashraf, M., Shahzad, S.M., Imtiaz, M., Rizwan, M.S. 2018. Salinity effects on nitrogen metabolism in plants–focusing on the activities of nitrogen metabolizing enzymes: A review. *Journal of Plant Nutrition*, 41(8), 1065–1081.

Bashir, S., Amir, M., Bashir, F., Javed, M., Hussain, A., Fatima, S., Parveen, R., Shahzadi, A. K., Afzal, S., Raza, S., Horain, T., Iqbal, A., Pervaiz, A., Rehman, A., Ayyaz, A., Zafar, Z. U., Athar, H.-R. 2021. Structural and functional stability of photosystem-II in "Moringa oleifera" under salt stress. *Australian Journal of Crop Science*, 15(5), 676–682.

Brengi S.H., Khedr A.E.M., Abouelsaad I.A. 2022. Effect of melatonin or cobalt on growth, yield and physiological responses of cucumber (*Cucumis sativus* L.) plants under salt stress. *Journal of the Saudi Society of Agricultural Sciences*, 21, 1, 51–60.

Breś W, Kleiber T, Markiewicz B, Mieloszyk E, Mieloch M. 2022. The effect of NaCl stress on the response of lettuce (*Lactuca sativa* L.). *Agronomy*, 12(2), 244.

Cai, Z., Wang, C., Chen, C., Chen, H., Yang, R., Chen, J., Chen, J., Tan, M., Mei, Y., Wei, L., Liu, X. 2021. Omics map of bioactive constituents in *Lonicera japonica* flowers under salt stress. *Industrial Crops and Products*, 167, 113526.

Cao, X., Zhong, C., Zhu, C., Zhu, L., Zhang, J., Wu, L., Jin, Q. 2018. Ammonium uptake and metabolism alleviate PEG-induced water stress in rice seedlings. *Plant Physiology and Biochemistry*, 132, 128–137.

de la Torre-González, A., Navarro-León, E., Blasco, B. and Ruiz, J.M. 2020. Nitrogen and photorespiration pathways, salt stress genotypic tolerance effects in tomato plants (*Solanum lycopersicum* L.). *Acta Physiologiae Plantarum*, 42(1), 1–8.

ElSayed, A. I., Rafudeen, M. S., Gomaa, A. M., and Hasanuzzaman, M. 2021. Exogenous melatonin enhances the reactive oxygen species metabolism, antioxidant defense-related gene expression, and photosynthetic capacity of *Phaseolus vulgaris* L. to confer salt stress tolerance. *Physiologia Plantarum*, 173(4), 1369–1381.

El-Taher, A.M., Abd El-Raouf, H.S., Osman, N.A., Azoz, S.N., Omar, M.A., Elkelish, A., Abd El-Hady, M.A.M. 2022. Effect of salt stress and foliar application of salicylic acid on morphological, biochemical, anatomical, and productivity characteristics of cowpea (*Vigna unguiculata* L.) Plants. *Plants*, 11(1), 115.

Epstein, E., Bloom, A.J. (Eds.). 2005. Mineral nutrition of plants. Principles and perspectives. Sinauer Associates, Inc., Sunderland.

Farhangi-Abriz, S., Torabian, S. 2018. Biochar improved nodulation and nitrogen metabolism of soybean under salt stress. *Symbiosis*, 74(3), 215–223.

Feng, H., Fan, X., Miller, A.J. and Xu, G. 2020. Plant nitrogen uptake and assimilation: Regulation of cellular pH homeostasis. *Journal of Experimental Botany*, 71(15), 4380–4392.

Foyer, C.H., Bloom, A.J., Queval, G., Noctor, G. 2009. Photorespiratory metabolism: Genes, mutants, energetics, and redox signaling. *Annual Review of Plant Biology*, 60, 455–484.

Geilfus, C.M. 2018. Chloride: From nutrient to toxicant. *Plant and Cell Physiology*, 59(5), 877–886.

Ghosh, U.K., Islam, M.N., Siddiqui, M.N., Cao, X., Khan, M. 2021. Proline, a multifaceted signalling molecule in plant responses to abiotic stress: understanding the physiological mechanisms. *Plant Biology* (Stuttgart, Germany). DOI:10.1111/plb.13363.

Guo, Q., Liu, L. Barkla, B.J. 2019. Membrane lipid remodeling in response to salinity. *International Journal of Molecular Sciences*, 20(17), 4264.

Hameed, A., Ahmed, M.Z., Hussain, T., Aziz, I., Ahmad, N., Gul, B., Nielsen, B.L. 2021. Effects of salinity stress on chloroplast structure and function. *Cells*, 10, 2022.

Huang, J., Zhu, C., Hussain, S., Huang, J., Liang, Q., Zhu, L., Cao, X., Kong, Y., Li, Y., Wang, L. and Li, J. 2020. Effects of nitric oxide on nitrogen metabolism and the salt resistance of rice (*Oryza sativa* L.) seedlings with different salt tolerances. *Plant Physiology and Biochemistry*, 155, 374–383.

Kim, B., Lee, H., Song, Y.H., Kim, H. 2021. Effect of salt stress on the growth, mineral contents, and metabolite profiles of spinach. *The Journal of the Science of Food and Agriculture*, 101(9), 3787–3794.

Kishorekumar, R., Bulle, M., Wany, A. Gupta, K.J. 2020. An overview of important enzymes involved in nitrogen assimilation of plants. *Nitrogen Metabolism in Plants*, (Clifton, N.J.), 2057, 1–13.

Kubi, H.A.A., Khan, M.A., Adhikari, A., Imran, M., Kang, S.M., Hamayun, M., Lee, I.J. 2021. Silicon and plant growth-promoting rhizobacteria *Pseudomonas psychrotolerans* CS51 mitigates salt stress in *Zea mays* L. *Agriculture*, 11, 272.

Kumar, S., Li, G., Yang, J., Huang, X., Ji, Q., Liu, Z., Ke, W., Hou, H. 2021. Effect of salt stress on growth, physiological parameters, and ionic concentration of water dropwort (*Oenanthe javanica*) cultivars. *Frontiers in Plant Science*, 12, 660409.

Li, S., Li, Y., He, X., Li, Q., Liu, B., Ai, X., Zhang, D. 2019. Response of water balance and nitrogen assimilation in cucumber seedlings to CO_2 enrichment and salt stress. *Plant Physiology and Biochemistry*, 139, 256–263.

Liao, L., Dong, T., Liu, X., Dong, Z., Qiu, X., Rong, Y., Sun, G., Wang, Z. 2019. Effect of nitrogen supply on nitrogen metabolism in the citrus cultivar 'Huangguogan'. *Plos One*, 14(3), e0213874.

Liao, Q., Gu, S., Kang, S., Du, T., Tong, L., Wood, J.D., Ding, R. 2022. Mild water and salt stress improve water use efficiency by decreasing stomatal conductance via osmotic adjustment in field maize. *Science of the Total Environment*, 805, 150364.

Liu, X., Hu, B., Chu, C. 2021. Nitrogen assimilation in plants: Current status and future prospects. *Journal of Genetics and Genomics*, S1673-8527(21)00376-3. Advance online publication.

Ma, Y., Dias, M.C., Freitas, H., 2020. Drought and salinity stress responses and microbe-induced tolerance in plants. *Frontiers in Plant Science*, 11, 1750.

Meng, S., Su, L., Li, Y., Wang, Y., Zhang, C., Zhao, Z. 2016. Nitrate and ammonium contribute to the distinct nitrogen metabolism of Populus simonii during moderate salt stress. *PloS One*, 11(3), e0150354.

Nadeem, M., Li, J., Yahya, M., Wang, M., Ali, A., Cheng, A., Wang, X., Ma, C. 2019. Grain legumes and fear of salt stress: Focus on mechanisms and management strategies. *International Journal of Molecular Sciences*, 20(4), 799.

Nigam, B., Dubey, R.S., Rathore, D. 2022. Protective role of exogenously supplied salicylic acid and PGPB (*Stenotrophomonas* sp.) on spinach and soybean cultivars grown under salt stress. 2022. *Scientia Horticulturae*, 293, 110654.

Park, H.J., Kim, W.Y., Yun, D.J. 2016. A new insight of salt stress signaling in plant. *Molecules and Cells*, 39(6), 447–459.

Patel, A., Tiwari, S. Prasad, S.M. 2021. Effect of time interval on arsenic toxicity to paddy field cyanobacteria as evident by nitrogen metabolism, biochemical constituent, and exopolysaccharide content. *Biological Trace Element Research*, 199(5), 2031–2046.

Pfannschmidt T, Brautigam K, Wagner R, Dietzel L, Schroter Y, Steiner S, Nykytenko A. 2009. Potential regulation of gene expression in photosynthetic cells by redox and energy state: approaches towards better understanding. *Annals of Botany*, 103, 599–607.

Raddatz, N., Morales De Los Ríos, L., Lindahl, M., Quintero, F.J., Pardo, J.M. 2020. Coordinated transport of nitrate, potassium, and sodium. *Frontiers in Plant Science*, 11, 247.

Sathee, L., Jha, S.K., Rajput, O.S., Singh, D., Kumar, S. and Kumar, A. 2021. Expression dynamics of genes encoding nitrate and ammonium assimilation enzymes in rice genotypes exposed to reproductive stage salinity stress. *Plant Physiology and Biochemistry*, 165, 161–172.

Seleiman, M.F Aslam, M.T., Alhammad, B.A., Hassan, M.U., Rizwan, M., et al. 2022. Salinity stress in wheat: Effects, mechanisms and management strategies. *Phyton-International Journal of Experimental Botany*, 91, 4, 667–694.

Selem, E., Hassan, A.A.S.A., Awad, M.F., Mansour, E., Desoky, E.S.M. 2022. Impact of exogenously sprayed antioxidants on physio-biochemical, agronomic, and quality parameters of potato in salt-affected soil. *Plants*, 11(2), 210.

Sikder, R.K., Wang, X., Zhang, H., Gui, H., Dong, Q., Jin, D., Song, M. 2020. Nitrogen enhances salt tolerance by modulating the antioxidant defense system and osmoregulation substance content in *Gossypium hirsutum*. *Plants*, 9(4), 450.

Tang, D., Liu, M.Y., Zhang, Q., Ma, L., Shi, Y., Ruan, J. 2020. Preferential assimilation of NH_4^+ over NO_3^- in tea plant associated with genes involved in nitrogen transportation, utilization and catechins biosynthesis. *Plant Science*, 291, 110369.

Tian, J., Pang, Y., Zhao, Z. 2021. Drought, salinity, and low nitrogen differentially affect the growth and nitrogen metabolism of *Sophora japonica* (L.) in a semi-hydroponic phenotyping platform. *Frontiers in Plant Science*, 12, 715456.

Uçarlı, C. 2020. Effects of salinity on seed germination and early seedling stage, abiotic stress in plants. In Fahad, S., Saud, S., Chen, Y., Wu, C., Wang, D. IntechOpen, DOI: 10.5772/intechopen.93647.

Ullah, A., Li, M., Noor, J., Tariq, A., Liu, Y., Shi, L. 2019. Effects of salinity on photosynthetic traits, ion homeostasis and nitrogen metabolism in wild and cultivated soybean. *Peer Journal*, 7, e8191.

Wang, H., Zhang, M., Guo, R., Shi, D., Liu, B., Lin, X., Yang, C. 2012. Effects of salt stress on ion balance and nitrogen metabolism of old and young leaves in rice (*Oryza sativa* L.). *BMC Plant Biology*, 12(1), 1–11.

Wang, Y.Y., Cheng, Y.H., Chen, K.E., Tsay, Y.F. 2018. Nitrate transport, signaling, and use efficiency. *Annual Review of Plant Biology*, 69, 85–122.

Yousuf, P.Y., Ahmad, A., Aref, I.M., Ozturk, M., Hemant, Ganie, A.H., Iqbal, M. 2016. Salt-stress-responsive chloroplast proteins in Brassica juncea genotypes with contrasting salt tolerance and their quantitative PCR analysis. *Protoplasma*, 253(6), 1565–1575.

Yousuf, P.Y., Ahmad, A., Ganie, A.H., Iqbal, M. 2016. Salt stress-induced modulations in the shoot proteome of Brassica juncea genotypes. *Environmental Science and Pollution Research International*, 23(3), 2391–2401.

Yousuf, P.Y., Ahmad, A., Ganie, A.H., Sareer, O., Krishnapriya, V., Aref, I.M., Iqbal, M. 2017. Antioxidant response and proteomic modulations in Indian mustard grown under salt stress. *Plant Growth Regulation*, 81, 31–50.

Yousuf, P.Y., Hakeem, K.U.R., Chandna, R., Ahmad, P. 2012. Role of glutathione reductase in plant abiotic stress. *In Abiotic Stress Responses in Plants*, Ahmad, P., Prasad, M.N.V., Eds., Springer: New York, NY, USA, pp. 149–158.

Yousuf, P.Y., Shabir, P.A., Hakeem, K.R. 2021. miRNAomic approach to plant nitrogen starvation. *International Journal of Genomics*, 8560323. https://doi.org/10.1155/2021/8560323

Zhang, Y., Zhang, L., Hu, X.H. 2014. Exogenous spermidine-induced changes at physiological and biochemical parameters levels in tomato seedling grown in saline-alkaline condition. *Botanical Studies*, 55(1), 1–8.

Zhao, Y., Wang, X., Wang, Y., Jiang, Z., Ma, X., Inyang, A.I., Cheng, H. 2019. Effects of salt on root aeration, nitrification, and nitrogen uptake in mangroves. *Forests*, 10(12), 1131.

Zhou, Y., Kishchenko, O., Stepanenko, A., Chen, G., Wang, W., Zhou, J., Pan, C., Borisjuk, N. 2022. The dynamics of NO_3^- and NH_4^+ uptake in duckweed are coordinated with the expression of major nitrogen assimilation genes. *Plants*, 11(1), 11.

16 Reactive Nitrogen Species in Plants

Urfi Jahan*, Uzma Kafeel, Fareed Ahmad Khan, and Afrin Jahan
Plant Ecology and Environment Laboratory, Department of Botany,
Aligarh Muslim University, Aligarh, Uttar Pradesh, India.
*Corresponding author. E-mail address: urfijahan111@gmail.com

CONTENTS

16.1 Introduction 203
16.2 Source of RNS in Plants 204
    16.2.1 Nitric Oxide (NO) 204
    16.2.2 Peroxynitrite (OONO⁻) 207
    16.2.3 Nitrosothiols 208
16.3 Conclusion and Future Outlook 208
References 209

16.1 INTRODUCTION

Nitric oxide(NO)-related compounds, such as S-nitrosothiols, S-nitrosoglutathione, and peroxynitrite, have been designated as reactive nitrogen species (RNS) that play essential roles in numerous physiological processes in plant cells (Halliwell and Gutteridge, 2007). These molecules can react with a broad spectrum of biomolecules, and they may act as reservoirs and transporters for NO in a wide range of plant cell signaling affairs. For promoting the interest of phytologists in RNS, this chapter will attempt to summarize the present understanding of RNS in plants, including their characteristics and physiological roles. The revelation that plant cells can create the free radical NO has opened new avenues of investigation. In physiological and phytopathological processes, several RNS have been implicated as signal molecules. Because they are involved directly or indirectly in many metabolic pathways, RNS metabolism appears to play a vital role in plant development. RNS have been involved in post-translational modification (nitration and *S*-nitrosylation) of protein, nitration of fatty acids and nucleic acids, and regulating reactive oxygen species metabolism. RNS are regulatory molecules in plants that control various physiological processes such as seed development, maturation, dormancy, and germination. Signaling mediated by RNS is primarily concerned with plant hormones that influence acclimatization, pathophysiology, and normal ontogeny (Figure 16.1). RNS also operate as molecular signals and are involved in oxidative damage and tissue dysfunction. They are toxic to plants and eventually kill them by peroxidizing lipid proteins and nucleic acids. RNS include both radicals like nitric oxide (NO), nitric dioxide (NO_2), as well as non-radicals like nitrous acid (HNO_2) and dinitrogen tetroxide (N_2O_4), among others, and they also play a significant role as signaling molecules in response to environmental stress. Similarly, when plants respond to pathogen attacks, NO is a critical mediator working in tandem with reactive oxygen species (ROS). As a result, the importance of RNS metabolism in higher plant physiology under ideal conditions has come to be thoroughly characterized. Seed dormancy and germination,

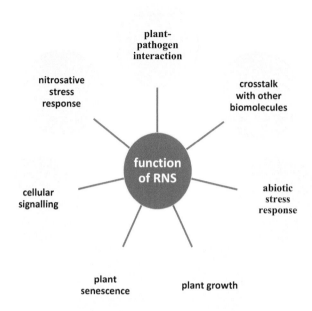

FIGURE 16.1 Function of RNS in plant system.

plant growth and development, stomatal motions, reproduction and pollen tube growth, aging, response to stress conditions, and senescence are all affected by RNS (Del Río, 2015; F. J. Corpas et al., 2008; Farnese et al., 2016).

16.2 SOURCE OF RNS IN PLANTS

RNS are produced in plants by various chemical events that occur in diverse compartments, such as respiration in mitochondria, photosynthesis in chloroplasts, oxidation-reduction reactions in the cytosol, and cellular photorespiration in peroxisomes (Kapoor et al., 2019). The earliest evidence of NOS activity in peroxisomes was found in plant tissues (del Río et al., 2006). Other cell organelles where NO production has been demonstrated, in addition to peroxisomes, are mitochondria and chloroplasts. (Kapuganti J. Gupta and Kaiser, 2010; Jasid et al., 2006). There are both non-enzymatic and enzymatic mechanisms in plants that produce NO. Moreover, plant peroxisomes have been shown to include L-arginine-dependent nitric oxide synthase (NOS) activity and produce RNS such as NO in recent years. S-nitrosoglutathione (GSNO), an intercellular and intracellular NO carrier, can be formed inside peroxisomes, and this RNS has been found in peroxisomes from numerous plant species. Plant peroxisomes can operate as subcellular sensors of plant stress by releasing RNS into the cytoplasm and causing particular changes in defense gene expression (stress signaling) (Del Río, 2015). Peroxisomes in plants have an RNS-mediated metabolic function in leaf senescence and certain types of abiotic stress, and can play a dual role in cells as both oxidative stress producers and RNS signal molecules (del Río et al., 2006) (Figure 16.2).

16.2.1 Nitric Oxide (NO)

Nitric oxide (NO) represents the family of RNS. RNS present in plants comprise mainly NO. It plays a crucial role in plant growth and development as an intercellular and intracellular signaling

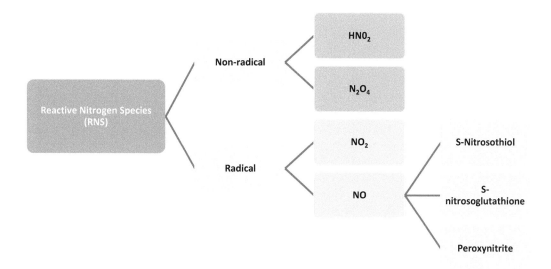

FIGURE 16.2 Understanding RNS.

molecule. NO regulates different processes by stimulating gene transcription or activating secondary messengers (Procházková, Wilhelmová, and Pavlík, 2015). Seed germination, pollen tube growth, cell wall lignification, root organogenesis, establishment and functioning of the legume–Rhizobium symbiosis, flowering, fruit ripening and senescence, and biotic and abiotic stress are among the many functions of NO in plant physiological and pathological processes (Kapoor et al., 2019). Stress causes an increase in ROS and RNS production in plants. Disturbance in metalloid homeostasis of plants alters endogenous NO levels. Plants respond to metalloids, such as boron (B), silicon (Si), selenium (Se), arsenic (As), and antimony (Sb), through complicated signaling pathways primarily mediated by NO. The stress-relieving effect of Si is amplified by NO, whereas the accumulation of NO or RNS adds to toxicity in Se, As, and Sb (Kolbert and Ördög, 2021). The reactive diatomic gas (NO) has been widely researched in plants. NO is a signaling chemical involved in a range of biological processes, according to this research. The discovery of nitric oxide synthases (NOS) and NO-forming nitrite reductase in algae, as well as the first functional studies of these enzymes, has resulted in more knowledge on NO biosynthesis in photosynthetic organisms (Astier et al., 2021). Huang et al. (2021) reported that NO, an endogenous signaling molecule in plants, can improve resilience to abiotic stressors. Studies also showed that nitrogen metabolism is essential for rice stress tolerance (Huang et al., 2021). Like NO and other metabolites, the gas transmitters play a dynamic role in root formation in higher plants, determining cell fate in the primary embryonic meristem. Furthermore, NO along with auxin, indole-3-acetic acid (IAA), ethylene, jasmonic acid (JA), strigolactones, alkamides, ROS, hydrogen sulfide (H2S), and melatonin affects root growth and architecture. Deciphering these interactions could be a useful biotechnological tool for crop management in soils, especially in the face of harsh environmental circumstances (Mukherjee and Corpas, 2020). Both enzymatic and non-enzymatic mechanisms are known to be involved in NO production in plant organs. The most reliable enzymatic sources of NO production in higher plants are currently nitrate reductase and NO synthase-like activity. In addition, NO is produced by inorganic and organic nitrite sources. The reactivity of nitrogen oxides with plant metabolites, polyamines, or the mitochondrial electron transport chain are non-enzymatic routes for NO production in plants (Astier, Gross, and Durner, 2018; Zs Kolbert et al., 2019). Plant roots alter NO production in response to diverse external stimuli, biotic stress, and soil nitrate levels. The involvement of NO and NO-derived compounds in root growth is critical (Sánchez-Vicente, Fernández-Espinosa, and Lorenzo, 2019; Correa-Aragunde, Graziano, and Lamattina, 2004). The temporal synchronization of root development and NO production by L-arginine-dependent NO synthase-like activity has been reported in several studies (F. J. Corpas and Barroso, 2017). Furthermore, the existence of different NO conjugates and RNS (S-nitrosoglutathione, peroxynitrite, N-nitrosomelatonin, etc.) in roots demonstrates that they have an active NO metabolism. Developmental and stress-regulated phases of root growth affect NO biosynthesis and accumulation, as well as other biomolecules. Protonation of nitrite to generate nitrous acid (HNO2), which yields NO and nitrogen dioxide (NO2), is a non-enzymatic NO source. This type of reaction is less likely to occur in the cytoplasm of plant cells since it occurs at a low pH (Begara-Morales et al., 2018).

However, the mechanism of NO formation, on the other hand, is most likely to be found in plant roots' apoplast. The apoplast NO generation is influenced by various factors such as low pH, nitrite permeable transporters, and nitrite. However, the apoplastic nitrite level of root tissues can fluctuate, and depends partly on the rhizosphere's nitrogen turnover rate. Furthermore, phenolics have enhanced NO production in apoplastic areas (Bethke, Badger, and Jones, 2004). Furthermore, root plasma membranes have NR activity, implying that apoplastic NO can be generated via enzymatic and non-enzymatic routes (Domingos et al., 2015). The rate of nitrate assimilation is determined by the range of root-NO levels found in different plant species (Chamizo-Ampudia et al., 2017). Moreover, plant roots create apoplastic and symplastic NO, likely to spread into the rhizosphere. The apical meristematic zone of the root apex, which is covered by the root cap, will most likely create a transient NO flux. Nitrogen availability in the soil, on the other hand, is likely to influence NO generation in this part of the roots (Ma et al., 2020). As a result of soil nitrate levels, nitrate-induced lateral root development occurs simultaneously as cellular NO production (Sun, Tao, et al., 2017). Plant roots generate a high amount of apoplastic NO when under anoxic or biotic stress. This NO production outside the cell shields cellular biomolecules from NO toxicity (Mukherjee and Corpas, 2020). Under physiological and stress situations, NO causes the synthesis of peroxynitrite (ONOO-) and NO conjugates such as S-nitrosoglutathione (GSNO) in the roots of many plant species (Chaki et al., 2009). NO also regulates cellulose and lignin biosynthesis and root tip elongation, lateral root proliferation, and adventitious rooting (F. Corpas and Barroso, 2015). Furthermore, NO can influence the activity of many proteins (endochitinase, alcohol dehydrogenase, fructose-bisphosphate aldolase, peroxidase, and NADP-isocitrate dehydrogenase) involved in redox, carbon, and nitrogen metabolism by post-translational modifications (Begara-Morales et al., 2013). However, accumulating evidence reveals that (Sharma et al., 2019) NO regulates root growth along with different biomolecules like auxin, ethylene, jasmonic acid, and strigolactones (Y. Li et al., 2020; Wei et al., 2020). The NO-ethylene interaction, for example, controls genes involved in Fe acquisition (García et al., 2018). During salt and metal stress, NO-melatonin crosstalk is linked to antioxidant system regulation. (Campos et al., 2019; Zhang et al., 2019). NO also influences lignin composition in *Vigna radiata* roots (Sharma et al., 2019). By changing the lignin composition in cell walls, NO availability influences root architecture and growth. Root growth response mediated by NO is modulated by adverse environmental circumstances such as heavy metal stress and salt stress. NO outbursts and nitro-oxidative stress are likely triggered by cadmium, zinc, and arsenic stress in root tips (Piacentini et al., 2020). External environmental stressors, such as anoxia stress, are expected to increase root cytosolic NR activity, resulting in nitrite depletion. NR activity is primarily responsible for hypoxia or anoxia-induced NO production in roots. Hence, low oxygen levels in roots trigger NO generation by a NOS-independent pathway (Fukao et al., 2019; Kapuganti Jagadis Gupta et al., 2020). The NO level is significant in programmed cell death of plant tissues (Domingos et al., 2015). Also, NO appears to be a positive regulator of root aerenchyma development (Mur et al., 2013). As a result, auxin-NO crosstalk controls root architecture and maintains a balance of ROS levels. In Arabidopsis, a higher NO concentration decreases root meristem development and cell division and changes the auxin efflux protein gradient (Yuan and Huang, 2016; Mukherjee and Corpas, 2020). A significant target of biotechnological applications in organogenesis and adventitious roots appears to be the NO-auxin crosstalk. Higher auxin and NO levels, on the other hand, may restrict primary root growth while promoting lateral root induction (Pagnussat et al., 2002; Sun, Feng, et al., 2017). Auxin stimulates the production of adventitious roots in tea plants via NO and H_2O_2-dependent pathways (WEI et al., 2018). Moreover, strigolactones control the form and architecture of roots, and NO inhibits strigolactone biosynthesis in sunflower roots (Bharti and Bhatla, 2015).

Additionally, Brassinolide also aided the establishment of adventitious roots by stimulating endogenous NO generation in cucumber via NO synthase and NR activity (Y. Li et al., 2020). Because of its single electron, NO functions as a free radical and generates the nitrosonium ion, leading to various NO-metal complexes (Hartsfield, 2002). The inhibitory effect of osmotic stress on seed germination in wheat seeds is reduced by CO–NO crosstalk (Liu et al., 2010). Furthermore, crosstalk between CO and NO in Arabidopsis roots aids in iron deficiency (Yang et al., 2016). The combination of H_2S (nucleophilic) and NO (electrophilic) forms a new intermediate called persulfide, which regulates ROS and RNS concentration in the cell (Mukherjee and Corpas, 2020). After decades of intensive work, NO has been identified as an important regulator of plant development and stress responses. Its route of action as an inducer of posttranslational changes of critical target proteins via cysteine S-nitrosylation and tyrosine nitration is determined by its reactivity as a free radical.

The recurrent synthesis of NO and ROS prevents programmed cell death. ROS and NO play an essential role in this course (Wang, Loake, and Chu, 2013). Further, both in vivo and in vitro approaches can detect NO in plants. It has been quantified by the Griess reagent method and the DAF-FMDA Fluorescence method (Z. G. Li, 2019). NO is vital for intracellular and intercellular signaling and plant growth and development (Kapoor et al., 2019). The development of precise noninvasive technologies for measuring endogenous NO and the implementation of genetic approaches that avoid the use of deceptive pharmacological studies would be crucial for getting significant advances in better knowledge of NO homeostasis and regulatory actions in plants (León and Costa-Broseta, 2020). Also, NO is engaged in long-distance signaling, which promotes abiotic stress adaptation. Plants produce NO quickly in response to a variety of biotic and abiotic stressors. Plant cells respond to various stressors, including pathogen exposure, by producing RNS, such as NO. NO is a free radical that has a variety of roles in plants, including stress physiology (Procházková, Wilhelmová, and Pavlík, 2015).

16.2.2 Peroxynitrite ($OONO^-$)

At physiological pH and temperature, peroxynitrite (OONO-) is a very short-lived RNS that may easily move through biological membranes and interact with target molecules in surrounding cells within a radius of one or two cells (Procházková, Wilhelmová, and Pavlík, 2015). The most potent RNS is OONO- (Del Río, 2015). Low quantities of OONO- are believed to be produced continually in photosynthesizing chloroplasts in plants, but greater levels are likely to be produced under stress, which induces NO production (Vandelle and Delledonne, 2011). Plants have been shown to produce OONO- in response to biotic stress. (Bellin, Delledonne, and Vandelle, 2016). Also, under abiotic stress, OONO- production in Arabidopsis when exposed to cadmium has been reported (F. J. Corpas and Barroso, 2014). OONO- interacts with target molecules in one of two ways. First, the OONO- anion or peroxynitrous acid might directly react with a specific target molecule (e.g., thiol oxidation). Second, peroxynitrous acid can homolyze, releasing nitrogen dioxide and hydroxyl radicals that react with the target molecule. In the absence of targets, the significant result of OONO- decay is nitrate, while secondary radical reactions can also produce nitrite and dioxygen, especially at alkaline pH. In a dose-dependent manner, ONOO- oxidizes chloroplastic proteins and reduces both oxygen evolution and photosystem (PS) II fluorescence production (Procházková, Wilhelmová, and Pavlík, 2015). Similarly, the nitration of tyrosine has been documented as a critical redox signaling process (Bartesaghi and Radi, 2018). ONOO- transduces the NO signal during the hypersensitive defense response by changing protein activity via tyrosine nitration (Frederickson Matika and Loake, 2014). Nitrotyrosine, like lipid peroxidation or protein carbonylation, is frequently utilized in plants as a marker of nitrosative stress during abiotic stress, such as salinity stress (Tanou et al., 2012; Mata-Pérez et al., 2016). The reaction between NO produces RNS OONO-, which has been identified as an effective regulator of cell signaling in plants via the tyrosine nitration process. Tyrosine nitration, which involves attaching a NO2 group to a tyrosine residue, can modify reduced nicotinamide-dinucleotide phosphate (NADPH)-generating systems and hence affect NADPH levels, which is an essential cofactor in cellular redox equilibrium (Begara-Morales et al., 2019). At the same time, OONO- is a nitrating/oxidant species synthesized by the robust reaction between NO and O2 and most commonly found in the organelles like peroxisomes (Arasimowicz-Jelonek and Floryszak-Wieczorek, 2011; F. J. Corpas et al., 2017). Under stress and normal conditions, such as protein nitration and S-nitrosylation, the production of ONOO- in plant tissue causes post-translational modification. The produced ONOO-molecules cause nitrosative changes in plants by synthesizing tyrosine nitration. Furthermore, endogenous nitration at the basal level could serve as a regulator. Several investigations found nitrated proteins in the chloroplast, cytosol, mitochondria, and peroxisomes of leaf cells (subcellular level). (Barroso, Valderrama, and Corpas, 2013; F. J. Corpas et al., 2015).

16.2.3 Nitrosothiols

Nitrosonium and nitroxyl ions are two other RNS described as mediating S-nitrosothiol production. S-nitrosylation appears to have a regulating role in plant physiology, according to a growing body of research. It can, for example, inhibit ethylene biosynthesis by S-nitrosylating S-adenosylmethionine, which is an ethylene precursor (Lindermayr, Saalbach, and Durner, 2005; Moreau et al., 2010). The GSNO family is one of the most common low molecular mass S-nitrosothiols (Broniowska, Diers, and Hogg, 2013). GSNO appears to be a crucial chemical in plant responses to abiotic and biotic stressors. Undertreatment of heavy metals, for example, resulted in a decrease in GSNO level in pea and Arabidopsis (Leterrier et al., 2011; Barroso, Valderrama, and Corpas, 2013). Nitrosoglutathione reductase is a critical enzyme in the regulation of GSNO pools (GSNOR). GSNOR degrades GSNO to glutathione disulfide and ammonia. Due to the widespread presence of GSNOR, it has been proposed that this enzyme works as a nitrosative stress protector rather than a cell signaling factor. (Procházková, Wilhelmová, and Pavlík, 2015). It has been shown that in the presence of O2, NO combines with glutathione via the S-nitrosylation reaction, forming GSNO (S-nitroglutathione), a key reservoir of NO bioactivity mainly found in plant species (Jedelská, Luhová, and Petřivalský, 2020).

16.3 CONCLUSION AND FUTURE OUTLOOK

The importance of RNS in plant biology has been proven. RNS metabolism appears to play an essential function in the growth of plants. Generally, it has been found that when plants are in ecological or abiotic stress, overproduction of RNS results. Different RNSs are derived from NO, and NO is the most extensively studied molecule. RNS has been shown to have an impact on stress reduction, in addition to their essential role in signaling. It also helps plants to develop more sustainably by reducing the adverse effects of various organic contaminants. It acts as a signal for plant growth and development as well as stress defense at low concentrations. RNS production is critical for the regulation of stress acclimation in plants. RNS has gotten much attention in recent years because of its critical function in systems that control plant growth and development in both standard and stressful conditions. Still, there is a scope of RNS metabolism that needs to be better explored. Also, RNS have a crucial role in intercellular and intracellular molecular communication. However, there are many things to learn about the interaction between the pathways mediated by RNS, ROS, cellular redox changes, calcium signaling, hormones, and other messenger molecules.

REFERENCES

Arasimowicz-Jelonek, Magdalena, and Jolanta Floryszak-Wieczorek. 2011. "Understanding the Fate of Peroxynitrite in Plant Cells – From Physiology to Pathophysiology." *Phytochemistry* 72 (8): 681–88. https://doi.org/10.1016/j.phytochem.2011.02.025.

Astier, Jeremy, Inonge Gross, and Jörg Durner. 2018. "Nitric Oxide Production in Plants: An Update." *Journal of Experimental Botany* 69 (14): 3401–11. https://doi.org/10.1093/jxb/erx420.

Astier, Jeremy, Jordan Rossi, Pauline Chatelain, Agnès Klinguer, Angélique Besson-Bard, Claire Rosnoblet, Sylvain Jeandroz, Valérie Nicolas-Francès, and David Wendehenne. 2021. "Nitric Oxide Production and Signalling in Algae." Edited by Zsuzsanna Kolbert. *Journal of Experimental Botany* 72 (3): 781–92. https://doi.org/10.1093/jxb/eraa421.

Barroso, Juan B., Raquel Valderrama, and Francisco J. Corpas. 2013. "Immunolocalization of S-Nitrosoglutathione, S-Nitrosoglutathione Reductase and Tyrosine Nitration in Pea Leaf Organelles." *Acta Physiologiae Plantarum* 35 (8): 2635–40. https://doi.org/10.1007/s11738-013-1291-0.

Bartesaghi, Silvina, and Rafael Radi. 2018. "Fundamentals on the Biochemistry of Peroxynitrite and Protein Tyrosine Nitration." *Redox Biology* 14 (April): 618–25. https://doi.org/10.1016/j.redox.2017.09.009.

Begara-Morales, Juan C., Mounira Chaki, Beatriz Sánchez-Calvo, Capilla Mata-Pérez, Marina Leterrier, José M. Palma, Juan B. Barroso, and Francisco J. Corpas. 2013. "Protein Tyrosine Nitration in Pea Roots during Development and Senescence." *Journal of Experimental Botany* 64 (4): 1121–34. https://doi.org/10.1093/jxb/ert006.

Begara-Morales, Juan C., Mounira Chaki, Raquel Valderrama, Beatriz Sánchez-Calvo, Capilla Mata-Pérez, María N. Padilla, Francisco J. Corpas, and Juan B. Barroso. 2018. "Nitric Oxide Buffering and Conditional Nitric Oxide Release in Stress Response." *Journal of Experimental Botany* 69 (14): 3425–38. https://doi.org/10.1093/jxb/ery072.

Begara-Morales, Sánchez-Calvo, Gómez-Rodríguez, Chaki, Valderrama, Mata-Pérez, López-Jaramillo, Corpas, and Barroso. 2019. "Short-Term Low Temperature Induces Nitro-Oxidative Stress That Deregulates the NADP-Malic Enzyme Function by Tyrosine Nitration in Arabidopsis Thaliana." *Antioxidants* 8 (10): 448. https://doi.org/10.3390/antiox8100448.

Bellin, Diana, Massimo Delledonne, and Elodie Vandelle. 2016. "Detection of Peroxynitrite in Plants Exposed to Bacterial Infection." In *Methods in Molecular Biology*, 1424:191–200. https://doi.org/10.1007/978-1-4939-3600-7_16.

Bethke, Paul C., Murray R. Badger, and Russell L. Jones. 2004. "Apoplastic Synthesis of Nitric Oxide by Plant Tissues." *Plant Cell* 16 (2): 332–41. https://doi.org/10.1105/tpc.017822.

Bharti, Niharika, and Satish C. Bhatla. 2015. "Nitric Oxide Mediates Strigolactone Signaling in Auxin and Ethylene-Sensitive Lateral Root Formation in Sunflower Seedlings." *Plant Signaling & Behavior* 10 (8): e1054087. https://doi.org/10.1080/15592324.2015.1054087.

Broniowska, Katarzyna A., Anne R. Diers, and Neil Hogg. 2013. "S-Nitrosoglutathione." *Biochimica et Biophysica Acta (BBA) – General Subjects* 1830 (5): 3173–81. https://doi.org/10.1016/j.bbagen.2013.02.004.

Campos, Fernanda V., Juraci A. Oliveira, Mayara G. Pereira, and Fernanda S. Farnese. 2019. "Nitric Oxide and Phytohormone Interactions in the Response of Lactuca Sativa to Salinity Stress." *Planta* 250 (5): 1475–89. https://doi.org/10.1007/s00425-019-03236-w.

Chaki, Mounira, Ana M. Fernández-Ocaña, Raquel Valderrama, Alfonso Carreras, Francisco J. Esteban, Francisco Luque, María V. Gómez-Rodríguez, Juan C. Begara-Morales, Francisco J. Corpas, and Juan B. Barroso. 2009. "Involvement of Reactive Nitrogen and Oxygen Species (RNS and ROS) in Sunflower-Mildew Interaction." *Plant and Cell Physiology* 50 (2): 265–79. https://doi.org/10.1093/pcp/pcn196.

Chamizo-Ampudia, Alejandro, Emanuel Sanz-Luque, Angel Llamas, Aurora Galvan, and Emilio Fernandez. 2017. "Nitrate Reductase Regulates Plant Nitric Oxide Homeostasis." *Trends in Plant Science* 22 (2): 163–74. https://doi.org/10.1016/j.tplants.2016.12.001.

Corpas, Francisco J., and Juan B. Barroso. 2014. "Peroxynitrite (ONOO–) Is Endogenously Produced in Arabidopsis Peroxisomes and Is Overproduced under Cadmium Stress." *Annals of Botany* 113 (1): 87–96. https://doi.org/10.1093/aob/mct260.

———. 2017. "Nitric Oxide Synthase-like Activity in Higher Plants." *Nitric Oxide* 68 (August): 5–6. https://doi.org/10.1016/j.niox.2016.10.009.

Corpas, Francisco J., Juan B. Barroso, José M. Palma, and Marta Rodriguez-Ruiz. 2017. "Plant Peroxisomes: A Nitro-Oxidative Cocktail." *Redox Biology* 11 (April): 535–42. https://doi.org/10.1016/j.redox.2016.12.033.

Corpas, Francisco J., Juan C Begara-Morales, Beatriz Sánchez-Calvo, Mounira Chaki, and Juan B. Barroso. 2015. "Nitration and S-Nitrosylation: Two Post-Translational Modifications (PTMs) Mediated by Reactive Nitrogen Species (RNS) and Their Role in Signalling Processes of Plant Cells." In Kapuganti Gupta and Abir Igamberdiev (eds), *Reactive Oxygen and Nitrogen Species Signaling and Communication in Plants. Signaling and Communication in Plants*, vol. 23, 267–81. Cham: Springer. https://doi.org/10.1007/978-3-319-10079-1_13

Corpas, Francisco J., Mounira Chaki, Ana Fernández-Ocaña, Raquel Valderrama, José M. Palma, Alfonso Carreras, Juan C. Begara-Morales, Morad Airaki, Luis A. Del Río, and Juan B. Barroso. 2008. "Metabolism of Reactive Nitrogen Species in Pea Plants under Abiotic Stress Conditions." *Plant and Cell Physiology* 49 (11): 1711–22. https://doi.org/10.1093/pcp/pcn144

Corpas, Francisco, and Juan Barroso. 2015. "Functions of Nitric Oxide (NO) in Roots during Development and under Adverse Stress Conditions." *Plants* 4 (2): 240–52. https://doi.org/10.3390/plants4020240.

Correa-Aragunde, Natalia, Magdalena Graziano, and Lorenzo Lamattina. 2004. "Nitric Oxide Plays a Central Role in Determining Lateral Root Development in Tomato." *Planta* 218 (6): 900–905. https://doi.org/10.1007/s00425-003-1172-7

Domingos, Patricia, Ana Margarida Prado, Aloysius Wong, Christoph Gehring, and Jose A. Feijo. 2015. "Nitric Oxide: A Multitasked Signaling Gas in Plants." *Molecular Plant* 8 (4): 506–20. https://doi.org/10.1016/j.molp.2014.12.010

Farnese, Fernanda S., Paulo E. Menezes-Silva, Grasielle S. Gusman, and Juraci A. Oliveira. 2016. "When Bad Guys Become Good Ones: The Key Role of Reactive Oxygen Species and Nitric Oxide in the Plant Responses to Abiotic Stress." *Frontiers in Plant Science* 7 (APR2016). https://doi.org/10.3389/fpls.2016.00471

Frederickson Matika, Debra E., and Gary J. Loake. 2014. "Redox Regulation in Plant Immune Function." *Antioxidants & Redox Signaling* 21 (9): 1373–88. https://doi.org/10.1089/ars.2013.5679

Fukao, Takeshi, Blanca Estela Barrera-Figueroa, Piyada Juntawong, and Julián Mario Peña-Castro. 2019. "Submergence and Waterlogging Stress in Plants: A Review Highlighting Research Opportunities and Understudied Aspects." *Frontiers in Plant Science* 10 (March). https://doi.org/10.3389/fpls.2019.00340

García, María J., Francisco J. Corpas, Carlos Lucena, Esteban Alcántara, Rafael Pérez-Vicente, Ángel M. Zamarreño, Eva Bacaicoa, José M. García-Mina, Petra Bauer, and Francisco J. Romera. 2018. "A Shoot Fe Signaling Pathway Requiring the Opt3 Transporter Controls Gsno Reductase and Ethylene in Arabidopsis Thaliana Roots." *Frontiers in Plant Science* 9. https://doi.org/10.3389/fpls.2018.01325

Gupta, Kapuganti J., and Werner M. Kaiser. 2010. "Production and Scavenging of Nitric Oxide by Barley Root Mitochondria." *Plant and Cell Physiology* 51 (4): 576–84. https://doi.org/10.1093/pcp/pcq022

Gupta, Kapuganti Jagadis, Luis A.J. Mur, Aakanksha Wany, Aprajita Kumari, Alisdair R. Fernie, and R. George Ratcliffe. 2020. "The Role of Nitrite and Nitric Oxide under Low Oxygen Conditions in Plants." *New Phytologist* 225 (3): 1143–51. https://doi.org/10.1111/nph.15969

Halliwell, Barry, and John M. C. Gutteridge. 2007. *Free Radicals in Biology and Medicine. Clarendon Press, Oxford.* Vol. 26. Oxford University Press. https://doi.org/10.1093/acprof:oso/9780198717478.001.0001

Hartsfield, Cynthia L. 2002. "Cross Talk Between Carbon Monoxide and Nitric Oxide." *Antioxidants & Redox Signaling* 4 (2): 301–7. https://doi.org/10.1089/152308602753666352

Huang, Dandan, Guangqin Jing, Lili Zhang, Changbao Chen, and Shuhua Zhu. 2021. "Interplay Among Hydrogen Sulfide, Nitric Oxide, Reactive Oxygen Species, and Mitochondrial DNA Oxidative Damage." *Frontiers in Plant Science* 12 (August). https://doi.org/10.3389/fpls.2021.701681

Jasid, Sebastián, Marcela Simontacchi, Carlos G. Bartoli, and Susana Puntarulo. 2006. "Chloroplasts as a Nitric Oxide Cellular Source. Effect of Reactive Nitrogen Species on Chloroplastic Lipids and Proteins." *Plant Physiology* 142 (3): 1246–55. https://doi.org/10.1104/pp.106.086918

Jedelská, Tereza, Lenka Luhová, and Marek Petřivalský. 2020. "Thioredoxins: Emerging Players in the Regulation of Protein s-Nitrosation in Plants." *Plants* 9 (11): 1–16. https://doi.org/10.3390/plants9111426

Kapoor, Dhriti, Simranjeet Singh, Vijay Kumar, Romina Romero, Ram Prasad, and Joginder Singh. 2019. "Antioxidant Enzymes Regulation in Plants in Reference to Reactive Oxygen Species (ROS) and

Reactive Nitrogen Species (RNS)." *Plant Gene* 19 (September): 100182. https://doi.org/10.1016/j.plgene.2019.100182

Kolbert, Zs, J.B. Barroso, R. Brouquisse, F.J. Corpas, K.J. Gupta, C. Lindermayr, G.J. Loake, et al., 2019. "A Forty Year Journey: The Generation and Roles of NO in Plants." *Nitric Oxide* 93 (December): 53–70. https://doi.org/10.1016/j.niox.2019.09.006

Kolbert, Zsuzsanna, and Attila Ördög. 2021. "Involvement of Nitric Oxide (NO) in Plant Responses to Metalloids." *Journal of Hazardous Materials* 420 (October): 126606. https://doi.org/10.1016/j.jhazmat.2021.126606

León, José, and Álvaro Costa-Broseta. 2020. "Present Knowledge and Controversies, Deficiencies, and Misconceptions on Nitric Oxide Synthesis, Sensing, and Signaling in Plants." *Plant, Cell & Environment* 43 (1): 1–15. https://doi.org/10.1111/pce.13617

Leterrier, Marina, Mounira Chaki, Morad Airaki, Raquel Valderrama, José M. Palma, Juan B. Barroso, and Francisco J. Corpas. 2011. "Function of S-Nitrosoglutathione Reductase (GSNOR) in Plant Development and under Biotic/Abiotic Stress." *Plant Signaling & Behavior* 6 (6): 789–93. https://doi.org/10.4161/psb.6.6.15161

Li, Yutong, Yue Wu, Weibiao Liao, Linli Hu, Mohammed Mujitaba Dawuda, Xin Jin, Zhongqi Tang, Jianjun Yang, and Jihua Yu. 2020. "Nitric Oxide Is Involved in the Brassinolide-Induced Adventitious Root Development in Cucumber." *BMC Plant Biology* 20 (1). https://doi.org/10.1186/s12870-020-2320-y

Li, Zhong Guang. 2019. "Measurement of Signaling Molecules Calcium Ion, Reactive Sulfur Species, Reactive Carbonyl Species, Reactive Nitrogen Species, and Reactive Oxygen Species in Plants." In *Plant Signaling Molecules: Role and Regulation under Stressful Environments*, 83–103. https://doi.org/10.1016/B978-0-12-816451-8.00005-8

Lindermayr, Christian, Gerhard Saalbach, and Jörg Durner. 2005. "Proteomic Identification of S-Nitrosylated Proteins in Arabidopsis." *Plant Physiology* 137 (3): 921–30. https://doi.org/10.1104/pp.104.058719

Liu, Yahui, Sheng Xu, Tengfang Ling, Langlai Xu, and Wenbiao Shen. 2010. "Heme Oxygenase/Carbon Monoxide System Participates in Regulating Wheat Seed Germination under Osmotic Stress Involving the Nitric Oxide Pathway." *Journal of Plant Physiology* 167 (16): 1371–79. https://doi.org/10.1016/j.jplph.2010.05.021

Ma, Ming, David Wendehenne, Laurent Philippot, Robert Hänsch, Emmanouil Flemetakis, Bin Hu, and Heinz Rennenberg. 2020. "Physiological Significance of Pedospheric Nitric Oxide for Root Growth, Development and Organismic Interactions." *Plant Cell and Environment* 43 (10): 2336–54. https://doi.org/10.1111/pce.13850

Mata-Pérez, Capilla, Juan C. Begara-Morales, Mounira Chaki, Beatriz Sánchez-Calvo, Raquel Valderrama, María N. Padilla, Francisco J. Corpas, and Juan B. Barroso. 2016. "Protein Tyrosine Nitration during Development and Abiotic Stress Response in Plants." *Frontiers in Plant Science* 7 (NOVEMBER2016). https://doi.org/10.3389/fpls.2016.01699

Moreau, Magali, Christian Lindermayr, Jörg Durner, and Daniel F. Klessig. 2010. "NO Synthesis and Signaling in Plants – Where Do We Stand?" *Physiologia Plantarum* 138 (4): 372–83. https://doi.org/10.1111/j.1399-3054.2009.01308.x

Mukherjee, Soumya, and Francisco J. Corpas. 2020. "Crosstalk among Hydrogen Sulfide (H2S), Nitric Oxide (NO) and Carbon Monoxide (CO) in Root-System Development and Its Rhizosphere Interactions: A Gaseous Interactome." *Plant Physiology and Biochemistry* 155 (October): 800–814. https://doi.org/10.1016/j.plaphy.2020.08.020

Mur, Luis A.J., Julien Mandon, Stefan Persijn, Simona M. Cristescu, Igor E. Moshkov, Galina V. Novikova, Michael A. Hall, Frans J.M. Harren, Kim H. Hebelstrup, and Kapuganti J. Gupta. 2013. "Nitric Oxide in Plants: An Assessment of the Current State of Knowledge." *AoB PLANTS* 5. https://doi.org/10.1093/aobpla/pls052

Pagnussat, Gabriela Carolina, Marcela Simontacchi, Susana Puntarulo, and Lorenzo Lamattina. 2002. "Nitric Oxide Is Required for Root Organogenesis." *Plant Physiology* 129 (3): 954–56. https://doi.org/10.1104/pp.004036

Piacentini, Diego, Federica Della Rovere, Adriano Sofo, Laura Fattorini, Giuseppina Falasca, and Maria Maddalena Altamura. 2020. "Nitric Oxide Cooperates With Auxin to Mitigate the Alterations in the Root System Caused by Cadmium and Arsenic." *Frontiers in Plant Science* 11 (August). https://doi.org/10.3389/fpls.2020.01182

Procházková, Dagmar, Naďa Wilhelmová, and Milan Pavlík. 2015. "Reactive Nitrogen Species and Nitric Oxide." In *Nitric Oxide Action in Abiotic Stress Responses in Plants*, 3–19. Cham: Springer International Publishing. https://doi.org/10.1007/978-3-319-17804-2_1

Río, Luis A. del, Luisa M. Sandalio, Francisco J. Corpas, José M. Palma, and Juan B. Barroso. 2006. "Reactive Oxygen Species and Reactive Nitrogen Species in Peroxisomes. Production, Scavenging, and Role in Cell Signaling." *Plant Physiology* 141 (2): 330–35. https://doi.org/10.1104/pp.106.078204

Río, Luis A. Del. 2015. "ROS and RNS in Plant Physiology: An Overview." *Journal of Experimental Botany* 66 (10): 2827–37. https://doi.org/10.1093/jxb/erv099

Sánchez-Vicente, Inmaculada, María Guadalupe Fernández-Espinosa, and Oscar Lorenzo. 2019. "Nitric Oxide Molecular Targets: Reprogramming Plant Development upon Stress." *Journal of Experimental Botany* 70 (17): 4441–60. https://doi.org/10.1093/jxb/erz339

Sharma, Sangeeta, Harminder Pal Singh, Daizy Rani Batish, and Ravinder Kumar Kohli. 2019. "Nitric Oxide Induced Modulations in Adventitious Root Growth, Lignin Content and Lignin Synthesizing Enzymes in the Hypocotyls of Vigna Radiata." *Plant Physiology and Biochemistry* 141: 225–30. https://doi.org/10.1016/j.plaphy.2019.05.028

Sun, Huwei, Fan Feng, Juan Liu, and Quanzhi Zhao. 2017. "The Interaction between Auxin and Nitric Oxide Regulates Root Growth in Response to Iron Deficiency in Rice." *Frontiers in Plant Science* 8. https://doi.org/10.3389/fpls.2017.02169

Sun, Huwei, Jinyuan Tao, Quanzhi Zhao, Guohua Xu, and Yali Zhang. 2017. "Multiple Roles of Nitric Oxide in Root Development and Nitrogen Uptake." *Plant Signaling & Behavior* 12 (1): e1274480. https://doi.org/10.1080/15592324.2016.1274480

Tanou, Georgia, Panagiota Filippou, Maya Belghazi, Dominique Job, Grigorios Diamantidis, Vasileios Fotopoulos, and Athanassios Molassiotis. 2012. "Oxidative and Nitrosative-Based Signaling and Associated Post-Translational Modifications Orchestrate the Acclimation of Citrus Plants to Salinity Stress." *Plant Journal* 72 (4): 585–99. https://doi.org/10.1111/j.1365-313X.2012.05100.x

Vandelle, Elodie, and Massimo Delledonne. 2011. "Peroxynitrite Formation and Function in Plants." *Plant Science* 181 (5): 534–39. https://doi.org/10.1016/j.plantsci.2011.05.002

Wang, Yiqin, Gary J. Loake, and Chengcai Chu. 2013. "Cross-Talk of Nitric Oxide and Reactive Oxygen Species in Plant Programed Cell Death." *Frontiers in Plant Science* 4 (Aug). https://doi.org/10.3389/fpls.2013.00314

WEI, Kang, Li-yuan WANG, Li RUAN, Cheng-cai ZHANG, Li-yun WU, Hai-lin LI, and Hao CHENG. 2018. "Endogenous Nitric Oxide and Hydrogen Peroxide Detection in Indole-3-Butyric Acid-Induced Adventitious Root Formation in Camellia Sinensis." *Journal of Integrative Agriculture* 17 (10): 2273–80. https://doi.org/10.1016/S2095-3119(18)62059-3

Wei, Lijuan, Meiling Zhang, Shouhui Wei, Jing Zhang, Chunlei Wang, and Weibiao Liao. 2020. "Roles of Nitric Oxide in Heavy Metal Stress in Plants: Cross-Talk with Phytohormones and Protein S-Nitrosylation." *Environmental Pollution* 259. https://doi.org/10.1016/j.envpol.2020.113943

Yang, Liming, Jianhui Ji, Hongliang Wang, Karen R. Harris-Shultz, Elsayed F. Abd_Allah, Yuming Luo, Yanlong Guan, and Xiangyang Hu. 2016. "Carbon Monoxide Interacts with Auxin and Nitric Oxide to Cope with Iron Deficiency in Arabidopsis." *Frontiers in Plant Science* 7 (Mar 2016). https://doi.org/10.3389/fpls.2016.00112

Yuan, Hong Mei, and Xi Huang. 2016. "Inhibition of Root Meristem Growth by Cadmium Involves Nitric Oxide-Mediated Repression of Auxin Accumulation and Signalling in Arabidopsis." *Plant Cell and Environment* 39 (1): 120–35. https://doi.org/10.1111/pce.12597

Zhang, Jiarong, Dongxu Li, Jian Wei, Wenna Ma, Xiangying Kong, Zed Rengel, and Qi Chen. 2019. "Melatonin Alleviates Aluminum-Induced Root Growth Inhibition by Interfering with Nitric Oxide Production in Arabidopsis." *Environmental and Experimental Botany* 161 (May): 157–65. https://doi.org/10.1016/j.envexpbot.2018.08.014

17 Nitrogen Toxicity in Plants, Symptoms, and Safeguards

Summia Rehman, Humara Fayaz, Ishfaq Ul Rehmaan, Kausar Rashid, and Sufiya Rashid

1Department of Botany, University of Kashmir, Srinagar, J&K, India
*Email: sumaiyarehman348@gmail.com

CONTENTS

17.1 Introduction 213
17.2 Nitrogen Deficiencies 214
17.3 Nitrogen Toxicity 214
    17.3.1 Toxicity of Excess N 214
17.4 Toxic Effects of Different Nitrogen Forms on Plants 215
    17.4.1 NH_4^+ Toxicity 215
        17.4.1.1 Effects of NH_4^+ on Plant Development 215
        17.4.1.2 Biochemical and Physiological Changes during NH_4^+ Toxicity 217
    17.4.2 NO_2^- Toxicity 219
    17.4.3 NO_3 Toxicity 220
17.5 Alleviation of NH4+ Toxicity 220
References 221

17.1 INTRODUCTION

Nitrogen (N) is among the foremost copious fundamental nutrients on Earth (Cesco et al., 2010), and is a key limiting factor in plant growth due to its limted availability (Graham and Vance, 2000; Hussain et al., 2016). Plants can obtain N from different molecules (such as nitrates, ammonium, urea, and amino acids) and are used for different metabolic purposes, including protein, nucleic acid production, and storage and signaling molecules (McAllister et al., 2012). Nitrate nitrogen (No_3) and ammonium nitrogen (NH4+) are the main nitrogen sources of plants, and the optimal conditions for enzyme activity involved in the conversion of inorganic N to organic N are critical to plant biomass accumulation, growth, and final productivity. In plants, easy-to-move NO_3 can be stored in vacuoles and is the main source of nitrogen in aerated aerobic soil conditions. However, it must be reduced to NH4+, to synthesize proteins and other organic compounds in plants (Garnett et al., 2009). Nitrate reductase (NR) converts NO_3 to nitrite in the non-organelle portion of the cytoplasm. All living plant cells can reduce NO_3 nitrite, use energy and reducing agents (NADH, NADPH) for photosynthesis and/or respiration in green tissue, and respiration in non-green tissue and roots. Ammonium is the only reduced form of N available for plant assimilation into N-carrying amino acids such as glutamic acid (Glu), glutamine (Gln), aspartic acid, and asparagine (Dadhich and Meena, 2014). For the biosynthesis of these N-amino acids, various enzymes, such as Glu synthase, Gln synthase (GS), glutamic acid dehydrogenase (GDH), alanine aminotransferase, aspartic acid aminotransferase, and asparagine synthase are important (Garnett et al.,2009).

17.2 NITROGEN DEFICIENCIES

The initial visual symptoms of nitrogen deficiency are due to limited chlorophyll synthesis, and the leaves are pale and yellow-green. Although chlorophyll is an indicator of nitrogen deficiency, the lack of chlorophyll itself is not necessarily harmful to plant growth. According to reports, soy mutant (glycine max L. Merr.) chlorophyll-deficient plants can maintain a normal plant equivalent rate of photosynthesis (Petrigrew et al., 1989). Since N is a mobile nutrient (Tucker, 1984), symptoms of nitrogen deficiency are most noticeable on older leaves. In addition to visual color symptoms, nitrogen deficiency leads to the formation of leaflets and reduced branching, tillering, and new shoots in various plant species. Total nitrogen analysis best reflects the long-term nitrogen state of plants, while the analysis of NO_3-plant part or wood pulp provides a reasonable indicator of the current nitrogen availability. In addition to being actively fixed N2 symbiosis some species use NHl as the main nitrogen form in legumes. N participation in many cellular components determines that the lack will have a profound impact on the growth and yield of crop plants.

17.3 NITROGEN TOXICITY

Of the inorganic nitrogen forms that may be encountered at roots in planting systems, NH_4^+ is most toxic to many plants when it is the only nitrogen source (Goyal & Huffaker, 1984). Although NHl is an intermediate of mineral in assimilation, it is toxic to many plants because the absorption may exceed the rate of assimilation, leading to the accumulation of NH_4^+ concentrations that are usually not present in plant cells. This accumulation may be due to NHl entering the cytoplasm after absorption, while the primary NH_4^+ assimilation capacity in the plastid. It is often difficult to distinguish the toxic effects of NHl concentrations from changes induced by the pH of the medium. In general, NHl is more harmful to the growth of roots than the growth of shoots (Haynes & Goh, 1977). This is because very little N is converted to NH_4^+, so the concentration of the roots is higher than that of the buds. Ammonium toxicity in the field conditions may not be as severe as seen from solution culture experiments. This is because the soil can buffer the pH changes that occur in NHl nutrition and most soils have some NO^{3-}, which can modify NHl assimilation. Some plant species prefer NHl nutrition NO^{3-}, including rice (Oryza sativa L.) (Wahhab & Bhatti, 1957) and many plants of the Ericaceae family, such as cranberries and blueberries (Haynes & Goh, 1978). Current literature suggests that more monocotyledonous plant species than dicotyledonous plant species are tolerant of NH_4^+ as the only N source, although this still involves only a limited number of species. When plant roots encounter strip anhydrous NH_3, ammonium is most likely to be toxic to field crops; however, this usually does not cause damage to crop production, because the roots cannot penetrate these bands, until due to microbial nitrification caused by NH_3 concentration decreases. Although NHl toxicity can be shown under controlled conditions, in the field conditions are not considered a serious problem. An exception is when the seedling line is immediately planted on a shallow belt of anhydrous NH_3. Most plants tolerate high levels of NO^{3-} without any physiological disorders, and nO^{3-} toxicity is rarely encountered in most agricultural conditions. If NO^{3-} metabolism has been blocked, for example, for plant mutants selected for nitrate reductase (NR) deficiency (Feenstra & Jacobsen, 1980), NO^{3-} can accumulate to toxic levels. It has been reported that the higher toxicity of nitrates is visible in herbs (Fragariaananassa Ouch.) and plants characterized by the browning of leaf margins and purple shades (Jackson, 1972). Symptoms of iron deficiency are associated with NO_3 toxicity that is due to Fe chelation produced by excess organic acid anions to NO_3 nutrition.

17.3.1 Toxicity of Excess N

The effects of excess N on most plants are well known. Excess nitrogen can cause many plants to grow vigorously dark green, but there are some characteristic changes in the development pattern, such as elongation of nutrition, slow maturation, elongation of the entire plant life cycle, increased fleshy. In food crops, plants can grow unusually tall, which increases the risk of food loss due to lodging. Plants may become more vulnerable to diseases and pests due to increased fleshy quality. In fruit plants, vegetative growth may be stimulated rather than flower or fruit set and development (Mills & Jones, 1979). Excess N may also change the biochemistry of plants. Beets grown at high levels of N accumulate less sugar (Mills & Jones, 1979). Excess N may induce a lack of S in some plants (Beaton et al., 1971). In coffee plants, an excess of N can lead to an imbalance of sand Mn, which in turn adversely affects protein synthesis (Muller, 1966). The level of N that may be excessive depends largely on the species. For some crops, the level of N in plant tissue has been defined, beyond the level where it may be excessive. A total N level of apples of 2.2% can reduce fruit quality and yield (Boynton, 1966). Embleton and Jones (1966) found that the most productive range of avocados is the change in leaf N content from 1.6 to 2.00/0. Relative growth of spinach plants declined sharply when the leaf NO_3-N content rose by more than 0.9% on a dry weight basis (Maynard & Barker, 1971). Some potato varieties are less resistant to nitrogen because, under the influence of high nitrogen, they tend to produce more transport and relatively less tuber growth (de Geus, 1973). The application of nitrogen from 40 to 45 kg/ha to horse gram and some other tropical legumes causes vegetation to grow very exuberant, but significantly reduces the setting of flowers and pods (personal observation, S. Goyal). The "critical concentration concept" (Ulrich, 1952) is a guide to N fertilization, which will help ensure maximum yield and avoid over-fertilization.

17.4 TOXIC EFFECTS OF DIFFERENT NITROGEN FORMS ON PLANTS

17.4.1 NH_4^+ TOXICITY

In the planting system, of all the inorganic nitrogen forms that the plant root system may encounter, NH_4^+ is probably most toxic to many plants. The influence of literature on NH_4^+ nutrition for plant growth and processes is significant. Although NH_4^+ is an intermediate for the assimilation of mineral N, it is toxic to many plants when it is the only exogenous source. Ammonium provided as an exogenous substrate presents a different situation than NH_4^+ being generated internally from NO_3^- reduction and light respiration. Using NO_3^- as a substrate, the steady-state concentration of NH_4^+ in cells is quite low (Goyal & Huffaker, 1981), as NH_4^+ is utilized almost as fast as it is formed. When NH_4^+ is applied to cells at concentrations that are usually not present, a large amount of NH_4^+ can be absorbed (Goyal, 1974). Many physiological and biochemical effects of NH_4^+ toxicity have been reported, but the actual cause of NH_4^+ toxicity is not yet known.

17.4.1.1 Effects of NH_4^+ on Plant Development

Growth may best indicate the overall performance of plants in their environment. Plant species vary widely in terms of growth due to nutrition. The effect of NH_4 on plant growth can usually be characterized as "very harmful" to "necessary," depending on the plant species. However, most crop species react negatively to NH_4^+ nutrition, and exhibit reduced growth rate and foliar damage. NH_4^+ as the only source of N is harmful to the growth of beans, sweet corn, cucumbers, peas (Maynard & Barker, 1979), and radish plants (Goyal, 1974). The dry weight of all plant parts of lima beans grown in solution culture is always low when provided in the form of NH_4^+ about 250 (McElhannon & Mills, 1978). In addition, the dry weight of southern pea roots, stems, leaves, pods, seeds, and total plant characters were significantly reduced in three different stages of development when grown in solution culture where NH_4^+ is the only source of N. In fact, 75% of N must be provided as NH_4^+ for maximum growth (Sasseville & Mills, 1979). Dirr et al. (1973) reported that when cultured on NH_4^+ the number and length of Leucothoe catesbaei shoots were lower than the number and length when cultured on NO_3^-. Ammonia nitrogen reduces growth at the top and roots of tomatoes compared to NO_3^- N (Torres De Claassen & Wilcox, 1974). The fresh and dry weight of pea and cucumber shoots grown in NH_4^+ N was significantly lower than that of shoots grown at equivalent concentrations of NO_3^- N (Barker & Maynard, 1972). In addition, Hohlt et al. (1970), during trials of several cultivated species in the family Solanaceae for NH^+ tolerance, found that tobacco is most tolerant to $(NH_4)_2SO$ applications, with tomato being the susceptible one. Barker et al. (1970) reported that the germination of cucumber seeds was lower on O.I N $(NH_4)_2SO$, compared with the same concentration of K_2SO4, even if the percentage of the emergence of seed primary roots was not affected. This indicates that NH_4^+ is toxic only after being physiologically absorbed by the root. Gaseous NH_3 released from the fertilizer strip can also lead to NH_3 toxicity (Bennett & Adams, 1970b). Maize seeds are continuously exposed to the partial vapor pressure of NH_3 as low as 0.063 mm Hg at the initial stage of germination, greatly compromising the early physiological development of maize; when exposure time increases NH_3 is toxic at lower levels (Allred & Ohlrogge, 1964). When NH_4^+ replaces $N0_3^-$ at equivalent concentrations, the medium does not support the growth of excised tomato roots (Robbins & Schmidt, 1938) or groundsel roots (Skinner & Street, 1954). In general, NH_4^+ seems to be more harmful to root growth than shooting (Haynes & Goh, 1977). Ammonium ions are ineffective as an N source for callus growth (Heller, 1954). From a practical point of view, the toxicity of NH_4^+ in field conditions may not be as severe as it seems from solution culture experiments. There are many possible reasons for this as the soil can act as a buffer for pH changes; therefore, root media acidification due to NH_4^+ absorption may be small, and NH_4^+ nitrification may occur, which constantly reduces NH_4^+ concentrations, and most of the soil at any time has a certain $N0_3$, which can change the assimilation of NH_4^+ and the root can choose to avoid NH_4^+ concentrated area and can be close to it after the start of nitrification. While NH_4^+ nutrition is toxic to many higher plants, it provides N better than $N0_3$ for many other plants. Ammonium is superior to $N0_3^-$ as an N source for rice (Wahhab & Bhatti, 1957), rhododendrons (Colgrove & Roberts, 1956), and blueberries (Cain, 1952). Greidanus et al. (1972) report that NH_4^+ is "necessary" for cranberry growth; the plant is unable to absorb $N0_3^-$ from soil or nutrient cultures and lacks the genes needed to induce nitrogen reductase enzyme. Many plants in the family Ericaceae (e.g., blueberries, cranberries) grow mainly in acidic soils, known as "calcification" ("acid-loving"), growing better at $NH4^+$ than $N0_3^-$ (Haynes & Goh, 1978). Another group of plants, apparently able to grow for a long time in flooded conditions (e.g., rice), also seems to prefer NH_4^+ as compared to $N0_3^-$. Wheat and ryegrass initially grew better and absorbed more N when fertilized with NH_4^+ than when fertilized with $N0_3^-$ (Spratt & Gasser, 1970). When growth and nitrogen uptake are fastest, wheat grows faster when fertilized with NH_4^+ than when fertilized with $N0_3^-$. In general, naturally grown plants that occur with little or no nitrification seem to prefer (or at least tolerate) NH_4^+ over $N0_3^-$. Since these plants grow naturally under acidic or flooded conditions, it is logical to assume that plants evolve under "NO_3^- free" conditions and therefore adapt well to the N form NH_4^+ that dominates these soils. Since inhibition of NH_4^+ nitrification is now feasible in many regions, future genetic manipulation of species that are currently not suitable for NH_4^+ nutrition to improve the ability to absorb NH_4^+ may be an important means of more efficient use of fertilizer nitrogen. The report shows that NH_4^+ as a good N source is not uncommon. Blair et al. (1970) found no difference in yield at the top of roots of maize grown in a low (2 mm) concentration of NO_3^- or NH_4^+ solution. The fresh and dry weight of molasses plants cultured on $N03^-$ or $NH4^+$ were very similar (Breteler, 1973). Morris and Giddens (1963) reported that there was no significant difference in plant weight due to NH_4^+ or $N0_3^-$ nutrition of cotton, corn, grain sorghum, and Coastal Dog root grown in soil. Whether NH_4^+ or $N0_3^-$ as the only source of N, the leaf body of *Spirodela oligorrhiza* grows equally well (Ferguson & Bollard, 1969). Many reports show that NH_4^+ together with $N0_3^-$ provides a more beneficial effect than any form alone. Mohanty and Fletcher (1976) reported that "Paul's Scarlet" Rose cells in suspended culture increased twice in a medium containing 25 mm $N0_3^-$. Wheat yield grown in NH_4^+ plus $N0_3^-$ is higher than any source grown alone in a continuous flow culture system (Cox & Reisenauer, 1973). Weissman (1964) reports that dry weight, total protein content, protein concentration, and protein percentage of total N are higher in sunflower leaves grown on

NH_4^+ plus NO_3^- than any one source alone. Gamborg (1970) reported that soy cells in suspended cultures did not grow on NO_3^- unless supplemented with NH_4^+ or glutamine. Current literature suggests that more monocotyledonous plant species can tolerate NH_4^+ than dicotyledonous plants. But we have not seen reports of successful cultivation of dicotyledonous plants in NH_4^+ as the only source of N (except calciferous bacteria). The reason why monocotyledonous plants are more tolerant of NH_4^+ than dicotyledonous plants is not clear.

17.4.1.2 Biochemical and Physiological Changes during NH_4^+ Toxicity

17.4.1.2.1 Photosynthesis

The physiological and biochemical effects on plants contrast with the effects caused by NO_3^-. The increase in NH_4^+ concentration in plant tissue is one of the first and essential elements of NH_4^+ toxicity (Goyar, 1974). The consequences of this are multifaceted. Ammonium ions act as a decoupling agent for photophosphorylation in isolated chloroplasts (Avron, 1960), increasing the low-energy forms of adenine nucleotides (AMP and ADP) and a simultaneous reduction in the amount of TP (Losada et al., 1973). No significant grana were found in tomato chloroplasts after 4 weeks of NH_4^+ nutrition, although mitochondria in the same cells seemed normal (Puritch & Barker, 1967). However, recent findings suggest that NH_4^+ increases photosynthetic CO_2 fixation in spinach isolated cells and Papaver somniferum (Paul et aI., 1978) and complete spinach chloroplasts. NH_4^+ stimulation of photosynthesis is described as activating RuBP carboxylase (Benedetti et al., 1976). However, Heath and Leech (1978) explained that this is caused by a change in the pH of the matrix, as increasing the pH of the medium from 7.6 to 8.2 reduces NH_4^+ induced photosynthetic stimulation. As the external pH value increases further, NH_4^+ inhibits CO_2 fixation. In some studies, NH_4^+ does not stimulate photosynthetic CO_2 fixation; for example, in isolated cotton cells, NH_4^+ in the presence of 5 mm $Ca(NO_3)_2$ and 7.5 H does not affect the total photosynthetic CO_2 fixation (Rehfeld and Jensen, 1973). In addition, Platt et al. (1977) found that NH_4^+ does not affect the total CO_2 fixation in the leaf disc of alfalfa plants grown on Hoagland solution. In studies where NH_4^+ stimulated CO_2 fixation, there were no other sources of reduced or reducible N, such as glutamine. Stimulated CO_2 fixation may be due to the creation of additional sinks for C and the response to reduced N because stimulation occurs only when there is no other reduced N source available. However, Heath and Leech (1978) showed that NH_4^+-induced stimulation of isolated CO_2 fixation in chloroplasts can be replicated by chlorinating NH_4Cl with methylamine.

17.4.1.2.2 Breathing

If NH_4^+ adversely affects photosynthesis, breathing can also be a predictable change. In general, the addition of N (via NH_4^+ or NO_3^- salt) to N hungry algae or higher plants is reported to cause an acceleration of respiratory rate (Hattori, 1958). Reports comparing respiration rates during NH_4^+ or NO_3^- assimilation are contradictory and often confusing. Both the gaseous NH_4 and the undissociated NH_3 inhibit respiration of barley root (Vines & Wedding, 1960) and are considered effective inhibitory forms. NH_4^+ ammonia does not inhibit NADH oxidation in red beetroot homogenate (Wedding & Vines, 1959). In addition, NH_3 can specifically inhibit the oxidation of NADH, thereby blocking the respiratory electron transport chain, which is considered to be part of the cause of NH_3 toxicity (Vines & Wedding, 1960). Some authors dispute the findings of NH_4^+ inhibition of respiration. Burkhart (1938) concluded that several yellowing seedlings absorb and utilize NH_4^+ until the available carbohydrates are exhausted, accelerating carbohydrate decomposition due to enhanced respiration during rapid NH_4^+ assimilation (Syrett, 1956a). Wakiuchi et al. (1971) reported that the respiration rate of cucumber leaves grown in 14.3 mm NH_4Cl was higher than that of cucumber leaves grown in 1.43 mm NH_4Cl, which was not due to mitochondrial decoupling (Matsumoto et al., 1971b). Ammonium also slightly increases respiration in separating barley leaves (Berner, 1971). Accelerating respiration and carbohydrate catabolism during NH_4^+ toxicity may require the supply of organic acids, especially α-Kg, to counteract NH_4^+ (Givan, 1979). In contrast, the respiration rate of dark beans (Barker et al., 1965) and radish leaves (Goyal, 1974) from plants grown in NO_3^- or NH_4^+ is the same.

17.4.1.2.3 Compounds Containing Nitrogen

The plant that grows in NH_4^+ always contains higher levels of free NH_4^+ and amide N ratio as compared to plants grown in NO_3^- (Kato, 1980). Amide N contributes 71% of total N in the secretions of sunflower plants grown in NH_4^+ and 19% NO_3^- (Weissman, 1964). Weissmann also reported that NH_4^+ nutrition favored alanine, arginine, leucine, serine, and valine in secretions, while NO_3^- nutrition favored, γ- aminobutyric acid, aspartic acid, glutamic acid, and lysine. Grown in suspension culture and provided with NO_3^- plus NH tobacco cells contained glutamine and alanine 50–100 times more than contained in NO_3^- alone (Bergmann et al., 1976). De Kock and Kirkby (1969) report that NH_4^+-fed buckwheat and mustard plants contain more free amino acids than NO_3^-, although different N forms do not cause differences in total N content. Hoff et al. (1974) also found elevated NH_4^+ nutrition levels of most free amino acids in the tomato plant part, with particularly high glutamic acid and aspartic acid and their amides. Ammonium nutrition enhances the production of asparagine and hinders its conversion to arginine in young apple trees (Tromp & Ovaa, 1979). However, Schrader et al. (1972) did not find any difference in the amino acid composition and total protein content of corn leaves due to NH_4^+ or NO_3^--nutrition. Alfalfa discs floating on buffer containing NH.Cl photosynthesizing with $^{14}CO_2$ produces more labels in glutamine, glutamic acid, aspartic acid, alanine, glycine, serine, and UDPG then did discs floating on buffer without NH.Cl (Pratt and Al., 1977). NH^+ nutrition is used to explain the rise in free amino acid levels. It was suggested that NH_4^+ plays a regulatory role in the transfer of C flow from carbohydrate biosynthesis to amino acid synthesis, presumably by activating pyruvate kinase and PEP carboxylase in the anaplerotic pathway (Paul et al., 1978).

17.4.1.2.4 Biocatalytic Activity

The activity of enzymes may change due to significant changes in the biochemical composition of plants when NH_4^+ is the source of N. The accumulation and depletion of free sugars (glucose and UDPG) and starches in cucumber leaves during NH_4^+ toxicity were linked to the reduction in the in vitro activity of granule-bound starch synthetase (Matsumoto et al., 1971a). When UDPG is a glucose donor, ammonium added to the reaction medium also inhibits starch synthase activity. In in vitro NH_4^+ injured cucumber plants, the enzyme GDH and all enzymes operating in glycolysis, TCA circulation, and mitochondrial respiratory chain were higher in activity (except aldolase and ATPase) (Matsumoto et al., 1971b). The highest increase (five-fold) was found in phosphate fructose kinase. These researchers concluded that when toxic NH_4^+ is applied to cucumber plants, starch synthesis is impaired, but carbohydrate catabolism accelerates to meet the higher demand for C skeleton to detoxify NH_4^+. However, they did not explain how NH_4^+ -injured plants are expected to have additional glucose converted to starch under increased respiratory activity unless photosynthesis increases accordingly. Although NH_4^+ inhibits starch synthesis, this does not seem to explain the NH_4^+ toxicity of cucumber plants. Apple enzyme activity is 3–4 times higher than the activity of the enzyme in NO_3^- fed cells alone (Bergmann et al., 1976). This may explain the reported lower levels of malic acid due to NH_4^+ nutrition. In suspension culture containing NO_3^- and NH_4^+ GDH and Glu synthase activity of "Paul's Scarlet" Rose cells were higher than the activity in those growing with NO_3^- alone. However, the opposite is true of gin synthase (Mohanty &Fletcher, 1980). The enzymatic potential of gin synthase and GDH in vivo greatly exceeds the actual rate of N assimilation. Therefore, it was concluded that these enzymes do not limit nitrogen assimilation, rather Glu synthase may have limited nitrogen assimilation, by transferring N to GDH. The activity of GDH in plants grown in NH_4^+ nutrition is higher than that of plants grown in NO_3^- (Goyar, 1974). Gamborg and Shyluk (1970) report that icdh and gin synthase activity in soy cells grown in NH_4^+ is lower than in cells grown in NO_3^-, but taking into account other observations, they conclude that lower levels of these enzymes do not limit the utilization of NH_4^+. The results of studies on ICDH in radish plants are also similar (Goyal, 1974). Sunflower and soybean roots that grew at NH_4^+ have higher and lower activity, respectively, than those grown at NO_3^- gin synthase (Weissman, 1972). G-6-PDH activity of soybeans and sunflower roots grown in NO_3^- is higher than the g-6-PDH activity of soybeans and sunflower roots grown in NH_4^+. However, the leaves of both species exhibit the same gin synthase and G-6-PDH activity in either form of N (Weissman, 1972). In summary, NH_4^+ nutrition alters the activity of many enzymes, and the effects vary from species to species. There is no conclusive evidence, however, that the activity of a particular enzyme limits NH_4^+ utilization when it is toxic.

17.4.2 NO_2^- TOXICITY

Nitrite is an intermediate product of NH_4^+ conversion to NO_3^- and NO_3^- conversion to N2 in soil, NO_3^- conversion to NH_4^+ in plants. It usually does not accumulate significantly in soil or plants. Some soils accumulate a large amount of NO_2^- especially if their pH is neutral to alkaline and fertilized in large quantities with urea or ammonia fertilizer (Chapman & Liebig, 1952), or if they are modified by sewage sludge (Yoneyama and Yoshida, 1978). Some NO_2^- may also accumulate in the soil during anaerobic processes. Therefore, the exposure of plants to the NO_2^- in the environment is not entirely theoretical. Although NO_2^- is generally considered harmful to higher plants, its toxicity is a function of many factors, and is Ph and concentration-dependent. A higher concentration of NO_2^- is toxic to avocado and citrus seedlings (Curtis, 1949) and tomato and barley plant growth (Bingham et al., 1954). According to Bingham et al. (1954), NO_2^- -N concentration of more than 50 mg/kg must dominate the root zone to damage plants. In tomatoes, NO_2^- concentration required to produce poisonous symptoms and the sternness of toxic symptoms are strictly related to the root supply of Mg_2^+, Fe_2^+ and air (Phipps & Cornforth, 1970). Rice plants are very delicate to NO_2^-, while cucumbers and wheat are tolerant. In the same way, avocados are more vulnerable than citrus (Curtis, 1949). $NaNO_2$ (40 mg/kg) is as suitable for the growth of guinea grass as $(NH_4)_2SO_4$ in the absence of lime (Oke, 1966). Sahulka (1973) reports that the normal dimension and fresh weight of pea roots grown on 2mM KNO_2 are significantly lower than the average length and fresh weight of pea roots grown on 2mM KNO_3. Generally, circumstances of greater acidity, little O_2 supply to the root, and greater environmental NO_2 concentrations appear to support NO_2 toxicity. Recently, Lee (1979) reported that a low O_2 supply to the roots increases the sensitivity of barley and corn plants to $NO2^-$, but when there is enough O_2, the change produced by NO_2 is not significant. Associated nitrous acid (HNO_2) (its concentration increases with a decrease in pH) rather than unassociated NO_2^- may be linked to growth inhibition (Lee, 1979). Jackson et al. (1974) showed that NO_3^- uptake in wheat seedlings is inhibited by NO_2^-. In N-hungry wheat seedlings, the absorption of K^+ and NO_2^- from KNO_2 was considerably lower than the absorption of k^+ and NO_3^- from equimolar KNO_3 (Jackson et al., 1974). In kidney beans, corn and sunflower roots absorb NO_2^- faster than NO_3^- and convert it to amino N (Yoneyama et al., 1980). Many authors have revealed that plants can assimilate and absorb NO_2^-. In the roots of most plants the complete reduction and assimilation pathway of NO_2^- seems to be active. While little or no NO_2^- can be derived from the root to the bud, in light as well as in dark, leaves of many species efficiently reduce and metabolize exogenously provided NO_2^-. This proposes that metabolites can enter the chloroplasts and reduce NADP, which may then reduce ferredoxin through NADP reductase, thereby providing electrons to nitrite reductase. However, the reduced light dependence of NO_2^- in barley leaves (Canvin & Atkins, 1974) and sunflower (Ito & Kumazawa, 1978) has also been reported. Vanecko and Varner (1955) showed that the reduction of NO_2^- during the light is dependent on chlorophyll.

17.4.3 NO$_3$ TOXICITY

Most plants tolerate high levels of N03- without any physiological disorders, but excess N03- nutrition can be toxic. But how it can be toxic is unknown (Barker & Mills, 1980). Nitrates are mainly toxic to plants, where NH4+ proved to be a superior nitrogen source compared to N03- such as members of the family, some species of the genus Zizhu, and rice. High levels of N03-can cause iron deficiency and produce symptoms of the well-known "lime-induced" chlorosis. Cain (1952) concluded that the toxicity of NO$_3^-$ does not affect the absorption of Fe, but affects internal functions. Colgrove and Roberts (1956) suggested that Fe is inactivated at higher pH levels promoted by NO$_3^-$ in plant tissues. On the other hand, in "Sitka" spruce (Picea sitchensis) and "Scottish" pine (camphor pine), greenery loss due to NO$_3^-$ nutrition is associated with higher organic anions (Nelson & Selby, 1974). Chlorosis caused by NO$_3^-$ seems to be best explained by the competitive chelation rationalism, which suggests that Fe activity in plant tissues is reduced by excess organic anions. Higher levels of organic acid anions due to NO$_3^-$ nutrition were discussed. Greidanus et al. (1972) reported that cranberry plants grown in NO$_3^-$ are N-defect because the plant lacks NO$_3^-$ absorption and reduction mechanism. Maynard and Barker (1971) showed that the dry matter yield of spinach decreased significantly when the NO$_3^-$ provided in sand culture was 24 and 48 mg/L instead of 12 mg/l. They concluded that the toxic effects of Ca (N03)$_2$ at high concentrations can be attributed to a specific NO$_3^-$ effect, as spinach is classified as a saline plant. Nitrate toxicity in strawberry plants is characterized by a change in leaf margin color to dark brown with a purple hue (Jackson, 1972). In severe cases, interventional organizations are also affected. The whiptail of cauliflower, a symptom of Mo deficiency, can be caused by the accumulation of a large number of NO$_3^-$ in leaves (Agarwala, 1952). Nitrate nutrition facilitates the accumulation of higher levels of NO$_3^-$ and organic acids such as malate and oxalate. Excessive consumption of NO$_3^-$ and oxalates can be toxic to humans and animals. As a result, NO$_3^-$ nutrition may reduce the food quality of some plants, especially those with fresh leaves to eat.

17.5 ALLEVIATION OF NH4+ TOXICITY

NH4+ toxicity can be alleviated by buffering the external pH, thereby offsetting rhizosphere acidification associated with ammonium absorption. Maintaining a neutral to weak alkaline pH also prevents a sharp decline in malic acid in cells that are usually associated with the provision of ammonium (Goodchild and Givan, 1990). In addition, optimizing the light state to avoid high light effects for ammonium-grown plants is more important than plants grown using nitrates or organic N. When NH4+ is used as the only nitrogen source, maintaining a high level of cations known to be suppressed in plant tissue in the nutrient solution is also very important. In particular, the supply level of K+ has been shown to ease the toxicity of solution in culture experiments and fields. Presently, it is not known whether the normal homeostasis of potassium controls cytoplasmic concentrations, or only vacuoles (Walker et al., 1996). In the case of calcium, it is interesting to speculate whether this widespread signaling ion (Berridge, 1997), NH4+ nutrition, lower multi-depressed vacuole, or any other intracellular pool may lead to inhibited Ca$_2^+$ spikes in response to various stimuli. One of the most fascinating aspects of NH4+ nutrition is that when NH4+ is provided alone, toxicity is observed in many species, and can be alleviated by co-providing nitrates (Schortemeyer et al., 1997). In addition, co-provision induces a synergistic growth reaction that can exceed the maximum growth rate of any N source in solution culture by up to 40 to 70% (Heberer, 1989). Interestingly, synergistic reactions are observed even in species such as conifers, where nitrate uptake is very small (Kronzucker et al., 1997). However, in a few cases, such as in some lianas, synergistic reactions are not present, and some plants even undergo growth inhibition of nitrates (Dijk and Eck, 1995). Some suggestions are put forward to try to explain the phenomenon of nitrate-ammonium synergistic effect. Many of these keys are the possible role of nitrates as a signal to stimulate a variety of biochemical reactions (Tischner, 2000). One possibility is to maximize cytokinin synthesis when NO_3^- and NH_4^+ are provided together (Chen et al., 1998). Another reason is that the rhizosphere alkalizing effect of plants on nitrate absorption may help limit acidification associated with NH_4^+ nutrition (Marschner, 1995). However, this effect may be partial at best, or require a high NO_3^- : NH_4^+ ratio in the nutrient solution, because ammonium inhibition of NO_3^- absorption is usually as high as 50% (Kronzucker et al., 2007), while NH_4 uptake can be moderately stimulated by nitrates (Kronzucker et al., 1999). Given that nitrogen efflux is also significantly reduced with co-provisions, the net result of the plant's use of two separate nitrogen sources together is that the total nitrogen uptake can be significantly (up to 75%) higher than the same nitrogen concentration presented in either N-source separate form (Kronzucker et al., 1999). An interesting aspect of this analysis is that, in rice, a 50/50 mixture of NO_3^- and NH_4^+ leads to more or less equal concentrations of NO_3^- and NH_4^+ in root cell solute (Kronzucker et al., 1999), attenuating any N source, at least in the requirements of cytosol charge balance. The most important synergistic reaction of co-provision of NO_3^- and NH_4^+ may lie in the enhanced transport of nitrogen to the branches. This is an issue of high agronomic importance because nitrogen stored in the bud tissue can be re-mobilized in a critical period of food requirement and fruit development, when the N transfer through the roots may be damaged due to aging (Mae et al., 1985). Several studies support the enhanced root assimilation in the presence of nitrates (Ota and Yamamoto, 1989) and can be mechanically explained by the nitrate-induced gs-GOGAT pathway, specifically targeting proplastids of roots (Redinbaugh and Campbell, 1993), opening up a pathway where ammonium assimilation cannot be performed without nitrates. In addition to these significant effects, the presence of nitrates may help reduce NH_4^+ toxicity, despite its reduced capacity in buds, slowing the difference in carbon emissions between roots and buds, and improving the flow of electrons between photosystems I and II. The synergistic reaction of NO_3^- and NH_4^+ jointly provided in addition to providing a promising way for agronomic improvement, but also on the mechanism of ammonium toxicity produced insights, is an area that needs further exploration.

REFERENCES

Agarwala, S. C. (1952). Relation of nitrogen supply to the molybdenum requirement of cauliflower grown in sand culture. *Nature*, *169*(4313), 1099–1099.

Allred, S. E., & Ohlrogge, A. J. (1964). Principles of nutrient uptake from fertilizer bands. VI. Germination and emergence of corn as affected by ammonia and ammonium phosphate 1. *Agronomy Journal*, *56*(3), 309–313.

Avron, M. (1960). Photophosphorylation by swiss-chard chloroplasts. *Biochimica et biophysica acta*, *40*, 257–272.

Barker, A. V., & Maynard, D. N. (1972). Cation and nitrate accumulation in pea and cucumber plants as influenced by nitrogen nutrition. *Journal of the American Society for Horticultural Science*, 97, 27–30.

Barker, A. V., & Mills, H. A. (1980). Ammonium and nitrate nutrition of horticultural crops. *Horticultural Reviews*, 2, 395–423.

Barker, A. V., Maynard, D. N., Mioduchowska, B., & Buch, A. (1970). Ammonium and salt inhibition of some physiological processes associated with seed germination. *Physiologia Plantarum*, *23*(5), 898–907.

Barker, A. V., Volk, R. J., & Jackson, W. A. (1965). Effects of Ammonium and Nitrate Nutrition on Dark Respiration of Excised Bean Leaves 1. *Crop Science*, *5*(5), 439–444.

Beaton, J. D., Tisdale, S. L., & Platou, J. (1971). Crop responses to sulphur in North America. Sulfur Inst. Tech. Bull. 18, 1–18.

Bennedetti, E. D., Forti, G., Garlaschi, F. M., & Rosa, L. (1976). On the mechanism of ammonium stimulation of photosynthesis in isolated chloroplasts. *Plant Science Letters*, *7*, 85–90.

Bennett, A. C., & Adams, F. (1970). Concentration of NH3 (aq) required for incipient NH3 toxicity to seedlings. *Soil Science Society of America Journal*, *34*(2), 259–263.

Bergmann, L., Grosse, W., & Koth, P. (1976). Influences of ammonium and nitrate on n-metabolism, malate accumulation and malic enzyme-activity in suspension cultures of nicotiana-tabacum var samsum. *zeitschrift fur pflanzenphysiologie*, *80*(1), 60–70.

Berner, E. (1971). Studies on the nitrogen metabolism of barley leaves. II. The effect of nitrate and ammonium on respiration and photosynthesis. *Plant Physiology Supplement*, *6*, 46–56.

Berridge, M. J. (1997). The AM and FM of calcium signalling. *Nature*, *386*(6627), 759–760.

Bingham, F. T., Chapman, H. D., & Pugh, A. L. (1954). Solution-culture studies of nitrite toxicity to plants. *Soil Science Society of America Journal*, *18*(3), 305–308.

Blair, G. J., Miller, M. H., & Mitchell, W. A. (1970). Nitrate and Ammonium as Sources of Nitrogen for Corn and Their Influence on the Uptake of Other Ions 1. *Agronomy Journal*, *62*(4), 530–532.

Boynoton, D. (1966). Apple nutrition. In: N. F. Childers (ed.), *Nutrition of Fruit Crops*. Somerset Press, Inc., Somerville, NJ, pp. 1–50.

Breteler, H. (1973). A comparison between ammonium and nitrate nutrition of young sugar-beet plants grown in nutrient solutions at constant acidity. 1. Production of dry matter, ionic balance and chemical composition. *NJAS Wageningen Journal of Life Sciences*, *21*(3), 227–244.

Cain, J. C. (1952). A comparison of ammonium and nitrate nitrogen for blueberries. *Proceedings of American Society of Horticultural Science*, *59*, 161–166.

Canvin, D. T., & Atkins, C. A. (1974). Nitrate, nitrite and ammonia assimilation by leaves: effect of light, carbon dioxide and oxygen. *Planta*, *116*(3), 207–224.

Cesco, S., Neumann, G., Tomasi, N., Pinton, R., & Weisskopf, L. (2010). Release of plant-borne flavonoids into the rhizosphere and their role in plant nutrition. *Plant and Soil*, *329*(1–2), 1–25.

Chapman, H. D., & Liebig, Jr., G. F. (1952). Field and laboratory studies of nitrite accumulation in soils. *Soil Science Society of America Journal*, *16*(3), 276–282.

Chen, J. G., Cheng, S. H., Cao, W., & Zhou, X. (1998). Involvement of endogenous plant hormones in the effect of mixed nitrogen source on growth and tillering of wheat. *Journal of Plant Nutrition*, *21*(1), 87–97.

Colgrove, M. S., Jr., & Roberts, A. N. (1956). Growth of azalea as influenced by ammonium and nitrate nitrogen. *Proceedings of American Society of Horticultural Science*, *68*, 522–536.

Cox, W. J., & Reisenauer, H. M. (1973). Growth and ion uptake by wheat supplied nitrogen as nitrate, or ammonium, or both. *Plant and Soil*, *38*(2), 363–380.

Curtis, D. S. (1949). Nitrite Injury on Avocado and Citrus Seedlings in Nutrient Solution1. *Soil Science*, *68*(6), 441–450.

Dadhich, R. K., & Meena, R. S. (2014). Performance of Indian mustard (*Brassica juncea* L.) in response to foliar spray of thiourea and thioglycollic acid under different irrigation levels. *Indian Journal of Ecology*, *41*(2), 376–378.

De Geus, J. G. (1973). *Fertilizer Guide for the Tropics and Subtropics*. Centre d'Etude de l'Azote, Zurich (Ed. 2), pp. 727–774.

De Kock, P. C., & Kirkby, E. A. (1969). Uptake by plants of various forms of nitrogen and effects on plant composition. *Tech. Bull. Minist. Agric. Fish. Food (GB).*, *15*, 7–14.

Dijk, E., & Eck, N. (1995). Ammonium toxicity and nitrate response of axenically grown Dactylorhiza incarnata seedlings. *New Phytologist*, *131*(3), 361–367.

Dirr, M. A., Barker, A. V., & Maynard, D. N. (1973). Growth and development of Leucothoe and Rhododendron under different nitrogen and pH regimes. *Horticultural Science*, *8*, 131–132.

Embleton, T. W., & Jones, W. W. (1966). Avocado and Mango Nutrition. *Nutrition of Fruit Crops: Tropical, Subtropical, Temperate Tree and Small Fruits*. Horticulture Publication Rutgers University New Brunswick, NJ, pp. 51–76.

Feenstra, W. J., & Jacobsen, E. (1980). Isolation of a nitrate reductase deficient mutant of Pisum sativum by means of selection for chlorate resistance. *Theoretical and Applied Genetics*, *58*(1), 39–42.

Ferguson, A. R., and E. G. Bollard. 1969. Nitrogen metabolism of *Spirodela oligorrhiza*. Utilization of ammonium nitrate, and nitrite. *Planta*, 88:344–352.

Gamborg, O. L. (1970). The effects of amino acids and ammonium on the growth of plant cells in suspension culture. *Plant Physiology*, *45*(4), 372–375.

Gamborg, O. L., & Shyluk, J. P. (1970). The culture of plant cells with ammonium salts as the sole nitrogen source. *Plant Physiology*, *45*, 598–600.

Garnett, T., Conn, V., & Kaiser, B. N. (2009). Root based approaches to improving nitrogen use efficiency in plants. *Plant, Cell & Environment*, *32*(9), 1272–1283.

Givan, C. V. (1979). Review. Metabolic detoxification of ammonia in tissues of higher plants. *Phytochemistry*, *18*, 375–382.

Goodchild, J. A., & Givan, C. V. (1990). Influence of ammonium and extracellular pH on the amino and organic acid contents of suspension culture cells of Acer pseudoplatanus. *Physiologia Plantarum*, *78*(1), 29–37.

Goyal, S. S. (1974). *Studies on the inhibitory effects of ammoniacal nitrogen on growth of radish plants (Raphanus sativus) and its reversal by nitrate*. University of California, Davis.

Goyal, S.S., & Huffaker, R. C. (1984). Nitrogen toxicity in plants. In R.D. Hauck (ed.), Nitrogen in Crop Production. ASA, CSSA, and SSSA, Madison, WI, pp. 97–118.

Graham, P. H., & Vance, C. P. (2000). Nitrogen fixation in perspective: an overview of research and extension needs. *Field Crops Research*, *65*(2–3), 93–106.

Greidanus, T., Peterson, L. A., Schrader, L. E., & Dana, M. N. (1972). Essentiality of ammonium for cranberry nutrition. *Journal of American Society of Horticultural Science*, *97*(2), 212–271.

Hattori, A. (1958). Studies on the metabolism of urea and other nitrogenous compounds in *Chlorella ellipsoidea*. II. Changes in levels of amino acids and amides during the assimilation of ammonia and urea by nitrogen-starved cells. *The Journal of Biochemistry*, 45, 57–64.

Haynes, R. J., & Goh, K. M. (1977). Evaluation of potting media for commercial nursery production of container-grown plants: II. Effects of media, fertiliser nitrogen, and a nitrification inhibitor on yield and nitrogen uptake of Callistephus chinensis (L.) Nees "Pink princess." *New Zealand journal of Agricultural Research*, *20*(3), 371–381.

Haynes, R. J., & Goh, K. M. (1978). Ammonium and nitrate nutrition of plants. *Biological Reviews*, *53*(4), 465–510.

Heath, R. L., & Leech, R. M. (1978). The stimulation of CO2-supported O2 evolution in intact spinach chloroplasts by ammonium ion. *Archives of Biochemistry and Biophysics*, *190*(1), 221–226.

Heberer, J. A., & Below, F. E. (1989). Mixed nitrogen nutrition and productivity of wheat grown in hydroponics. *Annals of Botany*, *63*(6), 643–649.

Heller, R. (1954). Les besoiusmineraux des tissus en culture. *AnneeBiol.*, *30*, 361–374.

Hoff, J. E., Wilcox, G. E., & Jones, C. M. (1974). The effect of nitrate and ammonium nitrogen on the free amino acid composition of tomato plants and tomato fruits. *Journal of the American Society for Horticultural Science,* 99, 27–30.

Hohlt, H. E., Maynard, D. N., & Barker, A. V. (1970). Studies on the ammonium tolerance of some cultivated Solanaceae. *Journal of the American Society of Horticultural Science*, *95*(3), 345–8.

Hussain, S., Khan, F & Cao W (2016) Seed priming alters the production and detoxification of reactive oxygen intermediates in rice seedlings grown under sub-optimal temperature and nutrient supply. Frontiers in Plant Science, *7*. https://doi.org/10.3389/fpls.2016.00439}

Ito, O., & Kumazawa, K. (1978). Amino acid metabolism in plant leaf: III. The effect of light on the exchange of 15N-labeled nitrogen among several amino acids in sunflower discs. *Soil Science and Plant Nutrition*, *24*(3), 327–336.

Jackson, D. C. (1972). Research note: nitrate toxicity in strawberries. *Agrochemophysica*, *4*(2), 45.

Jackson, W. A., Johnson, R. E., & Volk, R. J. (1974). Nitrite Uptake by Nitrogen-Depleted Wheat Seedlings. *Physiologia Plantarum*, *32*(1), 37–42.

Kato, T. (1980). Nitrogen assimilation in citrus trees: 1. Ammonium and nitrate assimilation by intact roots, leaves and fruits. *Physiologia Plantarum*, *48*(3), 416–420.

Kronzucker, H. J., Siddiqi, M. Y., & Glass, A. D. (1997). Conifer root discrimination against soil nitrate and the ecology of forest succession. *Nature*, *385*(6611), 59–61.

Kronzucker, H. J., Siddiqi, M. Y., Glass, A. D., & Kirk, G. J. (1999). Nitrate-ammonium synergism in rice. A subcellular flux analysis. *Plant Physiology*, *119*(3), 1041–1046.

Lee, R. B. (1979). The effect of nitrite on root growth of barley and maize. *New Phytologist*, *83*(3), 615–622.

Losada, M., Herrera, J., Maldonado, J. M., & Paneque, A. (1973). Mechanism of nitrate reductase reversible inactivation by ammonia in Chlamydomonas. *Plant Science Letters*, *1*(1), 31–37.

Mae, T., Hoshino, T., Ohira, K. (1985). Proteinase activities and loss of nitrogen in the senescing leaves of field-grown rice (*Oryza sativa* L.). *Soil Science Plant Nutrition*, *31*, 589–600.

Marschner, H. (1995). *Mineral Nutrition of Higher Plants*. 2nd Edn. Academic Press, London. pp. 889.

Matsumoto, H., Wakiuchi, N., & Takahashi, E. (1971a). Changes of starch synthetase activity of cucumber leaves during ammonium toxicity. *Physiologia Plantarum*, *24*(1), 102–10

Matsumoto, H., Wakiuchi, N., & Takahashi, E. (1971b). Changes of some mitochondrial enzyme activities of cucumber leaves during ammonium toxicity. *Physiologia Plantarum*, *25*(3), 353–357.

Maynard, D. N., & Barker, A. V. (1971). Critical nitrate levels for leaf lettuce, radish, and spinach plants. *Communications in Soil Science and Plant Analysis*, *2*(6), 461–470.

Maynard, D. N., & Barker, A. V. (1979). Regulation of nitrate accumulation in vegetables. *Acta Horticulturae*, *93*, 153–162.

McAllister, C. H., Beatty, P. H., & Good, A. G. (2012). Engineering nitrogen use efficient crop plants: the current status. *Plant biotechnology journal*, *10*(9), 1011–1025.

McElhannon, W. S., & Mills, H. A. (1978). Influence of Percent NO_3^-/NH_4^+ on growth, N absorption, and assimilation by lima beans in solution culture 1. *Agronomy Journal*, *70*(6), 1027–1032.

Mills, H. A., & Jones Jr, J. B. (1979). Nutrient deficiencies and toxicities in plants: Nitrogen. *Journal of Plant Nutrition*, *1*(2), 101–122.

Mohanty, B., & Fletcher, J. S. (1976). Ammonium influence on the growth and nitrate reductase activity of Paul's Scarlet rose suspension cultures. *Plant Physiology*, *58*(2), 152–155.

Mohanty, B., & Fletcher, J. S. (1980). Ammonium influence on nitrogen assimilating enzymes and protein accumulation in suspension cultures of Paul's Scarlet rose. *Physiologia Plantarum*, *48*(3), 453–459.

Morris, H. D., & Giddens, J. (1963). Response of Several Crops to Ammonium and Nitrate Forms of Nitrogen as Influenced by Soil Fumigation and Liming 1. *Agronomy Journal*, *55*(4), 372–374.

Muller, L. (1966). Coffee nutrition. In N. F. Childers (ed.), *Nutrition of Fruit Crops*. Somerset Press, Inc., Somerville, NJ, pp. 685–776.

Nelson, L. E., & Selby, R. (1974). The effect of nitrogen sources and iron levels on the growth and composition of Sitka spruce and Scots pine. *Plant and Soil*, *41*(3), 573–588.

Oke, O. L. (1966). Nitrite toxicity to plants. *Nature*, *212*(5061), 528–528.

Ota, K., & Yamamoto, Y. (1989). Promotion of assimilation of ammonium ions by simultaneous application of nitrate and ammonium ions in radish plants. *Plant and Cell Physiology*, *30*(3), 365–371.

Paul, J. S., Cornwell, K. L., & Bassham, J. A. (1978). Effects of ammonia on carbon metabolism in photosynthesizing isolated mesophyll cells from *Papaver somniferum* L. *Planta*, *142*(1), 49–54.

Paul, J. S., Cornwell, K. L., & Bassham, J. A. (1978). Effects of ammonia on carbon metabolism in photosynthesizing isolated mesophyll cells from Papaver somniferum L. *Planta*, *142*(1), 49–54.

Pettigrew, W. T., Hesketh, J. D., Peters, D. B., & Woolley, J. T. (1989). Characterization of Canopy Photosynthesis of Chlorophyll-Deficient Soybean Isolines. *Crop Science*, *29*(4), 1025–1029.

Phipps, R. H., & Cornforth, I. S. (1970). Factors effecting the toxicity of nitrite nitrogen to tomatoes. *Plant and Soil*, *33*(1–3), 457–466.

Platt, S. G., Plaut, Z., & Bassham, J. A. (1977). Ammonia regulation of carbon metabolism in photosynthesizing leaf discs. *Plant Physiology*, *60*, 739–742.

Puritch, G. S., & Barker, A. V. (1967). Structure and function of tomato leaf chloroplasts during ammonium toxicity. *Plant Physiology*, *42*(9), 1229–1238.

Redinbaugh, M. G., & Campbell, W. H. (1993). Glutamine synthetase and ferredoxin-dependent glutamate synthase expression in the maize (Zea mays) root primary response to nitrate (evidence for an organ-specific response). *Plant Physiology*, *101*(4), 1249–1255.

Rehfeld, D. W., & Jensen, R. G. (1973). Metabolism of separated leaf cells: III. Effects of calcium and ammonium on product distribution during photosynthesis with cotton cells. *Plant Physiology*, *52*(1), 17–22.

Robbins, W. J., & Schmidt, M. B. (1938). Growth of excised roots of the tomato. *Botanical Gazette*, *99*(4), 671–728.

Sahulka, J. (1973). The regulation of glutamate dehydrogenase, nitrite reductase, and nitrate reductase in excised pea roots by nitrite. *Biologia Plantarum*, *15*(4), 298–301.

Sasseville, D. N., & HA, M. (1979). N form and concentration: Effects on N absorption, growth, and total N accumulation with southernpeas. *Journal of American Society for Horticltural Science*, *104*, 586–591.

Schortemeyer, M., Stamp, P., & Feil, B. O. Y. (1997). Ammonium tolerance and carbohydrate status in maize cultivars. *Annals of Botany*, *79*(1), 25–30.

Schrader, L. E., Domska, D., Jung Jr, P. E., & Peterson, L. A. (1972). Uptake and Assimilation of Ammonium-N and Nitrate-N and Their Influence on the Growth of Corn (*Zea mays* L.) 1. *Agronomy Journal*, *64*(5), 690–695.

Skinner, J. C., & Street, H. E. (1954). Studies on the growth of excised roots. II. Observations on the growth of excised groundsel roots. *The New Phytologist*, *53*(1), 44–67.

Spratt, E. D., & Gasser, J. K. R. (1970). The effect of ammonium sulphate treated with a nitrification inhibitor, and calcium nitrate, on growth and N-uptake of spring wheat, ryegrass and kale. *The Journal of Agricultural Science*, *74*(1), 111–117.

Syrett, P. J. (1956a). The assimilation of ammonia and nitrate by nitrogen-starved cells of Chiarella vulgaris. II. The assimilation of large quantities of nitrogen. *Plant Physiology*, 9, 19–27.

Tischner, R. (2000). Nitrate uptake and reduction in higher and lower plants. *Plant, Cell & Environment*, *23*(10), 1005–1024.

Torres de Claassen, M. E., & Wilcox, G. E. (1974). Effect of nitrogen form on growth and composition of tomato and pea tissue. *Journal of the American Society for Horticultural Science*, *99*, 171–174.

Tromp, J., & Ovaa, J. C. (1979). Uptake and distribution of nitrogen in young apple trees after application of nitrate or ammonium, with special reference to asparagine and arginine. *Physiologia Plantarum*, *45*(1), 23–28.

Tucker, T.C. (1984). Diagnosis of nitrogen deficiency in plants. In R.D. Hauck (ed.) *Nitrogen in Crop Production*. ASA, CSSA, and SSSA, Madison, WI, pp. 249–262.

Ulrich, A. (1952). Physiological bases for assessing the nutritional requirements of plants. *Annual Review of Plant Physiology*, *3*(1), 207–228.

Vanecko, S., & Varner, J. E. (1955). Studies on nitrite metabolism in higher plants. *Plant Physiology*, *30*(4), 388.

Vines, H. M., & Wedding, R. T. (1960). Some effects of ammonia on plant metabolism and a possible mechanism for ammonia toxicity. *Plant Physiology*, *35*(6), 820.

Wahhab, A., & Bhatti, H. M. (1957). Effect of Various Sources of Nitrogen on Rice Paddy Yield 1. *Agronomy Journal*, *49*(3), 114–116.

Wakiuchi, N., Matsumoto, H. and Takahashi, E.. (1971). Changes of some enzyme activities of cucumber during ammonium toxicity. *Physiologia Plantarum*, 24:248–253.

Walker, D. J., Leigh, R. A., & Miller, A. J. (1996). Potassium homeostasis in vacuolate plant cells. *Proceedings of the National Academy of Sciences*, *93*(19), 10510–10514.

Wedding, R. T., & Vines, H. M. (1959). Inhibition of reduced diphosphopyridine nucleotide oxidation by ammonia. *Nature*, *184*(4694), 1226–1227.

Weissman, G. S. (1964). Effect of ammonium and nitrate nutrition on protein level and exudate composition. *Plant Physiology*, *39*(6), 947.

Weissman, G. S. (1972). Influence of ammonium and nitrate nutrition on enzymatic activity in soybean and sunflower. *Plant Physiology*, *49*(2), 138–141.

Yoneyama, T., and Yoshida, T. (1978). Nitrogen mineralization of sewage sludges in soil. *Soil Science and Plant Nutrition*, 24, 139–144.

Yoneyama, T., Iwata, E., and Yazaki, J. (1980). Nitrite utilization in the roots of higher plants. *Soil Science. Plant Nutrition*, 26, 9–23.

18 Nitrogen Metabolism Enzymes
Structure, Role, and Regulation

Shafia Liyaqat Nahvi[1*] *and Afiya Khurshid*[2]
[1]Department of Fruit Science, Faculty of Horticulture, Sher-e-Kashmir University of Agricultural Science and Technology, Shalimar, Kashmir, India
[2]Department of Vegetable Science, Faculty of Horticulture, Sher-e-Kashmir University of Agricultural Science and Technology, Shalimar, Kashmir, India
[*]Corresponding Author: Email: shafia.nahvi@gmail.com

CONTENTS

- 18.1 Introduction 228
- 18.2 Nitrate Reductase (NR) 228
  - 18.2.1 Structure 228
  - 18.2.2 Classes of NR 228
- 18.3 Environmental Factors Regulating Nitrate Reductase Enzyme 229
  - 18.3.1 Light and Photosynthesis 229
  - 18.3.2 Response of Leaf NR Activation to Nitrate Supply 229
  - 18.3.3 Modulation of NR in Roots: Response to Anoxia 230
  - 18.3.4 Response of NR to Salt Stress 230
  - 18.3.5 Response of NR to Water Stress 230
- 18.4 Glutamine Synthetase 230
  - 18.4.1 Structure of Glutamine Synthetase 230
  - 18.4.2 Factors Affecting the Activity of Glutamine Synthetase 231
    - 18.4.2.1 Light 231
    - 18.4.2.2 Nitrogen and Carbon Concentration 231
    - 18.4.2.3 Metabolites 231
- 18.5 Glutamate Synthase 232
  - 18.5.1 Structure 232
  - 18.5.2 External Factors Affecting GOGAT 232
    - 18.5.2.1 LIGHT 232
    - 18.5.2.2 Temperature 233
    - 18.5.2.3 Salinity Stress 233
    - 18.5.2.4 Water 233
- References 234

DOI: 10.1201/9781003248361-18

18.1 INTRODUCTION

Nitrogen (N) is critical for the growth and development of the plant as well as resistance to biotic and abiotic stresses since it plays a crucial role in energy metabolism and protein synthesis. Macronutrient nitrogen is required for the formation of chlorophyll as well as a variety of primary and secondary metabolites and hormones. Nitrogen is directly involved in the biosynthesis of amino acids, proteins, and nucleic acids. Nitrogen deficiency can result in decreased root growth, altered root framework, reduction in plant biomass, and lowered photosynthesis. As a result, it is critical to comprehend the structure and regulation of nitrogen metabolism. A variety of enzymes and intermediates are involved in nitrogen assimilation. The activity of these enzymes is dependent on internal factors such as the stage of the crop as well as the external factors like light, temperature, salinity, water, and absence of oxygen. In this chapter we discuss the structure and factors regulating the key enzymes involved in nitrogen metabolism, such as nitrate reductase, glutamine synthase, and GOGAT.

18.2 NITRATE REDUCTASE (NR)

18.2.1 Structure

In higher plants, algae, and fungus nitrate reductase is a homodimer cytosolic protein composed of two identical ~100-kD subunits with five domains in each monomer:

- Mo-MPT domain containing a single molybdopterin cofactor
- Dimer interface domain
- Cytochrome b domain
- NADH domain that interacts with a FAD domain to create the cytochrome b reductase fragment

A GPI (glycosylphosphatidylinositol)-anchored variant can be found on the plasma membrane's outer face. Its precise function is still unknown. Nitrate reductase belongs to the dimethyl sulphoxide reductase (DMSOR) family of molybdoenzymes. There are four types of eukaryotic assimilatory NR (Euk-NR) (Aylott et al.,1997; Bachmann et al.,1996a) and three different bacterial enzymes (i.e., cytoplasmic assimilatory nitrate reductases (Nas), membrane-bound respiratory nitrate reductases (Nar), and periplasmic dissimilatory nitrate reductases (Nap) (Bachmann et al.,1996b; Barber et al., 1997; Berks et al., 1995; Bruns and Karplus, 1995).

The occurrence of molybdenum (Mo) and molybdenum cofactor (Mo-co) in the active core of NR enzymes is a frequent feature. The active core of bacterial NR contains molybdopterin-guanine dinucleotide (MGD), whereas the active center of eukaryotic NR contains mononucleotide. Molybdenum ion is bound to the four thiolate ligands located in the two MGD halves. Molybdenum is also regulated by sulfur (–S), oxygen (–O), or selenium (–Se) bonds of cysteine, serine, or selenocysteine residues of the polypeptide chain, as well as available oxo (–O) and hydroxyl (–OH) groups or water in several molybdenum enzymes (Brunt et al., 1992; Campbell et al., 1997; Campbell, 1989; Campbell, 1999). The molybdenum is covalently bonded to the protein by a cysteine ligand in periplasmic dissimilatory nitrate reductases (Nap), and an aspartate in membrane-bound respiratory nitrate reductases (Nar) (Tavares et al., 2006).

18.2.2 Classes of NR

In bacteria two types of nitrate reductase have been found: a) NR dependent on ferredoxin or flavodoxin and b) NR dependent on NADPH. The structure of the cofactor is the basis of this classification.

1. Ferredoxin- or flavodoxin-dependent NR is found in bacteria and cyanobacteria and has been reported to contain Moco and (Fe-S) clusters in their active centers while lacking FAD and cytochromes. (Campbell and Kinghorn,1990; Cannon et al., 1993). NR present in *Azotobacter spp.* and *Plectonema spp.* has one 105 kD subunit that contains a (4Fe-4S) center and molybdenum (1 atom/molecule) and utilizes flavodoxin as an electron donor (Cannon et al., 1991). The amino acid sequences of these enzymes were examined, and it was observed that at the N-terminus residues of cysteine are present that bind the (4Fe-4S) center. *A. chroococcum, Clostridium perfringens,* and *Ectothiorhodospira shaposhnikovi* have all been found to have ferredoxin-dependent NR.
2. NR of Klebsiella oxytoca (pneumoniae) and Rhodobacter capsulatus are NADPH-dependent heterodimers made up of the diaphorase FAD-consisting 45 kD subunit and 95 kD catalytic subunit with molybdenum as a cofactor and N-terminal (4Fe-4S) core (Chen et al., 1995). Furthermore, NR of *Klebsiella oxytoca (pneumoniae)* consists of an extra (2Fe-2S) center attached to a C-terminal swarm of cysteine residues, which is identical, by its sequence of amino acid, to the NifU protein, the nifU gene product, which plays a role in the synthesis of (Fe-S) clusters in the nitrogenase active center.

18.3 ENVIRONMENTAL FACTORS REGULATING NITRATE REDUCTASE ENZYME

18.3.1 Light and Photosynthesis

In the light, NR synthesis is more, whereas, in the dark, it is less (Weiner and Kaiser, 1999). When triggered by a variety of artificial means, such as anoxia or acidification of tissue, or uncouplers of respiration like CCCP or DNP, NR protein becomes more stable in dark-colored leaves (Kaiser et al., 1999), which means the active form is more resistant to destruction than its inactive form (bound to 14-3-3s). When 35S-methionine was supplied, NR degradation was significantly quicker in the dark than in the light.

As previously stated, in the light, NR present in leaves is more active (70–90%), while in the dark, it is less active (10–30%). Surprisingly, despite the presence of a significant amount of active NR in the darkened leaves extract, green leaf nitrate reduction is almost non-existent in the dark (Kaiser et al., 2000). As a result, light is not a direct indication for activating NR. Despite the presence of constant bright light, NR becomes inactive when CO_2 is not present. Photosynthesis is hence necessary for NR activation. Assimilates are most likely translocated as signals from the chloroplast function. Indeed, by giving sugars to the leaves, NR may be triggered in the dark. Furthermore, dark inactivation of NR in a starchless mutant of *N. sylvestris*, which accumulates high amounts of sugar-Ps as a result of the block in starch synthesis, is significantly reduced compared to the wild type (Bachmann et al., 1995). Finally, removing foliage in the light causes photosynthates to build up in the leaf, resulting in NR "hyperactivation" in the light (Huber et al., 1992a, b). All these reactions suggest the sensitiveness of the activation state of NR to metabolites, which may be explained by the discovery that physiological amounts of hexose monophosphates block the NR protein kinase in vitro (Kaiser et al., 1999). This reaction to photosynthates suggests that carbon and nitrogen uptake must be coordinated, hence indicating sensitiveness of N reduction to stomatal resistance. As a result, when plants cover their stomata to conserve water during a drought, not only does photosynthesis decrease, but NR becomes less active.

18.3.2 Response of Leaf NR Activation to Nitrate Supply

Differences in nitrate supply do not affect the state of activation of NR. Plants deficient in nitrate may have a similar activation state of NR (high in the light, low in the dark) as plants with ample nitrate, but with lower NR protein levels and total NR activity in their leaves. The regulation was directly influenced only in very few cases when NR levels were relatively low because of prolonged deficiency of nitrate (Man et al., 1999) or when the Nia genes were mutated (Scheible et al., 1997b). For both scenarios, NR was significantly less dark-inactivated (Man et al., 1999). Once again, the underlying cause is not certain. Normally, in tobacco plants that have an ample nitrate supply, NR mRNA manufacturing in the foliage commences late in the night and peaks around noontime, but NR protein (activity) tends to increase in the early hours and continues to remain steady or slightly declines until nightfall. Intriguingly, when the supply of nitrate to barley plants was disrupted and the concentration of internal nitrate decreased to very low levels, NR protein and activity lowered significantly over the day, implying rapid NR breakdown (Man et al., 1999). In this case, NR remained active in the dark, even though NR destruction (e.g., in the dark) is typically followed by inactivation.

18.3.3 Modulation of NR in Roots: Response to Anoxia

The fact that NR is more active in anoxic plant tissues than in air has long been known, but the reason for this is uncertain. It's been shown that NR is stimulated within minutes of being exposed to anoxia, not only in darkened leaves but also in roots (Glaab and Kaiser, 1993; Botrel et al., 1996; Botrel and Kaiser, 1997). Anoxic cells' cytoplasm is known to be acidified, lowering its pH from 7.2 to 6.5. Darkened leaf discs or portions of roots if acidified artificially similarly activated NR, leading to the conclusion that anoxia-induced NR activation was most likely mediated by cytosolic acidity (Kaiser and Brendle-Behnisch, 1995). Other theories, on the other hand, may appear plausible.

18.3.4 Response of NR to Salt Stress

Changes in salinity affect NR activity. It has been demonstrated that when maize seedlings are subjected to moderate salinity stress, the activation status of NR remains relatively constant in both light and dark conditions (Baki et al., 2000). NR activity decreased in the leaves while increasing in the roots as salinity increased. Variations in leaf nitrate content were a contributing factor. Nitrate concentrations in salt-stressed seedling leaves plummeted, whereas chloride concentrations increased. NR-mRNA concentrations in leaves have also altered little throughout the day. NR-mRNA revealed a clear morning peak and a second, less noticeable peak in the late-night hours in control plants. With salinity, the latter maximum vanished, and the early light phase rise in NR protein and activity was postponed. As a result, rather than affecting post-translational NR regulation, salinity stress happens to influence the expression of NR. The half-life of NR is only a few hours. As a result, the concentration of NR protein present is determined not only by the rate of synthesis but also by the rate of breakdown.

18.3.5 Response of NR to Water Stress

After being subjected to water stress, the abundance of NR transcripts was found to rapidly decrease in maize (*Zea mays L.*) plants. Nitrate reductase mRNA was about 80% lower after 7 days of water deprivation than in water-replete plants; however, the nitrate reductase mRNA pool was restored within a day after water-deficit plants were rehydrated after 3 days. Water stress decreased total extractable foliar NR activity (Foyer et al., 1998).

18.4 GLUTAMINE SYNTHETASE

18.4.1 Structure of Glutamine Synthetase

Glutamine synthetase is made up of eight, ten, or twelve identical subunits arranged in two face-to-face rings. Bacterial GS are dodecamers that have 12 active sites in each monomer (Eisenberg et al.,

2000). Each active site forms a "tunnel," which has three different substrate-binding sites: nucleotide, ammonium ion, and amino acid (Ginsburg et al., 1970). ATP attaches to the top of the bifunnel that connects GS to the outside world. Glutamate binds to the active site of the bottom. Two divalent cation-binding sites (Mn^{+2} or Mg^{+2}) can be found in the center of the bifunnel. The first cation-binding site is crucial in the phosphorylation of glutamate by ATP, whereas the second stabilizes active GS and aids glutamate binding.

The two rings of GS are held together by hydrogen bonding and hydrophobic interactions. In its sequence, each subunit has a C-terminus and an N-terminus. By inserting into the hydrophobic area of the subunit across in the other ring, the C-terminus (helical thong) stabilizes the GS structure. The solvent is exposed to the N-terminus. In addition, six four-stranded sheets made up of anti-parallel loops from the 12 subunits form the core channel.

18.4.2 Factors Affecting the Activity of Glutamine Synthetase

18.4.2.1 Light

Studies on the impact of light on the expression of the GS genes provided the first insight into their reciprocal regulation. For example, a study of the expression of nuclear genes for the chloroplastic form of GS, GS2, in pea (Edward et al., 1989), Phaseolus (Cock et al., 1991), and Arabidopsis (Peterman et al., 1991) showed that light regulates the gene for this GS isoform in a manner regulated at least in part by phytochrome (Oliveira and Coruzzi, 1999). Light, on the other hand, has no effect on the levels of gene expression for cytosolic GS1. Because potential cis-acting light regulatory elements, which are essential for phytochrome-mediated control, have been discovered in GS2 promoters, light regulation of GS2 can be explained via transcriptional mechanisms (Tjaden et al., 1995). Plant gene expression can be influenced by light either directly through phytochrome activation or indirectly through changes in carbon metabolite levels. The activation of photosynthesis, which leads to a rise in carbon metabolite production, is one well-known indirect consequence of light. Indeed, an increase in the amounts of hexoses and/or non-hexose carbon metabolites can affect GS gene expression in Arabidopsis. For example, in the absence of light, sucrose administration can cause the increase of mRNA for chloroplastic GLN2 and cytosolic GLN1 in a time- and dose-dependent manner. Surprisingly, the non-hexose carbon source 2-oxoglutarate has a beneficial effect on GS expression in the cytosol. This finding could be explained by the fact that 2-oxoglutarate is most significant as a carbon-level signal in non-photosynthetic tissues like roots, where the cytosolic GS1 isoforms are most abundantly expressed (Oaks, 1992).

18.4.2.2 Nitrogen and Carbon Concentration

In nitrogen-deficient conditions, GS activity is roughly five times higher than in nitrogen-rich settings, while in a complete medium consisting of glucose, GS activity is around four times higher. Under carbon-limited environments, overall GS activity is reduced thrice. The energy requirements for this enzyme's route can be explained by its reduced/uninduced activity. Not only is ATP used to create each glutamate molecule, but the adenylation process also gets its AMP from ATP. As a result, when there is a carbon/energy shortage, GDH is preferred for glutamate synthesis (Schulz et al., 2001).

18.4.2.3 Metabolites

Amino acids can sometimes act as messengers, allowing plant gene expression to be regulated. Indeed, it has been discovered that different amino acids have varying degrees of sucrose antagonism for individual GS genes (Oliveira and Coruzzi, 1999). Asparagine, for example, was found to be the most efficient amino acid in reducing sucrose-induced GS mRNA increase. Aspartate, on the other hand, was the least effective. The following is an explanation for the amino acid differences in sucrose-induced GS gene expression: 1) Each amino acid may have various but partially overlapping mechanisms for eliciting its effects, 2) each amino acid may be taken up at different rates, and/or 3) different amino acids may be removed from the plant or accumulate in different places. The results observed in plants for GS control by carbon and amino acids are similar to a mechanism found in *E. coli* (Magasanik and Neidhardt, 1987). The relative internal levels of glutamine to 2-oxoglutarate control the absorption of inorganic nitrogen into glutamine in *E. coli*.

18.5 GLUTAMATE SYNTHASE

18.5.1 Structure

Glutamate synthase (GltS) is an enzyme that is also known as glutamine oxoglutarate aminotransferase and is commonly abbreviated as GOGAT. This enzyme, along with glutamine synthetase (abbreviated GS), is important in the control of nitrogen assimilation in photosynthetic eukaryotes and prokaryotes. Plants have two glutamate synthase enzymes: one that uses reduced ferredoxin as an electron donor (Fd GOGAT/Fd-GltS) and another that uses NADH as an electron donor (NADH-GOGAT/NADH-GltS). A third bacterial variant uses NADPH as an electron donor (NADPH GOGAT/NADPH-GltS) (Reitzer, 1996). In photosynthetic eukaryotes, GS and GltS isoenzymes are localized in the cytosol and chloroplast (Oliveira et al., 1997).

- Fd-Glts, also known as FD-GOGAT, is a 165 kDa polypeptide chain that contains flavine mononucleotide (FMN) and an iron-sulfur cluster. It is found in higher plants, cyanobacteria, and algae. The enzyme's reducing equivalents are supplied by ferredoxin. This enzyme is responsible for the absorption of ammonium (NH_4^+) produced by photorespiration, as well as the light-dependent reduction of nitrate (NO_3) (Sakakibara et al.,1991). This enzyme is found in chloroplasts and contributes more than 96% of total glutamate synthase activity (Suzuki and Rothstein, 1997). The chloroplast of the green alga Caulerpa simpliciuscula has been found to contain Fd-glutamate synthase (McKenzie et al., 1979). The plastid genome of red algae contains a glsF structural gene for Fd-glutamate synthase in *Antithamnion sp.* (Valentin et al., 1993) and gltB in Porphyra purpurea (Valentin et al., 1993; Reith and Munholland, 1993). Fd-glutamate synthase activity has also been demonstrated in *Syn echococcus sp.* PCC 6301 (Marqueś et al., 1992).
- NADH-GltS, which is found in fungi, lower animals, and non-green tissues of plants such as seeds and roots, is made up of a single polypeptide chain of 225–230 kDa that is thought to have evolved from gene fusion of the subunits of NADPH-GOGAT; however, NADH-GOGAT is extremely selective for NADH. It catalyzes the absorption of NH_4^+ from nitrogen-fixing bacteria, which is then incorporated into amino acids. Nitrogen fixation and NH_4^+ content is proportional to increased NADH-GOGAT gene expression (Suzuki and Knaff, 2005).
- NADPH-dependent GOGAT or NADPH-GltS is mostly found in eubacteria, whereas ferredoxin-dependent GOGAT or Fd-Glts is found in cyanobacteria and photosynthetic tissues of plants. NADPH-dependent GOGAT or NADPH-GltS is formed by two different subunits (α subunit or αGltS, ~150 kDa, and β subunit or βGltS, ~ 50 kDa) tightly linked to produce the catalytically active αβ protomer, whereas the ferredoxin-dependent GOGAT (Fd-GltS) is produced by a single polypeptide chain identical to αGltS. The enzyme, isolated from yeast and non-photosynthetic tissues of plants, is composed of a solitary polypeptide chain generated from the combination of the bacterial α and β subunits, and it has been demonstrated to be NADH-dependent (Suzuki and Knaff, 2005).

18.5.2 External Factors Affecting GOGAT

18.5.2.1 LIGHT

For GltS gene expression, plants exhibit cell-specific and organ-specific patterns, by sensing light and metabolite signals in the regulation of in vivo function of glutamate synthase isoforms, according to expression analysis (Edwards et al., 1990; Thum et al., 2003). Two distinct functional genes encode Fd-Glts in Arabidopsis, GLU1 and GLU2. GLU1 is expressed in leaves, whereas GLU2 is expressed more prominently in roots and etiolated tissues (Lam et al., 1996; Oliveira et al., 1997). Light positively regulates GLU1 and GLU2 gene induction; however, sucrose in the dark regulates GLU2 gene induction (Suzuki and Rothstein, 1997; Oliveira et al., 1997). Investigating the occurrence of NADH- and ferredoxin (Fd)-dependent glutamate synthases (GOGATs) in the organs (roots, hypocotyl, and cotyledons) of Scots pine seedlings, it was found that cytosolic NADH-GOGAT during the experimental period (4 to 12 days after sowing) was reduced to a low level and was not affected by light. In contrast, Fd-GOGAT of plastids increased significantly in response to light. In contrast to NADH-GOGAT, which was found in all organs in similar amounts, Fd-GOGAT was mainly found in cotyledons, even in the presence of nitrate (Elmlinger and Mohr, 1991).

18.5.2.2 Temperature

The activity of glutamate synthase has also been shown to be affected by temperature. The temperature has been reported to inversely regulate GOGAT activity. Liang et al. (2011) found GOGAT activities in rice grains changed during the grain-filling stage, increasing gradually to a peak value and then decreasing. The average activity of GltS in grains of Koshihikari variety of rice decreased significantly at high temperatures, and it decreased significantly more in grains of IR72 rice. When compared to the normal temperature, the average activity of GOGAT increased significantly in Koshihikari grains and significantly more in IR72 grains. Hungria and Kaschuk (2014) discovered contradictory results in different strains of common bean (*Phaseolus vulgaris L.*). At full flowering, high temperatures reduced the activity of all enzymes involved in N metabolism. For the NADH-Glts, activity at 28°C (174.5 mol NADH mg-1 protein min1) decreased by 31% when the temperature was raised to 34°C, and by 76% when the temperature was raised to 39°C. NADH-Glts is produced and regulated by plant metabolism, but the experiment revealed that differences in NADH-Glts activities are most likely caused by strain-specific differences. At 28°C, strains BR 814, BR 6010, and CNPAF 126 achieved rates of NADH-Glts activity one and half times higher than CIAT 899 and CFN 299, and six to seven times higher than BR 817 strain of common bean.

18.5.2.3 Salinity Stress

One of the most common abiotic stresses limiting GltS/GOGAT activity is salinity. Excess sodium and chlorine ions weaken the GOGAT/GS pathway while enhancing the GDH pathway. Rice has been reported to accumulate Na+ and Cl to toxic levels in old leaves when subjected to salt stress. Salt stress reduced the expression of OsNADH-GOGAT2 and OsFd-GOGAT in both young and old leaves, with expression levels of the two genes significantly lower in old leaves than in young leaves. Downregulation of OsGS2 and OsFd-GOGAT in old rice leaves may be a harmful response to Na+ and Cl excesses (Wang et al., 2012). When exposed to salt stress, GOGAT activity varies between strains of the same species. The "100/1B" barley strain was found to support nitrogen assimilation when exposed to salinity by enhancing the GS and GOGAT cycle under high nitrogen conditions and stimulating the GDH pathway under low nitrogen conditions. Salinity inhibited the GS/GOGAT cycle while increasing GDH activity in "Barley medenine" (Ben Azaiez et al., 2020).

18.5.2.4 Water

The activity of the GOGAT enzyme has been reported to decrease under water stress. The effect of water-deficit stress on GOGAT activities of Pisum sativum nodules has been reported by González et al. (1998). Under more severe water-deficit stress (> -1.5 nodule Ψw (MPa) over 14 days) there was a more pronounced reduction of GOGAT. This contrasts with the mild water-deficit stress (-1.0 nodule Ψw (MPa) for 14 days) where GOGAT activity did not show any significant decline. Reduced GOGAT activity upon water-deficit stress has previously been documented for lucerne (Becana et al., 1984) and soybean (Gordon et al., 1997).

REFERENCES

Aylott JW, Richardson DJ, Russell DA. 1997. Optical biosensing of nitrate ions using a sol-gel immobilized nitrate reductase. Analyst, 122:77–80.

Bachmann M, Huber JL, Liao PC, Gage DA, Huber SC. 1996a. The inhibitor protein of phosphorylated nitrate reductase from spinach. Spinacia oleracea leaves is a 14-3-3 protein. FEBS Letters 387:127–31.

Bachmann M, Shiraishi N, Campbell WH, Yoo BC, Harmon AC, et al. 1996b. Identification of the major regulatory site as Ser-543 in spinach leaf nitrate reductase and its phosphorylation by a Ca^{2+}-dependent protein kinase in vitro. Plant Cell, 8:505–17.

Bachmann, M, McMichael Jr, RW, Huber, JL, Kaiser, WM, Huber, SC, 1995. Partial purification and characterization of a calcium-dependent protein kinase and an inhibitor protein required for inactivation of spinach leaf nitrate reductase. Plant Physiology, 108(3): 1083–91.

Baki, GAE, Siefritz, F, Man, HM, Weiner, H, Kaldenhoff, R, Kaiser, WM, 2000. Nitrate reductase in Zea mays L. under salinity. Plant, Cell & Environment, 23(5): 515–21.

Barber MJ, Trimboli AJ, Nomikos S, Smith ET. 1997. Direct electrochemistry of the flavin domain of assimilatory nitrate reductase: effects of NAD^+ and NAD^+ analogs. Archives of Biochemistry and Biophysics, 345:88–96.

Becana M, Aparicio-Tejo PM, Sánchez-Díaz M. 1984. Effects of water stress on enzymes of ammonia assimilation in root nodules of alfalfa (Medicago sativa). Physiologia Plantarum, 61, 653–7.

Ben Azaiez, FE, Ayadi, S, Capasso, G, Landi, S, Paradisone, V, Jallouli, S, Hammami, Z, Chamekh, Z, Zouari, I, Trifa, Y, Esposito, S. 2020. Salt stress induces differentiated nitrogen uptake and antioxidant responses in two contrasting barley landraces from MENA region. Agronomy, 10(9): 1426.

Berks BC, Ferguson SJ, Moir JW, Richardson DJ. 1995. Enzymes and associated electron transport systems that catalyse the respiratory reduction of nitrogen oxides and oxyanions. Biochimica et Biophysica Acta, 1232:97–173.

Botrel A, Magné C, Kaiser WM, 1996. Nitrate reduction, nitrite reduction and ammonium assimilation in barley roots in response to anoxia. Plant Physiology and Biochemistry (Paris), 34(5): 645–52.

Botrel, A, Kaiser, WM. 1997. Nitrate reductase activation state in barley roots in relation to the energy and carbohydrate status. Planta, 201(4): 496–501.

Bruns CM, Karplus PA. 1995. Refined crystal structure of spinach ferredoxin reductase at 1.7 A resolution: oxidized, reduced and 20-phospho-50-AMP bound states. Journal of Molecular Biology, 247:125–45.

Brunt CE, Cox MC, Thurgood AG, Moore GR, Reid GA, et al. 1992. Isolation and characterization of the cytochrome domain of flavocytochrome b2 expressed independently in *Escherichia coli*. Biochemical Journal, 283:87–90.

Campbell ER, Corrigan JS, Campbell WH. 1997. Field determination of nitrate using nitrate reductase. In *Proc. Symp. Field Anal. Methods Hazard. Wastes Toxic Chem.*, ed. E. Koglin, pp. 851–60. Pittsburgh: Air & Waste Manage. Assoc.

Campbell WH, Kinghorn JR. 1990. Functional domains of assimilatory nitrate reductases and nitrite reductases. Trends in Biochemical Sciences, 15:315–319.

Campbell WH. 1989. Structure and synthesis of higher plant nitrate reductase. See Ref. 102a, pp. 123–54.

Campbell WH. 1999. Nitrate reductase structure, function and regulation: bridging the gap between biochemistry and physiology. *Annual Review of Plant Biology*, 50(1): 277–303.

Cannons AC, Barber MJ, Solomonson LP. 1993. Expression and characterization of the heme-binding domain of Chlorella nitrate reductase. Journal of Biological Chemistry 268:3268–71.

Cannons AC, Iida N, Solomonson LP. 1991. Expression of a cDNA clone encoding the haem-binding domain of Chlorella nitrate reductase. Biochemical Journal, 278:203–9.

Chen R, Greer A, Dean AM. 1995. A highly active decarboxylating dehydrogenase with rationally inverted coenzyme specificity. Proceedings of the National Academy of Sciences, 92:11666–70.

Cock JM, Brock IW, Watson AT, Swarup R, Morby AP, Cullimore JV. 1991. Regulation of glutamine synthetase genes in leaves of Phaseolus vulgaris. Plant Molecular Biology, 17:761–71.

Edwards JW, Coruzzi GM. 1989. Photorespiration and light act in concert to regulate the expression of the nuclear gene for chloroplast glutamine synthetase. Plant Cell, 1:241–8.

Edwards JW, Walker EL, Coruzzi GM. 1990. Cell-specific expression in transgenic plants reveals nonoverlapping roles for chloroplast and cytosolic glutamine synthetase.. Proceedings of the National Academy of Sciences USA, 87: 3459–63.

Eisenberg D, Gill HS, Pfluegl GM, Rotstein SH. 2000. Structure-function relationship of glutamine synthetases. Biochimica et Biophysica Acta, (1–2): 122–45.

Elmlinger MW, Mohr H. 1991. Coaction of blue/ultraviolet-A light and light absorbed by phytochrome in controlling the appearance of ferredoxin-dependent glutamate synthase in the Scots pine (*Pinus sylvestris* L.) seedling. Planta, 183(3): 374–80.

Foyer CH, Valadier MH, Migge A, Becker TW. 1998. Drought-induced effects on nitrate reductase activity and mRNA and on the coordination of nitrogen and carbon metabolism in maize leaves. Plant Physiology, 117(1): 283–92.

Ginsburg A, Yeh J, Hennig SB, Denton MD. 1970. Some effects of adenylylation on the biosynthetic properties of the glutamine synthetase from Escherichia coli. Biochemistry, 3:633–49.

Glaab, J, Kaiser, W.M, 1993. Rapid modulation of nitrate reductase in pea roots. Planta, 191(2): 173–9.

González, E.M, Aparicio-Tejo, P.M, Gordon, A.J, Minchin, F.R, Royuela, M. and Arrese-Igor, C, 1998. Water-deficit effects on carbon and nitrogen metabolism of pea nodules. Journal of Experimental Botany, 49(327): 1705–14.

Gordon AJ, James CL. 1997. Enzymes of carbohydrate and amino acid metabolism in developing and mature nodules of white clover. Journal of Experimental Botany, 48:895–903.

Huber JL, Huber SC, Campbell WH, Redinbaugh MG. 1992a. Reversible light/dark modulation of spinach leaf nitrate reductase activity involves protein phosphorylation. Archives of Biochemistry and Biophysics, 296(1): 58–65.

Huber SC, Huber JL, Campbell WH, Redinbaugh MG. 1992b. Comparative studies of the light modulation of nitrate reductase and sucrose-phosphate synthase activities in spinach leaves. Plant Physio*logy*, 100(2):706–12.

Hungria M, Kaschuk G. 2014. Regulation of N2 fixation and NO3−/NH4+ assimilation in nodulated and N-fertilized Phaseolus vulgaris L. exposed to high temperature stress. Environmental and Experimental Botany, 98:32–39.

Kaiser WM, Brendle-Behnisch E. 1995. Acid-base-modulation of nitrate reductase in leaf tissues. Planta, 196(1): 1–6.

Kaiser WM, Kandlbinder A, Stoimenova M, Glaab J. 2000. Discrepancy between nitrate reduction rates in intact leaves and nitrate reductase activity in leaf extracts: what limits nitrate reduction in situ? Planta, 210(5): 801–7.

Kaiser WM, Weiner H, Huber SC. 1999. Nitrate reductase in higher plants: a case study for transduction of environmental stimuli into control of catalytic activity. Physiologia Plantarum, 105(2): 384–9.

Lam HM, Coschigano KT, Oliveira IC, MeloOliveira R, Coruzzi GM. 1996. The molecular-genetics of nitrogen assimilation into amino acids in higher plants. Annual Review of Plant Biology, 47: 569–93.

Liang CG, Chen LP, Yan WANG, Jia LIU, Xu GL, Tian LI. 2011. High temperature at grain-filling stage affects nitrogen metabolism enzyme activities in grains and grain nutritional quality in rice. Rice Science, 18(3): 210–16.

Magasanik B, Neidhardt, F.C. 1987. Regulation of carbon and nitrogen utilization. In: Neidhardt FC, Ingraham JL, Low KB, Magasanik B, Schaechter M and Umbarger HE (Eds.), *Escherichia coli and Salmonella typhimurium: Cellular and Molecular Biology*. American Society for Microbiology, Washington, DC, 1318–25.

Man HM, Abd-El Baki G.K, Stegmann P, Weiner H, Kaiser WM, 1999. The activation state of nitrate reductase is not always correlated with total nitrate reductase activity in leaves. Planta, 209(4): 462–8.

Marque´s S, Florencio FJ and Candau P (1992) Purification and characterization of the ferredoxin-glutamate synthase from the unicellular cyanobacterium Synechococcus sp. PCC 6301. European Journal of Biochemistry, 206: 69–77.

McKenzie GH, Ch'Ng AL, Gayler KR (1979) Glutamine synthetase/glutamine:a-ketoglutarate aminotransferase in chloroplasts from the marine alga Caulerpa simpliciuscula. Plant Physiology, 63: 578–2.

Oaks, A. 1992. A re-evaluation of nitrogen assimilation in roots. BioScience, 42: 103–11.

Okuhara H, Matsumura T, Fujita Y, Hase T. 1999. Cloning and inactivation of genes encoding ferredoxin- and NADHdependent glutamate synthases in the cyanobacterium Plectonema boryanum. Imbalances of nitrogen and carbon assimilations caused by deficiency of the ferredoxin-dependent enzyme. Plant Physiology, 120: 33–42.

Oliveira IC, Lam H-M, Coschigano K, Melo-Oliveira R, Coruzzi GM. 1997. Molecular-genetic dissection of ammonium assimilation in Arabidopsis thaliana. Plant Physiology and Biochemistry 35:185–98.

Oliveira IC, Coruzzi, G. 1999. Reciprocal regulation by carbon and nitrogen metabolites of genes for glutamine synthetase in Arabidopsis. Plant Physiology, 121: 301–9.

Oliver G, Gosset G, Sanchez-Pescador R, Lzoya E, Ku LM, Florez N, Becerril B, Valle F. 1987. Determination of the nucleotide sequence for the glutamate synthase structural genes of Escherichia coli K-12. Gene, 60: 1–11.

Pelanda R, Vanoni MA, Perego M, Piubelli L, Galizzi A, Curti B, Zanetti G. 1993. Glutamate synthase genes of the diazotroph Azospirillum brasilense. Cloning, sequencing, and analysis of functional domains. Journal of Biological Chemistry, 268: 3099–106.

Peterman TK, Goodman HM. 1991. The glutamine synthetase gene family of Arabidopsis thaliana: Light-regulation and differential expression in leaves, roots and seeds. Molecular and General Genetics, 230:145–54.

Reith M, Munholland J. 1993. A high-resolution gene map of the chloroplast genome of the red alga Porphyra purpurea. Plant Cell, 5: 465–75.

Reitzer LJ (1996) Ammonia assimilation and the biosynthesis of glutamine, glutamate, aspartate, asparagine, L-alanine and D-alanine. In: F.C. Neidhart (ed.), *Escherichia coli and Salmonella: Cellular and Molecular Biology*, pp. 391–407. ASM Press, Washington, DC.

Sakakibara H, Watanabe M, Hase T, Sugiyama T. 1991. Molecular cloning and characterization of complementary DNA encoding for ferredoxin-dependent glutamate synthase in maize leaf. Journal of Biological Chemistry, 266:2028–35.

Scheible WR, González-Fontes A, Morcuende R, Lauerer M, Geiger M, Glaab J, Gojon A, Schulze ED, Stitt M. 1997b. Tobacco mutants with a decreased number of functional nia genes compensate by modifying the diurnal regulation of transcription, post-translational modification and turnover of nitrate reductase. Planta, 203(3): 304–19.

Schulz AA, Collett HJ, Reid SJ. 2001. Nitrogen and carbon regulation of glutamine synthetase and glutamate synthase in *Corynebacterium glutamicum*. FEMS Microbiology Letters, 205:361–7.

Suzuki A, Rothstein S. 1997. Structure and regulation of ferredoxin-dependent glutamate synthase from Arabidopsis thaliana. Cloning of cDNA, expression in different tissues of wild-type and gltS mutant strains, and light induction. European Journal of Biochemistry, 243: 708–18.

Suzuki A, Rothstein S. 1997. Structure and regulation of ferredoxin-dependent glutamase synthase from Arabidopsis thaliana: cloning of cDNA, expression in different An Overview of Important Enzymes Involved in Nitrogen Assimilation of Plants 11 tissues of wild-type and gltS mutant strains, and light induction. European Journal of Biochemistry, 243:708–18.

Suzuki, A, Knaff, D.B. 2005. Glutamate synthase: structural, mechanistic and regulatory properties, and role in the amino acid metabolism. Photosynthesis Research, 83(2): 191–217.

Tavares P, Pereira AS, Moura JJ, Moura I. 2006. Metalloenzymes of the denitrification pathway. Journal of Inorganic Biochemistry, 100(12): 2087–100.

Thum KE, Shasha DE, Lajay LV, Coruzzi GM. 2003. Light- and carbon-signaling pathways. Modeling circuits of interactions. Plant Physiology, 132: 440–52.

Tjaden G, Edwards JW, Coruzzi GM. 1995. Cis elements and trans-acting factors affecting regulation of a non-photosynthetic light-regulated gene for chloroplast glutamine synthetase. Plant Physiology, 108: 1109–17.

Valentin K, Kostrzewa M, Zetsche K. 1993. Glutamate synthase is plastid-encoded in a red alga: implications for the evolution of glutamate synthases. Plant Molecular Biology, 23: 77–85.'

Wang H, Zhang M, Guo R, Shi, D, Liu, B, Lin, X, Yang, C. 2012. Effects of salt stress on ion balance and nitrogen metabolism of old and young leaves in rice (*Oryza sativa* L.). BMC Plant Biology, 12(1): 1–11.

Weiner H, Kaiser WM. 1999. 14-3-3 proteins control proteolysis of nitrate reductase in spinach leaves. FEBS Letters, 455(1–2): 75–8.

19 Regulatory RNAs and Their Role in Nitrogen Metabolism of Diazotrophs

Azra Quraishi[1*] and Mehreen Fatima[2]
[1]Knowledge and Research Support Services [KRSS], University of Management and Technology, Lahore, Pakistan
[2]Department of Life Sciences, University of Management and Technology, Lahore, Pakistan
*Corresponding author: E-mail: azra.quraishi@umt.edu.pk

CONTENTS

19.1 Introduction..237
19.2 Role of sRNAs in Nitrogen Metabolism..239
    19.2.1 Indirect Involvement of sRNAs in Nitrogen Metabolism 239
        19.2.1.1 Role of sRNAs in Cyanobacteria..239
        19.2.1.2 Role of sRNAS in Gammaproteobacteria......................................240
        19.2.1.3 Role of sRNA in Alphaproteobacteria...242
        19.2.1.4 Role of sRNA in Methanoarchae...242
    19.2.2 Direct Involvement of sRNAs in Nitrogen Metabolism...................................242
        19.2.2.1 sRNA NSiR4 in Synechocystis 6803..242
        19.2.2.2 sRNA NfiS in Pseudomonas stutzeri A1501.................................243
        19.2.2.3 $sRNA_{154}$ in M. mazei strain Gö1..243
19.3 Can Regulatory RNA Sense Nitrogen Status?..244
19.4 Conclusion 244
References..245

19.1 INTRODUCTION

The regulatory networks that regulate the metabolic processes in living organisms allow the microbes to survive conditions of stress and starvation due to changes in the environment (1–5). These metabolic regulatory networks control the metabolism of carbon energy sources. The regulatory systems are responsible for the acquisition of different nitrogen sources from the environment. These nitrogen sources are important for microbes to survive conditions of starvation and stress (6–9).

For hundreds of years, bacteria and archaea have developed various strategies for the utilization of nitrogen sources (e.g., ammonia, amino acids, and compounds containing inorganic nitrogen (9)). Diazotrophs utilize molecular nitrogen in the absence of available nitrogen sources. This process is present only in prokaryotes and seems to be absent in eukaryotic microorganisms.

Nitrogen uptake and utilization is a strictly regulated process as high-energy processes are triggered in the utilization of nitrogen to acquire energy (8). The regulatory process is mainly carried out at the transcriptional and post-translational levels. These regulatory processes may differ on different microorganisms (i.e., bacteria and archaea).

In bacteria, the transcriptional machinery involved in nitrogen metabolism is regulated by major global transcriptional regulators. These regulators are mostly constitutively expressed. Two main components of this regulatory system are NtrB/NtrC (10). They are present in proteobacteria and regulate transcriptional activation of σ^{54} promoters that are recognized by RNAP (alternative RNA polymerase). This polymerase contains an alternative σ^{54} sigma factor (RpoN).

The transcriptional regulation of nitrogen metabolism in these organisms is carried out by master regulator NtcA. This transcription factor belongs to the CRP family (i.e., cyclic AMP receptor protein). NtcA can form dimers, which in turn can bind to specific promoter motifs required for nitrogen metabolism. These promoters require nitrogen for activation and thus activation of these promoters coordinates as a cellular response to nitrogen availability at the transcriptional level (11).

Ammonia or ammonium is the nitrogen source of choice for nitrogen metabolism as it can incorporate directly to glutamate by enzyme glutamate dehydrogenase at the expenditure of very low energy, but the enzyme has a low-binding affinity to ammonium (12).

In conditions of limited nitrogen, sometimes GS (glutamine synthetase) has the binding affinity to ammonium and GOGAT system (glutamine oxoglutarate aminotransferase) are responsible for ammonium utilization. 1 ATP per ammonium is utilized for ammonium assimilation and thus this GS-GOGAT system is activated only under nitrogen limitation. GS is deactivated by covalent modification or direct protein interactions in case of an uplift in ammonium concentration (14).

The main proteins that are responsible for sensing nitrogen and its regulation are the PII-like proteins that contain highly conserved sequences for nitrogen status sensing by keeping in check three main nitrogen metabolites: glutamate, glutamine, and 2 oxoglutarate (8, 15). The PII-like proteins transmit the nitrogen status signal to target proteins, which results in post-translational modifications in response to nitrogen presence.

The ability of diazotrophs to convert molecular nitrogen to a nitrogen source is of great interest in the field of agriculture as this process is an important part of the nitrogen cycle and prevents loss of environmental nitrogen by denitrification (e.g., the cyanobacteria present in the ocean perform a crucial role of fixing atmospheric dinitrogen to be used as an energy source (16, 17)).

Nitrogenase is the main enzyme responsible for the process of nitrogen fixation (8). It is composed of two enzymes, dinitrogenase (encoded by nif D and nifK) and dinitrogenase reductase (encoded by nif H). The reduction of molecular nitrogen is an energy-driven process and thus is tightly regulated at transcriptional and post-translational levels (9). As compared to bacterial nitrogen fixation, very little knowledge is available on nitrogen fixation and nitrogen metabolism in Archae (18, 20). The Archae contain transcriptional and translational machinery similar to eukaryotic organisms and have less similarity in this aspect to the bacteria.

Methanococcus maripalidus and *Methanosarcina mazei* are the best-studied examples of nitrogen-fixing archae. Studies of these species indicate that the nitrogen regulation occurs at the transcriptional level by the NrpR regulator that regulates transcription of target genes through binding of operator site in case of nitrogen availability (20, 21). This leads to the blocking of RNA polymerase to bind at the promoter site. Upon onset of nitrogen-limiting conditions the intracellular 2 oxoglutarate concentration increases, which is the indicator of nitrogen depletion. 2 oxoglutarate binds to the NrpR, which results in conformational changes in the NrpR structure causing the release of the repressor from the target promoter site and initiation of nitrogen fixation process by activation of target genes (i.e., nif operon, gln A1, gln K1-amtB1operon, and nrpA (22, 26)).

Proteins showing homology to NrpR have been discovered in various methanoarchael genomes, which indicates that this protein might be common in regulating nitrogen metabolism in Archae (23, 28). A second protein NrpA has recently been discovered in *M. mazei*. This protein is under the direct control of NrpR. Under nitrogen starvation conditions, NrpA is expressed, which binds and activates the transcription of nif operon (29).

It has recently been discovered that apart from transcriptional and post-translational modification, a third method/process of regulating nitrogen metabolism is also present in microorganisms.

This process is in fine-tuning with the metabolic networks and holds great importance (30, 31). The small non-coding RNAs (sRNAs) have been identified to play a major role in post-transcriptional regulation. They are involved in metabolic and stress regulation as well as nitrogen metabolism. In the next section, we will discuss the direct and indirect involvement of sRNAs in nitrogen metabolism and regulation.

19.2 ROLE OF SRNAS IN NITROGEN METABOLISM

The involvement of sRNAs in the regulation of nitrogen fixation and response to nitrogen fluctuation due to stress and environmental change has been of keen interest in the past few years. These sRNAs act by mediating the expression of *nif* genes and proteins through a controlled process at the transcriptional and post-translational levels. However, there are also various examples of the indirect effect of sRNAs on nitrogen metabolism.

19.2.1 Indirect Involvement of sRNAs in Nitrogen Metabolism

19.2.1.1 Role of sRNAs in Cyanobacteria

In 2011 Mitschke et al. performed an RNA sequence analysis on the genome of the filamentous cyanobacteria *Anabaena sp.* PCC7120 and tested its response to nitrogen availability. Six hundred transcriptional start sites that correspond to sRNAs were identified (both cis and trans acting) (32). This suggested that sRNAs had a critical role in the nitrogen assimilation process in this organism. The transcoded sRNAs were further studied in detail (e.g., heterocyst-dependent sRNA NsiR1). It was seen that under low nitrogen conditions the cells that are normally responsible for the production of oxygen through photosynthesis were transformed into nitrogen-fixing cells called heterocysts (33). These cells provide the organism with fixed nitrogen. This nitrogen is then transported to all the non-differentiated cells present within the filament. This sRNA is highly conserved in various species of cyanobacteria (heterocyst generating). HetR is the protein that regulates the process of cell differentiation in *Anabaena*. Fusion studies of reporter region of filament growing and thriving in the presence of atmospheric nitrogen have indicated that NsiR1 synthesis is activated in both morphologically distinct heterocysts and heterocysts that show characteristics of vegetative cells. This suggests that the NsiR1 is an early marker of cellular differentiation in cyanobacteria (specifically filamentous cyanobacteria) when the colony is provided with nitrogen-limiting conditions (34, 35).

Another non-coding 6SRNA (i.e., one of the primarily characterized and studies sRNAs) has been seen to be involved in the regulation of the RNAP activity in the stationary growth phase of cyanobacteria (36). In *Synechocystis sp.* (PCC6803) numerous evidence indicated the involvement of 6SRNA in promoting the process of recovery in nitrogen-depleted conditions. Upon comparison of the *Synechocystis* strain with chromosomal deletion of 6SRNA gene with wild-type *Synechocystis* under nitrogen-limiting conditions, it was observed that the physiological and transcriptional response of the mutated strain was comparatively very slow as compared to the wild-type strain (e.g., photosynthetic pigment reassembly was delayed in the deleted strain as compared to the wild-type strain, which was seen as a more bleached phenotype when compared to the natural strain). This is due to the delay in the production of phycobilisome after the organism recovers from nitrogen limitation (37, 38). Furthermore, it was observed that the photosynthetic activity of the deleted strain was 15% less than the wild-type strain. There are two phases of the nitrogen recovery process. Upon onset of the first phase the basal cellular function is recovered including ATP synthetase expression. In the second phase, the process of glycogen degeneration and photosynthetic activity is increased (39, 42).

Heilmann and coworkers did similar research on *Synechocystis* to study the storage of carbon polymers under nitrogen scarcity and it was seen that carbon polymers were rapidly depleted in the wild-type strain when nitrogen concentration was shifted to a sufficient amount but delay in carbon polymers depletion was observed in 6SRNA-deleted strain (43, 48).

Genome-wide microarray to evaluate transcriptome was also carried out in various studies, which indicated that genes encoding ATP synthetase subunits and ribosomal subunits have upregulated level of transcription when nitrogen is added to the habitat in case of the wild-type strain. Therefore, it can be concluded that 6SRNA plays a vital role in the acclimation system of cyanobacteria (49). Some studies also indicate that 6SRNA has a role in sigma factor recruitment during nitrogen recovery. Protein pull-down studies using 6SRNA-deleted *Synechocystis* strain and wild strain indicated that sigma factor (sigA) is recruited into RNAP complex at decreased levels in 6SRNA-deleted strain as compared to the wild-type strain (50)

It was seen that sig B and sig C were recruited more quickly in the deleted 6S strain under conditions of nitrogen recovery as compared to the wild-type strain. This indicated that recruitment of sig B and sig C into RNAP core enzyme is inhibited by the 6SRNA. Sig C is theorized to act as a nitrogen metabolism regulator in the stationary growth phase (51). It is hypothesized that sigC prevents cells from leaving a stress adaptive state whereas sigB inhibits cells from being vulnerable to strass. In short, it can be deduced that 6SRNA is not critically essential for *Synechocystis*, but it has a crucial role in promoting the recovery of the cyanobacteria from nitrogen starvation. This is carried out by the regulation of two groups of sigma factors through transcriptional control (52, 49).

19.2.1.2 Role of sRNAS in Gammaproteobacteria

Numerous sRNAs have been found in gammaproteobacteria that are regulated upon the alteration in nitrogen status in the bacterial habitat. These sRNAs are mostly involved in the biosynthetic pathways and do not directly take part in nitrogen metabolism (53, 54). In 2012, while studying the genome of *Pseudomonas aureginosa* (PA01) (i.e., a cis-acting regulatory RNA element) was identified that had a nitrogen-dependent expression. NalA was initially described as a putative sRNA (55–58). The NalA was found in front of the *nirB* gene found in the *nas* operon having σ^{54}-dependent promoters and was named as nitrate assimilation leader A (NalA). This is because the *nas* operon genes are involved in the process of conversion of nitrate to nitrite by nitrogen assimilation to produce ammonium (54). The process of nitrogen assimilation in *P. aureginosa* was studied by Romeo and coworkers. The nitrate assimilation operon was studied and the NalA (cis-acting RNA) was studied in detail using genetic and biochemical studies and experiments (59).

The NalA has a σ^{54}-binding site present in its promoter sequence. Subsequently, its promoter sequence also has a binding site for the nitrogen response regulator NtrC. This indicates that NalA has nitrogen-dependent transcriptional upregulation upon shift to nitrogen-depleted habitat (57). Transcriptional fusion studies of the Nal A promoter to lacZ gene resulted in very low Nal A activity in σ^{54}-deleted strains as compared to wild strain. *nalA* promoter fusion was tested under different nitrogen sources and it was seen that NalA activity was very low in the presence of ammonium or glutamine whereas high growth conditions were seen in the presence of nitrite and nitrate or glutamate. These results indicate that NalA has an NtrC- and σ^{54}-dependent promoter (60–62).

Close phylogenetic linkage has been observed between *P. aureginosa* and *Azotobacter vinelandii* by analysis of several housekeeping proteins and conserved sequences in the two strains (61). The nitrite assimilation operons are highly conserved in *P. aureginosa, A. vinelandii*, and *Klebsiella pnemoniae*. Antitermination studies within the NalA sequence were carried out in *P. aureginosa* and *A. vinelandii* (54). The authors were deducing whether antitermination within the NalA sequence is necessary to complete nitrate assimilation operon expression (63). Transcription profiling of strains grown in minimal media in the absence of nitrate indicated that *nalA* and *nirB* cotranscription occurs only in nitrate presence. For full expression of the NalA, the entire sequence of *nalA* and promoter plus the first 12 nucleotides of *nirB* are required. If *nalA* is deleted then no expression of NalA is detected. Homology of NasT is present in *A. vinelandii* that acts as an antiterminator protein (64). It has a similar function in *P. aureginosa* PA01. All this data and studies suggest that nitrogen assimilation is regulated by a well-conserved process consisting of various metabolic processes in *P. aureginosa* and *A. vinelandii*. Thus, the strains are phylogenetically related (67).

In 2014, by utilization of bioinformatics tools, the intergenic regions of RpoN (σ^{54}) were analyzed to identify novel sRNA. A highly conserved sRNA with nitrogen-dependent characteristics was identified in *P. aureginosa* PA01 (68–71). This sRNA was named NrsZ. Lac Z fusion studies of this sRNA promoter with various nitrogen concentrations indicated that the promoter activation of *the nrsZ* gene is triggered by nitrogen starvation conditions (72). RNA deep sequencing indicated that NrsZ transcript is activated in conditions of nitrogen starvation or nitrogen depletion (73).

In *E. coli*, some sRNA have been identified that are regulated by nitrogen availability. CyaR is one such sRNA that is under the direct influence of the Crp (global regulator) that mediates the catabolic repression in *E. coli*. CyaR has shown direct inhibition of the essential NAD synthetase encoding gene *nad E*. NAD synthetase uses ammonia for the last step of NAD synthesis (75). Further study by DeLay and Gottesman suggested that the target sequence of CyaR is located exactly in front of the *nad E* translational site (74). It was proposed from their study and research data that Nad E downregulation is carried out with the assistance of CyaR and this downregulation helps to limit the use of ammonia in *E. coli*. (76).

Further studies indicated another sRNA that is not directly nitrogen regulated. GcvB acts as an indirect regulatory factor involved in nitrogen metabolism. It acts by repressing the transporters responsible for amino acid uptake. GcvB is a highly conserved sRNA present in gammaproteobacteria. It has been extensively studied in *Salmonella*. Studies indicate that GcvB controls the gene expression of genes in amino acid metabolism. This regulation is carried out at a post-transcriptional level. The expression of GcvB is high in nutrient-rich media. The regulation of gene by GcvB is carried out by a C/A rich single-stranded region present within the GcvB gene. The region is highly conserved and interacts with the 5' UTRs (5' untranslated regions) of many ABC transporters. As a result, the translation of these genes is blocked due to blockage of the ribosomal-binding site. Due to the inhibition of the specific ABC transporter expression the amino acid uptake is altered in the case of nitrogen availability (77). The inhibition/repression of the ABC transporters helps save energy used in unnecessary amino acid transport when a sufficient amount of nutrients is already present in the environment.

E. coli has a variety of sRNA. One such example is the Hfq-dependent sRNA named sdsN. It is transcribed in a σ^s-dependent manner and its expression is induced upon the onset of the stationary growth phase in *E. coli*. Crl (i.e., a σ^s-related assembly factor dependent on nitrogen) is involved in assisting the transcription of sdsN (80). Under nitrogen-limiting conditions the expression of Crl is reduced, which results in the reduction of sdsN expression. The NarP-NarQ regulatory system is responsible for the transcriptional regulation of nitrate/nitrite responsive genes under anaerobic conditions (81). The NarP-NarQ system is inhibited by the expression of sdsN. Furthermore, sdsN is responsible for the repression of genes involved in the metabolism of oxidized nitrogen compounds.

A highly conserved nitrogen-associated sRNA is present in *A. vinelandii*. This sRNA is named ArrF and is expressed under the condition of iron depletion. The sRNA is negatively regulated by Fur-Fe^{+2}. Fur-Fe^{+2} can be described as the iron metabolism master regulator that tightly controls iron storage and uptake. An "iron box" is present in the ArrF promoter region that binds to the Fur-Fe^{+2} and causes transcriptional repression of the sRNA under iron-rich conditions. Studies with fur-mutated and Δarrf-mutated strains were carried out in comparison to wild *A. vinelandii* strain to study the potential target of the ArrF. FeSII was identified as one of the targets of ArrF. It was observed that ArrF acts to repress FeSII expression in iron-limited conditions. FeSII plays a crucial role in nitrogen metabolism by defending the nitrogenase enzyme from oxygen inactivation. This is carried out by the formation of a protective complex with the nitrogenase enzyme (82). This suggests that ArrF might play a regulatory role in nitrogen metabolism. It was also observed that phbR expression was also regulated by ArrF. phbR is the transcriptional activator of the phbBAC operon. This operon carries out the synthesis of PHB (poly hydroxyl butyrate) at the post-transcriptional level.

PHB acts as an energy source in nutrient-limiting conditions in *A. vinelandii*. Upon iron depletion, it is seen that the phbBAC operon is activated by phbR. phbR is activated by ArrF under Fe-limiting conditions.

19.2.1.3 Role of sRNA in Alphaproteobacteria

MmgR is an sRNA present in *Sinorhizobium meliloti*. This sRNA acts as a bridge between central nitrogen and carbon metabolism (83). A study indicated that the transcription of MmgR is dependent on the cellular nitrogen conditions and the sRNA is only activated in nitrogen-limiting conditions. It has been suggested that NtrC plays a role in the transcription of MmgR by regulating the transcriptional start site of MmgR. NtrC is a component of the Nrt B/C regulatory system. In 2017, it was found that MmgR is involved in the regulation of PHB accumulation quite similarly to ArrF in *A. vinelandii*. MmgR post-transcriptionally regulates Phasin gene expression that encodes PHB granule-associated proteins 1 and 2 (phap 1 and phap 2) (84). This is done in response to carbon surplus growth and nitrogen depletion conditions. Upon deletion of the mmgR core region in *S. melilotii*, an imbalance of PHB accumulation was observed under C surplus. This resulted in an almost 20% increase in PHB granules leading to uncontrolled PHB accumulation that caused an increase in biomass. Upon complementation of the mmgR the PHB levels were restored to the corresponding PHB levels in the wild-type strain. Another study indicated that a conserved heptamer motif is present in the MmgR promoter that is involved in the transcription initiation under conditions of nitrogen limitation (84).

Many of the genes that are involved in nitrogen metabolism in *S. melilotii* (e.g., nifH) and several enterobacterial genes (e.g., glnA) are under the regulatory transcriptional control of NtrC (85). When the reporter region of the mmgR and gfp were fused in an ntrC-mutated strain, it was seen that the promoter activity of mmgR was decreased. This suggests that NtrC has direct activational capacity of MmgR transcription upon the onset of nitrogen-limiting conditions. MmgR plays an important role in both carbon and nitrogen metabolism by regulating genes under specific conditions (86).

19.2.1.4 Role of sRNA in Methanoarchae

In *M. mazei* Gö1, an sRNA ($sRNA_{41}$) is present that has a 100-fold induction in nitrogen-rich conditions as compared to nitrogen-limiting conditions. This $sRNA_{41}$ is highly conserved among the Methanosarcina species. Bioinformatics studies conducted to detect targets of this $sRNA_{41}$ revealed two probable targets (87). These targets are the ACDS (acetyl coenzyme A decarbonylase/synthetase complex) encoding homologous operons. Studies of ACDS levels via PCR in $sRNA_{41}$-deleted and overexpressed strains indicated negligible changes. This suggests that $sRNA_{41}$ carries out post-transcriptional regulation of the ACDS. It was observed that subunits of ACDS operon were upregulated in strains with deleted $sRNA_{41}$ by using a global proteome approach to study protein levels. During growth in the presence of methanol, ACDS is crucial for the production of acetyl coenzyme A (used to generate amino acids by the Wood-Ljungdahl pathway). The decrease in $sRNA_{41}$ causes an increase in the ACDS complex under nitrogen depletion (88). This in turn results in the production of large amounts of amino acids for nitrogenase synthesis.

19.2.2 DIRECT INVOLVEMENT OF sRNAs IN NITROGEN METABOLISM

19.2.2.1 sRNA NSiR4 in Synechocystis 6803

Several sRNAs are unregulated in cyanobacteria under conditions of nitrogen starvation (e.g., *Anabaena 7120* and *Synechocystis 6714 and 6803*) (32, 33). This indicates that these sRNAs are involved in nitrogen regulation in these organisms. The very first sRNA to be functionally characterized to determine its direct involvement in nitrogen metabolism is NsiR4 (nitrogen stress-induced RNA 4) found in *Synechocystis* 6803. This sRNA uses the enzyme GS to effect nitrogen assimilation (89, 91). Studies and research of NSiR4 indicated strong upregulation of the sRNA under conditions of nitrogen depletion. This indicates that it has an important regulatory role in nitrogen metabolism. The sequence of NSir4 is highly conserved at the genomic level. It is distributed over five distinct morphological sections. It was experimentally confirmed that the binding motif of Ntc A was present in the NSiR4 promoter. Ntc A is the global transcriptional regulator of nitrogen assimilation in cyanobacteria (92). This NtcA-binding site is present 48 to 35 nt. Upstream of NSiR4 transcriptional start site (92). This means that the binding site is present in an activating position near the NSiR4 start site in *Synechocystis* 6803. The proximity between the two sites indicates that NSiR4 transcription is induced in the case of nitrogen depletion. 2 oxoglutarate is responsible for the regulation of NtcA activity (92, 98). Insufficient nitrogen presence, 2 oxoglutarate is present in low levels in the cyanobacteria which cause inactivation of NtcA and has low affinity to its target promoter sites. Bioinformatics and experimental techniques were used to deduce the potential targets of NSiR4 (11). Microarray studies were carried out using RNA isolated from different mutated strains of nsiR4 (i.e., overexpressed, deleted, and compensatory strain of NSiR4). This experiment deduced that gifA mRNA was the most probable target of NSiR4.

The GS inactivating factor (IF7) is encoded by the gifA gene (10, 93). This gene is under direct negative control of NtcA. Therefore, under nitrogen depletion, the gifA transcription is downregulated. In strains with deleted NSiR4, it was observed that the mRNA levels of gifA were increased, whereas in NSiR4 overexpressed strains, reduced levels of the gifA were observed (94, 95). The interaction site of NSiR4 is located in the 5' UTR of gifA. Western blot analysis of synechocystis indicated that IF7 protein levels were decreased upon overexpression of NSiR4 as compared to wild strain, whereas in NSiR4-deleted strains 60% higher IF7 protein levels were observed. As compared to IF7, the regulation of IF17 is still unclear. Other sRNAs may be present for regulation that are still to be discovered (96, 97).

19.2.2.2 sRNA NfiS in Pseudomonas stutzeri A1501

Another sRNA was found in *P. stutzeri* A1501 that has a direct role in nitrogen metabolism. This sRNA is NfiS. *P.stutzeri* A1501 is a root-associated bacterium that can fix nitrogen due to the presence of *nif* operon acquired through time by horizontal gene transfer. The NifS has σ^{54}-dependent transcription and is under the direct control of the NtrC/NifA regulatory cascade (100–103). This results in a high induction level of the sRNA upon inducing sorbitol stress or nitrogen depletion. NfiS is highly conserved and is present only in *P. stutzeri* strains. The transcript staability of the sRNA is affected by Hfq. It was observed through experimentation that upon deletion of the hfq gene, NfiS levels were only detected in trace amounts in the bacteria as compared to the wild-type strain (89). The knockout of the NfiS gene results in an extreme decrease in nitrogenase activity in bacteria. However, overexpression of NfiS causes an almost 150% increase in nitrogenase activity. This indicates that NfiS plays a very critical role in the regulation of the nitrogen fixation process. It was seen that the proteins that are encoded by the nif-related genes and nif gene are greatly affected by the overexpression of mutant NfiS (increase in expression), whereas in deletion strain, these genes show a decrease in expression (104). Structure studies of NfiS have shown a stem-loop structure, and it is predicted that this structure can form base pair at the 5' end with the nifK mRNA. The nif K gene encodes a key subunit of the nitrogenase enzyme. It was seen that nif K mRNA degrades more rapidly in the absence of NfiS as compared to the presence of the sRNA. This suggests that the interaction of nifK mRNA with the NfiS helps stabilize the mRNA and increases its half-life. The secondary structure of nifK contains an inhibitory pin that is broken when nif K interacts with NfiS. This means that the mRNA becomes available for transcription upon interaction with NfiS. NfiS is located in the core genome of *P. stutzeri* A1501 and regulates components of *the nif* operon (104).

19.2.2.3 sRNA$_{154}$ in M. mazei strain Göl

sRNA$_{154}$ was the first reported sRNA in archaea that had a role in nitrogen regulation. Its function is the regulation of the expression of the Nif promoter-specific activator NrpA, nitrogenase, and GS.

It works by influencing transcript stability (87). A differential RNA sequence approach was used to identify this sRNA by comparing nitrogen-limiting conditions to nitrogen-sufficient conditions. In 2009, a regulatory function of $sRNA_{154}$ was already predicted. Having a highly conserved structure as well as sequence, the $sRNA_{154}$ has a transcriptional control on the NrpR global repressor of nitrogen. This sRNA is conserved to the methanosarcinales family (94, 105). The transcriptional control on NrpR indicates that $sRNA_{154}$ has a direct regulatory role in nitrogen metabolism. Bioinformatics studies indicated the interaction of several mRNAs with $sRNA_{154}$. These mRNAs encode nitrogen-associated enzymes. Secondary structure studies showed two conserved stem-loops in $sRNA_{154}$ that interact with the target mRNAs (106). Chromosomal deletion studies and differential RNA sequence studies were carried out to confirm this prediction. It was seen in these studies that the transcript levels of genes encoding PII like protein GlnK1, GS1, a transcriptional activator of nif operon, NrpA, and nitrogenase subunit NifH were all downregulated in the $sRNA_{154}$-deleted strain. Further studies indicated that the transcript stability of nifH, glnA1, and nrpA is highly reduced in the absence of the $sRNA_{154}$, whereas levels of glnA2 mRNA showed a slight increase in the absence of $sRNA_{154}$ (107). This indicates that $sRNA_{154}$ has a negative impact on glnA2 expression. Direct interactions between glnA1, A2, nrpA mRNAs, and $sRNA_{154}$ were seen in electrophoretic mobility shift assays, whereas no direct interactions were observed between nifH and $sRNA_{154}$. Deletion of $sRNA_{154}$ causes a decrease in NifH protein levels. This was confirmed by western blot analysis (94, 105).

From all the experimental data it was deduced that the loop structure of several $sRNA_{154}$ stabilize the polycistronic nif mRNA and inhibit endonucleolytic cleavage. This is done by masking the RNAse recognition sites (106).

19.3 CAN REGULATORY RNA SENSE NITROGEN STATUS?

It is important to sense the nitrogen status in the cell to respond correctly to nitrogen availability (12, 14). The sensing mechanism that has been studied descriptively involves the internal metabolic pools of glutamate, glutamine, and 2 oxoglutarate. Sometimes PII-like proteins also transmit a signal to the target proteins. It can be hypothesized that the regulatory RNAs may be able to bind the charged nitrogen metabolites directly (90, 99). As a result of this binding, the regulatory function of the RNA can be affected. Riboswitch-mediated RNA binding if nitrogen metabolites is an interesting mechanism to sense the internal nitrogen status directly (106, 112). In cyanobacteria and marine metagenomes, the naturally occurring aptamers of glutamine have been found upstream from the genes involved in nitrogen metabolism. The majority of the sRNA reported in archaea and bacteria are involved in nitrogen-dependent regulation at the post-transcriptional level (108). These sRNAs are expressed in a nitrogen-dependent manner and are induced or repressed by global transcriptional regulators that sense the nitrogen status (88). However, sRNAs might be able to sense the nitrogen status signal directly by binding to the nitrogen metabolites or by interacting with proteins that are produced or repressed by changes in nitrogen levels (110, 111). Further studies need to be carried out in this field to confirm or negate this theory (90).

19.4 CONCLUSION

The nitrogen metabolism is regulated at a molecular level in prokaryotes and sRNAs and along with finely tuned regulators play a critical role in directly or indirectly regulating the process of nitrogen metabolism. Many of the sRNAs discussed have been shown to play a crucial role in the regulation of both the nitrogen and carbon metabolic cycle in prokaryotes. However, the sRNA that plays a direct role in regulating nitrogen metabolism is scarce. The nitrogen cycle is one of the most important metabolic pathways, and numerous sRNAs may be involved in its regulation that are still undiscovered. Further research is crucial in this matter to identify these sRNAs and understand their role and significance in the regulation of the nitrogen cycle.

REFERENCES

1. Fischer HM. 1994. Genetic regulation of nitrogen fixation in rhizobia. Microbiol Rev 58:352–386.
2. Kessler PS, McLarnan J, Leigh JA. 1997. Nitrogenase phylogeny and the molybdenum dependence of nitrogen fixation in Methanococcus maripaludis. J Bacteriol 179:541–543. http://dx.doi.org/10.1128/jb.179.2
3. Smith DR, Doucette-Stamm LA, Deloughery C, Lee H, Dubois J, Aldredge T, Bashirzadeh R, Blakely D, Cook R, Gilbert K, Harrison D, Hoang L, Keagle P, Lumm W, Pothier B, Qiu D, Spadafora R, Vicaire R, Wang Y, Wierzbowski J, Gibson R, Jiwani N, Caruso A, Bush D, Reeve JN. 1997. Complete genome sequence of Methanobacterium thermoautotrophicum ΔH: functional analysis and comparative genomics. J Bacteriol 179:7135–7155. http://dx.doi.org/10.1128/jb.179.22.7135-7155.1997
4. Leigh JA. 2000. Nitrogen fixation in methanogens: the archaeal perspective. Curr Issues Mol Biol 2:125–131.
5. Dos Santos PC, Fang Z, Mason SW, Setubal JC, Dixon R. 2012. Distribution of nitrogen fixation and nitrogenase-like sequences amongst microbial genomes. BMC Genomics 13:162. http://dx.doi.org/10.1186/1471-2164-13-162
6. Moure VR, Costa FF, Cruz LM, Pedrosa FO, Souza EM, Li XD, Winkler F, Huergo LF. 2015. Regulation of nitrogenase by reversible mono-ADP-ribosylation. Curr Top Microbiol Immunol 384:89–106. http://dx.doi.org/10.1007/82_2014_380
7. Merrick M. 2015. Post-translational modification of P II signal transduction proteins. Front Microbiol 5:763. http://dx.doi.org/10.3389/fmicb.2014.00763
8. Dixon R, Kahn D. 2004. Genetic regulation of biological nitrogen fixation. Nat RevMicrobiol 2:621–631. http://dx.doi.org/10.1038/nrmicro954
9. Leigh JA, Dodsworth JA. 2007. Nitrogen regulation in Bacteria and Archaea. Annu Rev Microbiol 61:349–377. http://dx.doi.org/10.1146/annurev.micro.61.080706.093409. ASMscience
10. Muro-Pastor MI, Reyes JC, Florencio FJ. 2005. Ammonium assimilation in cyanobacteria. Photosynth Res 83:135–150. http://dx.doi.org/10.1007/s11120-004-2082-7
11. Herrero A, Muro-Pastor AM, Flores E. 2001. Nitrogen control in cyanobacteria. J Bacteriol 183:411–425. http://dx.doi.org/10.1128/JB.183.2.411-425.2001
12. van Heeswijk WC, Westerhoff HV, Boogerd FC. 2013. Nitrogen assimilation in Escherichia coli: putting molecular data into a systems perspective. Microbiol Mol Biol Rev 77:628–695. http://dx.doi.org/10.1128/MMBR.00025-13
13. Forchhammer K, Lüddecke J. 2016. Sensory properties of the PII signalling protein family. FEBS J 283:425–437. http://dx.doi.org/10.1111/febs.13584
14. Schumacher J, Behrends V, Pan Z, Brown DR, Heydenreich F, Lewis MR, BennettMH, Razzaghi B, KomorowskiM, Barahona M, Stumpf MP, Wigneshweraraj S, Bundy JG, BuckM. 2013. Nitrogen and carbon status are integrated at the transcriptional level by the nitrogen regulator NtrC in vivo. mBio 4:e00881-e13. http://dx.doi.org/10.1128/mBio.00881-13
15. Zehr JP. 2011. Nitrogen fixation by marine cyanobacteria. Trends Microbiol 19:162–173. http://dx.doi.org/10.1016/j.tim.2010.12.004
16. Hoffman BM, Lukoyanov D, Yang ZY, Dean DR, Seefeldt LC. 2014. Mechanism of nitrogen fixation by nitrogenase: the next stage. Chem Rev 114:4041–4062. http://dx.doi.org/10.1021/cr400641x
17. Seefeldt LC, Hoffman BM, Dean DR. 2009. Mechanism of Modependent nitrogenase. Annu Rev Biochem 78:701–722. http://dx.doi.org/10.1146/annurev.biochem.78.070907.103812
18. Huergo LF, Pedrosa FO, Muller-Santos M, Chubatsu LS, Monteiro RA, Merrick M, Souza EM. 2012. PII signal transduction proteins: pivotal players in post-translational control of nitrogenase activity. Microbiology 158:176–190. http://dx.doi.org/10.1099/mic.0.049783-0
19. Masepohl B, Hallenbeck PC. 2010. Nitrogen and molybdenum control of nitrogen fixation in the phototrophic bacterium Rhodobacter capsulatus. Adv Exp Med Biol 675:49–70. http://dx.doi.org/10.1007/978-1-4419-1528-3_4.
20. Dodsworth JA, Leigh JA. 2006. Regulation of nitrogenase by 2-oxoglutarate-reversible, direct binding of a PII-like nitrogen sensor protein to dinitrogenase. Proc Natl Acad Sci U S A 103:9779–9784. http://dx.doi.org/10.1073/pnas.0602278103
21. Leigh JA. 1999. Transcriptional regulation in Archaea. Curr Opin Microbiol 2:131–134. http://dx.doi.org/10.1016/S1369-5274(99)80023-X

22. Cohen-Kupiec R, Blank C, Leigh JA. 1997. Transcriptional regulation in Archaea: in vivo demonstration of a repressor binding site in a methanogen. Proc Natl Acad Sci U S A 94:1316–1320. http://dx.doi.org /10.1073/pnas.94.4.1316
23. Lie TJ, Leigh JA. 2003. A novel repressor of nif and glnA expression in the methanogenic archaeon Methanococcus maripaludis. Mol Microbiol 47:235–246. http://dx.doi.org/10.1046/j.1365-2958.2003.03293.x
24. Weidenbach K, Glöer J, Ehlers C, Sandman K, Reeve JN, Schmitz RA. 2008. Deletion of the archaeal histone in Methanosarcina mazei Göl results in reduced growth and genomic transcription. Mol Microbiol 67:662–671. http://dx.doi.org/10.1111/j.1365-2958.2007.06076.x
25. Lie TJ, Hendrickson EL, Niess UM, Moore BC, Haydock AK, Leigh JA. 2010. Overlapping repressor binding sites regulate expression of the Methanococcus maripaludisglnK1 operon. Mol Microbiol 75:755–762. http://dx.doi.org/10.1111/j.1365-2958.2009.07016.x
26. Weidenbach K, Ehlers C, Kock J, Schmitz RA. 2010. NrpRII mediates contacts between NrpRI and general transcription factors in the archaeon Methanosarcina mazei Göl. FEBS J 277:4398–4411. http://dx.doi.org /10.1111/j.1742-4658.2010.07821.x
27. Lie TJ, Wood GE, Leigh JA. 2005. Regulation of nif expression in Methanococcus maripaludis: roles of the euryarchaeal repressor NrpR, 2-oxoglutarate, and two operators. J Biol Chem 280:5236–5241. http:// dx.doi.org/10.1074/jbc.M411778200
28. Lie TJ, Leigh JA. 2007. Genetic screen for regulatory mutations in Methanococcus maripaludis and its use in identification of inductiondeficient mutants of the euryarchaeal repressor NrpR. Appl Environ Microbiol 73:6595–6600. http://dx.doi.org/10.1128/AEM.01324-07
29. Lie TJ, Dodsworth JA, Nickle DC, Leigh JA. 2007. Diverse homologues of the archaeal repressor NrpR function similarly in nitrogen regulation. FEMS Microbiol Lett 271:281–288. http://dx.doi.org/10.1111 /j.1574-6968.2007.00726.x
30. Weidenbach K, Ehlers C, Schmitz RA. 2014. The transcriptional activator NrpA is crucial for inducing nitrogen fixation in Methanosarcina mazei Göl under nitrogen-limited conditions. FEBS J 281:3507–3522. http://dx.doi.org/10.1111/febs.12876
31. Wagner EG, Romby P. 2015. Small RNAs in bacteria and archaea: who they are, what they do, and how they do it. Adv Genet 90:133–208. http://dx.doi.org/10.1016/bs.adgen.2015.05.001
32. Mitschke J, Vioque A, Haas F, Hess WR, Muro-Pastor AM. 2011. Dynamics of transcriptional start site selection during nitrogen stressinduced cell differentiation in Anabaena sp. PCC7120. Proc Natl Acad Sci U S A 108:20130–20135. http://dx.doi.org/10.1073/pnas.1112724108
33. Ionescu D, Voss B, Oren A, Hess WR, Muro-Pastor AM. 2010. Heterocyst-specific transcription of NsiR1, a non-coding RNA encoded in a tandem array of direct repeats in cyanobacteria. J Mol Biol 398: 177–188. http://dx.doi.org/10.1016/j.jmb.2010.03.010
34. Zhao J, Wolk CP. 2008. Developmental biology of heterocysts, 2006, p 397–418. In Whitworth DE (ed), *Myxobacteria: Multicellularity and Differentiation*. ASM Press, Washington, DC.
35. Muro-Pastor AM, Hess WR. 2012. Heterocyst differentiation: from single mutants to global approaches. Trends Microbiol 20:548–557. http://dx.doi.org/10.1016/j.tim.2012.07.005
36. Muro-Pastor AM. 2014. The heterocyst-specific NsiR1 small RNA is an early marker of cell differentiation in cyanobacterial filaments. mBio 5:e01079-e14. http://dx.doi.org/10.1128/mBio.01079-14
37. Olmedo-Verd E, Muro-Pastor AM, Flores E, Herrero A. 2006. Localized induction of the ntcA regulatory gene in developing heterocysts of Anabaena sp. strain PCC 7120. J Bacteriol 188:6694–6699. http://dx.doi.org/10.1128/JB.00509-06.
38. Rajagopalan R, Callahan SM. 2010. Temporal and spatial regulation of the four transcription start sites of hetR from Anabaena sp. strain PCC 7120. J Bacteriol 192:1088–1096. http://dx.doi.org/10.1128/JB.01297-09
39. Barrick JE, Sudarsan N, Weinberg Z, Ruzzo WL, Breaker RR. 2005. 6S RNA is a widespread regulator of eubacterial RNA polymerase that resembles an open promoter. RNA 11:774–784. http://dx.doi.org /10.1261/rna.7286705
40. Steuten B, Hoch PG, Damm K, Schneider S, Köhler K, Wagner R, Hartmann RK. 2014. Regulation of transcription by 6S RNAs: insights from the Escherichia coli and Bacillus subtilis model systems. RNA Biol 11:508–521. http://dx.doi.org/10.4161/rna.28827

41. Cavanagh AT, Wassarman KM. 2014. 6S RNA, a global regulator of transcription in Escherichia coli, Bacillus subtilis, and beyond. Annu Rev Microbiol 68:45–60. http://dx.doi.org/10.1146/annurev-micro-092611-150135
42. Burenina OY, Elkina DA, Hartmann RK, Oretskaya TS, Kubareva EA. 2015. Small noncoding 6S RNAs of bacteria. Biochemistry (Mosc) 80:1429–1446. http://dx.doi.org/10.1134/S0006297915110048
43. Wassarman KM, Storz G. 2000. 6S RNA regulates E. coli RNA polymerase activity. Cell 101:613–623. http://dx.doi.org/10.1016/S0092-8674(00)80873-9
44. Trotochaud AE, Wassarman KM. 2004. 6S RNA function enhances long-term cell survival. J Bacteriol 186:4978–4985. http://dx.doi.org/10.1128/JB.186.15.4978-4985.2004
45. Cavanagh AT, Klocko AD, Liu X, Wassarman KM. 2008. Promoter specificity for 6S RNA regulation of transcription is determined by core promoter sequences and competition for region 4.2 of σ70. Mol Microbiol 67:1242–1256. http://dx.doi.org/10.1111/j.1365-2958.2008.06117.x
46. Neusser T, Polen T, Geissen R, Wagner R. 2010. Depletion of the non-coding regulatory 6S RNA in E. coli causes a surprising reduction in the expression of the translation machinery. BMC Genomics 11:165. http://dx.doi.org/10.1186/1471-2164-11-165
47. Cavanagh AT, Sperger JM, Wassarman KM. 2012. Regulation of 6S RNA by pRNA synthesis is required for efficient recovery from stationary phase in E. coli and B. subtilis. Nucleic Acids Res 40:2234–2246. http://dx.doi.org/10.1093/nar/gkr1003
48. Cabrera-Ostertag IJ, Cavanagh AT, Wassarman KM. 2013. Initiating nucleotide identity determines efficiency of RNA synthesis from 6S RNA templates in Bacillus subtilis but not Escherichia coli. Nucleic Acids Res 41:7501–7511. http://dx.doi.org/10.1093/nar/gkt517
49. Heilmann B, Hakkila K, Georg J, Tyystjärvi T, Hess WR, Axmann IM, Dienst D. 2017. 6S RNA plays a role in recovery from nitrogen depletion in Synechocystis sp. PCC 6803. BMC Microbiol 17:229. http://dx.doi.org/10.1186/s12866-017-1137-9
50. Klotz A, Georg J, Bučinská L, Watanabe S, Reimann V, Januszewski W, Sobotka R, Jendrossek D, Hess WR, Forchhammer K. 2016. Awakening of a dormant cyanobacterium from nitrogen chlorosis reveals a genetically determined program. Curr Biol 26:2862–2872. http://dx.doi.org/10.1016/j.cub.2016.08.054
51. Asayama M, Imamura S, Yoshihara S, Miyazaki A, Yoshida N, Sazuka T, Kaneko T, Ohara O, Tabata S, Osanai T, Tanaka K, Takahashi H, Shirai M. 2004. SigC, the group 2 sigma factor of RNA polymerase, contributes to the late-stage gene expression and nitrogen promoter recognition in the cyanobacterium Synechocystis sp. strain PCC 6803. Biosci Biotechnol Biochem 68:477–487. http://dx.doi.org/10.1271/bbb.68.477.5
52. Tuominen I, Tyystjärvi E, Tyystjärvi T. 2003. Expression of primary sigma factor (PSF) and PSF-like sigma factors in the cyanobacterium Synechocystis sp. strain PCC 6803. J Bacteriol 185:1116–1119. http://dx.doi.org/10.1128/JB.185.3.1116-1119.2003
53. Livny J, Brencic A, Lory S, Waldor MK. 2006. Identification of 17 Pseudomonas aeruginosa sRNAs and prediction of sRNA-encoding genes in 10 diverse pathogens using the bioinformatic tool sRNAPredict2. Nucleic Acids Res 34:3484–3493. http://dx.doi.org/10.1093/nar/gkl453
54. Romeo A, Sonnleitner E, Sorger-Domenigg T, Nakano M, Eisenhaber B, Bläsi U. 2012. Transcriptional regulation of nitrate assimilation in Pseudomonas aeruginosa occurs via transcriptional antitermination within the nirBD-PA1779-cobA operon. Microbiology 158:1543–1552. http://dx.doi.org/10.1099/mic.0.053850-0
55. Moreno-Vivián C, Cabello P, Martínez-Luque M, Blasco R, Castillo F. 1999. Prokaryotic nitrate reduction: molecular properties and functional distinction among bacterial nitrate reductases. J Bacteriol 181: 6573–6584.
56. Richardson DJ. 2001. Introduction: nitrate reduction and the nitrogen cycle. Cell Mol Life Sci 58:163–164. http://dx.doi.org/10.1007/PL00000844
57. Lin JT, Stewart V. 1996. Nitrate and nitrite-mediated transcription antitermination control of nasF (nitrate assimilation) operon expression in Klebsiella pheumoniae M5al. J Mol Biol 256:423–435. http://dx.doi.org/10.1006/jmbi.1996.0098
58. Setubal JC, dos Santos P, Goldman BS, Ertesvåg H, Espin G, Rubio LM, Valla S, Almeida NF, Balasubramanian D, Cromes L, Curatti L, Du Z, Godsy E, Goodner B, Hellner-Burris K, Hernandez

JA, Houmiel K, Imperial J, Kennedy C, Larson TJ, Latreille P, Ligon LS, Lu J, Maerk M, Miller NM, Norton S, O'Carroll IP, Paulsen I, Raulfs EC, Roemer R, Rosser J, Segura D, Slater S, Stricklin SL, Studholme DJ, Sun J, Viana CJ, Wallin E, Wang B, Wheeler C, Zhu H, Dean DR, Dixon R, Wood D. 2009. Genome sequence of Azotobacter vinelandii, an obligate aerobe specialized to support diverse anaerobic metabolic processes. J Bacteriol 191:4534–4545. http://dx.doi.org/10.1128/JB.00504-09
59. Rediers H, Vanderleyden J, De Mot R. 2004. Azotobacter vinelandii: a Pseudomonas in disguise? Microbiology 150:1117–1119. http://dx.doi.org/10.1099/mic.0.27096-0
60. Chai W, Stewart V. 1998. NasR, a novel RNA-binding protein, mediates nitrate-responsive transcription antitermination of the Klebsiella oxytoca M5al nasF operon leader in vitro. J Mol Biol 283:339–351. http:// dx.doi.org/10.1006/jmbi.1998.2105
61. Gutierrez JC, Ramos F, Ortner L, Tortolero M. 1995. nasST, two genes involved in the induction of the assimilatory nitrite-nitrate reductase operon (nasAB) of Azotobacter vinelandii. Mol Microbiol 18:579–591. http://dx.doi.org/10.1111/j.1365-2958.1995.mmi_18030579.x
62. Stülke J. 2002. Control of transcription termination in bacteria by RNA-binding proteins that modulate RNA structures. Arch Microbiol 177:433–440. http://dx.doi.org/10.1007/s00203-002-0407-5
63. Wenner N, Maes A, Cotado-Sampayo M, Lapouge K. 2014. NrsZ: a novel, processed, nitrogen-dependent, small non-coding RNA that regulates Pseudomonas aeruginosa PAO1 virulence. Environ Microbiol 16: 1053–1068. http://dx.doi.org/10.1111/1462-2920.12272
64. Maier RM, Soberón-Chávez G. 2000. Pseudomonas aeruginosa rhamnolipids: biosynthesis and potential applications. Appl Microbiol Biotechnol 54:625–633. http://dx.doi.org/10.1007/s00253 0000443
65. Soberón-Chávez G, Lépine F, Déziel E. 2005. Production of rhamnolipids by Pseudomonas aeruginosa. Appl Microbiol Biotechnol 68:718–725. http://dx.doi.org/10.1007/s00253-005-0150-3
66. Köhler T, Curty LK, Barja F, van Delden C, Pechère JC. 2000. Swarming of Pseudomonas aeruginosa is dependent on cell-to-cell signaling and requires flagella and pili. J Bacteriol 182:5990–5996. http://dx.doi.org/10.1128/JB.182.21.5990-5996.2000
67. Déziel E, Lépine F, Milot S, Villemur R. 2003. rhlA is required for the production of a novel biosurfactant promoting swarming motility in Pseudomonas aeruginosa: 3-(3-hydroxyalkanoyloxy) alkanoic acids (HAAs), the precursors of rhamnolipids. Microbiology 149:2005–2013. http://dx.doi.org/10.1099/mic.0.26154-0
68. De Lay N, Gottesman S. 2009. The Crp-activated small noncoding regulatory RNA CyaR (RyeE) links nutritional status to group behavior. J Bacteriol 191:461–476. http://dx.doi.org/10.1128/JB.01157-08
69. Saier MH Jr. 1998. Multiple mechanisms controlling carbon metabolism in bacteria. Biotechnol Bioeng 58:170–174. http://dx.doi.org/10.1002/(SICI)1097-0290(19980420)58:2/3<170::AID-BIT9>3.0.CO;2-I
70. Görke B, Stülke J. 2008. Carbon catabolite repression in bacteria: many ways to make the most out of nutrients. Nat Rev Microbiol 6:613–624. http://dx.doi.org/10.1038/nrmicro1932
71. Deutscher J. 2008. The mechanisms of carbon catabolite repression in bacteria. Curr Opin Microbiol 11:87–93. http://dx.doi.org/10.1016 /j.mib.2008.02.007
72. Sharma CM, Darfeuille F, Plantinga TH, Vogel J. 2007. A small RNA regulates multiple ABC transporter mRNAs by targeting C/A-rich elements inside and upstream of ribosome-binding sites. Genes Dev 21: 2804–2817. http://dx.doi.org/10.1101/gad.447207
73. Sharma CM, Papenfort K, Pernitzsch SR, Mollenkopf HJ, Hinton JC, Vogel J. 2011. Pervasive post-transcriptional control of genes involved in amino acid metabolism by the Hfq-dependent GcvB small RNA. Mol Microbiol 81:1144–1165. http://dx.doi.org/10.1111/j.1365-2958.2011.07751.x
74. Hao Y, Updegrove TB, Livingston NN, Storz G. 2016. Protection against deleterious nitrogen compounds: role of σS-dependent small RNAs encoded adjacent to sdiA. Nucleic Acids Res 44:6935–6948. http://dx.doi.org/10.1093/nar/gkw404
75. Zafar MA, Carabetta VJ, Mandel MJ, Silhavy TJ. 2014. Transcriptional occlusion caused by overlapping promoters. Proc Natl Acad Sci U S A 111:1557–1561. http://dx.doi.org/10.1073/pnas.1323413111
76. Stewart V. 1994. Dual interacting two-component regulatory systems mediate nitrate- and nitrite-regulated gene expression in Escherichia coli. Res Microbiol 145:450–454. http://dx.doi.org/10.1016/0923-2508(94) 90093-0

77. Durand S, Braun F, Lioliou E, Romilly C, Helfer AC, Kuhn L, Quittot N, Nicolas P, Romby P, Condon C. 2015. A nitric oxide regulated small RNA controls expression of genes involved in redox homeostasis in Bacillus subtilis. PLoS Genet 11:e1004957. http://dx.doi.org/10.1371/journal.pgen.1004957
78. Bandyra KJ, Said N, Pfeiffer V, Górna MW, Vogel J, Luisi BF. 2012. The seed region of a small RNA drives the controlled destruction of the target mRNA by the endoribonuclease RNase E. Mol Cell 47:943–953. http://dx.doi.org/10.1016/j.molcel.2012.07.015
79. Papenfort K, Espinosa E, Casadesús J, Vogel J. 2015. Small RNAbased feedforward loop with AND-gate logic regulates extrachromosomal DNAtransfer in Salmonella. Proc Natl Acad Sci U S A 112:E4772–E4781. http://dx.doi.org/10.1073/pnas.1507825112
80. Jung YS, Kwon YM. 2008. Small RNA ArrF regulates the expression of sodB and feSII genes in Azotobacter vinelandii. Curr Microbiol 57: 593–597. http://dx.doi.org/10.1007/s00284-008-9248-z
81. Moshiri F, Kim JW, Fu C, Maier RJ. 1994. The FeSII protein of Azotobacter vinelandii is not essential for aerobic nitrogen fixation, but confers significant protection to oxygen-mediated inactivation of nitrogenase in vitro and in vivo. Mol Microbiol 14:101–114. http://dx.doi.org /10.1111/j.1365-2958.1994.tb01270.x
82. Muriel-Millán LF, Castellanos M, Hernandez-Eligio JA, Moreno S, Espín G. 2014. Posttranscriptional regulation of PhbR, the transcriptional activator of polyhydroxybutyrate synthesis, by iron and the sRNA ArrF in Azotobacter vinelandii. Appl Microbiol Biotechnol 98:2173–2182. http://dx.doi.org/10.1007/s00253-013-5407-7. 8
83. Ceizel Borella G, Lagares A Jr, Valverde C. 2016. Expression of the Sinorhizobium meliloti small RNA gene mmgR is controlled by the nitrogen source. FEMS Microbiol Lett 363:fnw069. http://dx.doi.org /10.1093/femsle/fnw069
84. Lagares A Jr, Ceizel Borella G, Linne U, Becker A, Valverde C. 2017. Regulation of polyhydroxybutyrate accumulation in Sinorhizobium meliloti by the trans-encoded small RNA MmgR. J Bacteriol 199:e00776–16. http://dx.doi.org/10.1128/JB.00776-16
85. Ow DW, Sundaresan V, Rothstein DM, Brown SE, Ausubel FM. 1983. Promoters regulated by the glnG (ntrC) and nifA gene products share a heptameric consensus sequence in the −15 region. Proc Natl Acad Sci U S A 80:2524–2528. http://dx.doi.org/10.1073/pnas.80.9.2524
86. Ceizel Borella G, Lagares A Jr, Valverde C. 2018. Expression of the small regulatory RNA gene mmgR is regulated negatively by AniA and positively by NtrC in Sinorhizobium meliloti 2011. Microbiology 164:88–98. http://dx.doi.org/10.1099/mic.0.000586
87. Jäger D, Sharma CM, Thomsen J, Ehlers C, Vogel J, Schmitz RA. 2009. Deep sequencing analysis of the Methanosarcina mazei Gö1 transcriptome in response to nitrogen availability. Proc Natl Acad Sci U S A 106:21878–21882. http://dx.doi.org/10.1073/pnas.0909051106
88. Buddeweg A, Sharma K, Urlaub H, Schmitz RA. 2018. sRNA41 affects ribosome binding sites within polycistronic mRNAs in Methanosarcina mazei Gö1. Mol Microbiol 107:595–609. http://dx.doi.org/10.1111/mmi.13900
89. Kopf M, Klähn S, Scholz I, Hess WR, Voß B. 2015. Variations in the non-coding transcriptome as a driver of inter-strain divergence and physiological adaptation in bacteria. Sci Rep 5:9560. http://dx.doi.org /10.1038/srep09560
90. Kopf M, Klähn S, Pade N, Weingärtner C, Hagemann M, Voß B, Hess WR. 2014. Comparative genome analysis of the closely related Synechocystis strains PCC 6714 and PCC 6803. DNA Res 21:255–266. http://dx.doi.org/10.1093/dnares/dst055
91. Giner-Lamia J, Robles-Rengel R, Hernández-Prieto MA, Muro-Pastor MI, Florencio FJ, Futschik ME. 2017. Identification of the direct regulon of NtcA during early acclimation to nitrogen starvation in the cyanobacterium Synechocystis sp. PCC 6803. Nucleic Acids Res 45:11800–11820. http://dx.doi.org/10.1093/nar/gkx860
92. Klähn S, Schaal C, Georg J, Baumgartner D, Knippen G, Hagemann M, Muro-Pastor AM, Hess WR. 2015. The sRNA NsiR4 is involved in nitrogen assimilation control in cyanobacteria by targeting glutamine synthetase inactivating factor IF7. Proc Natl Acad Sci U S A 112:E6243–E6252. http://dx.doi.org/10.1073/pnas.1508412112
93. Golden JW, Yoon HS. 2003. Heterocyst development in Anabaena. Curr Opin Microbiol 6:557–563. http://dx.doi.org/10.1016/j.mib.2003.10.004

94. Wright PR, Georg J, Mann M, Sorescu DA, Richter AS, Lott S, Kleinkauf R, Hess WR, Backofen R. 2014. CopraRNA and IntaRNA: predicting small RNA targets, networks and interaction domains. Nucleic Acids Res 42(Web Server issue):W119–W123. http://dx.doi.org/10.1093/nar/gku359
95. Wright PR, Richter AS, Papenfort K, Mann M, Vogel J, Hess WR, Backofen R, Georg J. 2013. Comparative genomics boosts target prediction for bacterial small RNAs. Proc Natl Acad Sci U S A 110:E3487–E3496. http://dx.doi.org/10.1073/pnas.1303248110
96. Urban JH, Vogel J. 2009. A green fluorescent protein (GFP)-based plasmid system to study post-transcriptional control of gene expression in vivo. Methods Mol Biol 540:301–319. http://dx.doi.org/10.1007/978-1-59745-558-9_22
97. Urban JH, Vogel J. 2007. Translational control and target recognition by Escherichia coli small RNAs in vivo. Nucleic Acids Res 35:1018–1037. http://dx.doi.org/10.1093/nar/gkl1040
98. Ames TD, Breaker RR. 2011. Bacterial aptamers that selectively bind glutamine. RNA Biol 8:82–89. http://dx.doi.org/10.4161/rna.8.1.13864
99. Ren A, Xue Y, Peselis A, Serganov A, Al-Hashimi HM, Patel DJ. 2015. Structural and dynamic basis for low-affinity, high-selectivity binding of l-glutamine by the glutamine riboswitch. Cell Rep 13:1800–1813. http://dx.doi.org/10.1016/j.celrep.2015.10.062
100. Yan Y, Yang J, Dou Y, Chen M, Ping S, Peng J, Lu W, Zhang W, Yao Z, Li H, Liu W, He S, Geng L, Zhang X, Yang F, Yu H, Zhan Y, Li D, Lin Z, Wang Y, Elmerich C, Lin M, Jin Q. 2008. Nitrogen fixation island and rhizosphere competence traits in the genome of root-associated Pseudomonas stutzeri A1501. Proc Natl Acad Sci U S A 105:7564–7569. http://dx.doi.org/10.1073/pnas.0801093105
101. Yu H, Yuan M, Lu W, Yang J, Dai S, Li Q, Yang Z, Dong J, Sun L, Deng Z, Zhang W, Chen M, Ping S, Han Y, Zhan Y, Yan Y, Jin Q, Lin M. 2011. Complete genome sequence of the nitrogen-fixing and rhizosphereassociated bacterium Pseudomonas stutzeri strain DSM4166. J Bacteriol 193:3422–3423. http://dx.doi.org/10.1128/JB.05039-11
102. Bentzon-Tilia M, Severin I, Hansen LH, Riemann L. 2015. Genomics and ecophysiology of heterotrophic nitrogen-fixing bacteria isolated from estuarine surface water. mBio 6:e00929. http://dx.doi.org/10.1128/mBio.00929-15
103. Yan Y, Lu W, Chen M, Wang J, Zhang W, Zhang Y, Ping S, Elmerich C, Lin M. 2013. Genome transcriptome analysis and functional characterization of a nitrogen-fixation island in root-associated Pseudomonas stutzeri, p 851–863. In de Bruijn FJ (ed), *Molecular Microbial Ecology of the Rhizosphere*. John Wiley & Sons, Inc, New York, NY. doi:10.1002/9781118297674.ch80
104. Zhan Y, Yan Y, Deng Z, Chen M, Lu W, Lu C, Shang L, Yang Z, Zhang W, Wang W, Li Y, Ke Q, Lu J, Xu Y, Zhang L, Xie Z, Cheng Q, Elmerich C, Lin M. 2016. The novel regulatory ncRNA, NfiS, optimizes nitrogen fixation via base pairing with the nitrogenase gene nifK mRNA in Pseudomonas stutzeri A1501. Proc Natl Acad Sci U S A 113:E4348–E4356. http://dx.doi.org/10.1073/pnas.1604514113
105. Busch A, Richter AS, Backofen R. 2008. IntaRNA: efficient prediction of bacterial sRNA targets incorporating target site accessibility and seed regions. Bioinformatics 24:2849–2856. http://dx.doi.org/10.1093/bioinformatics/btn544
106. Prasse D, Förstner KU, Jäger D, Backofen R, Schmitz RA. 2017. sRNA154 a newly identified regulator of nitrogen fixation in Methanosarcina mazei strain Gö1. RNA Biol 14:1544–1558. http://dx.doi.org/10.1080/15476286.2017.1306170
107. Sharma CM, Vogel J. 2014. Differential RNA-seq: the approach behind and the biological insight gained. Curr Opin Microbiol 19:97–105. http://dx.doi.org/10.1016/j.mib.2014.06.010
108. Hervás AB, Canosa I, Little R, Dixon R, Santero E. 2009. NtrCdependent regulatory network for nitrogen assimilation in Pseudomonas putida. J Bacteriol 191:6123–6135. http://dx.doi.org/10.1128/JB.00744-09
109. Baumgartner D, Kopf M, Klähn S, Steglich C, Hess WR. 2016. Small proteins in cyanobacteria provide a paradigm for the functional analysis of the bacterial micro-proteome. BMC Microbiol 16:285. http://dx.doi.org/10.1186/s12866-016-0896-z
110. Miseta A, Csutora P. 2000. Relationship between the occurrence of cysteine in proteins and the complexity of organisms. Mol Biol Evol 17:1232–1239. http://dx.doi.org/10.1093/oxfordjournals.molbev.a026406

111. Prasse D, Thomsen J, De Santis R, Muntel J, Becher D, Schmitz RA. 2015. First description of small proteins encoded by spRNAs in Methanosarcina mazei strain Gö1. Biochimie 117:138–148. http://dx.doi.org/10.1016/j.biochi.2015.04.007
112. Cassidy L, Prasse D, Linke D, Schmitz RA, Tholey A. 2016. Combination of bottom-up 2D-LC-MS and semi-top-down GelFree-LC-MS enhances coverage of proteome and low molecular weight short open reading frame encoded peptides of the archaeon Methanosarcina mazei. J Proteome Res 15:3773–3783. http://dx.doi.org/10.1021/acs.jproteomeb0056

Index

A

abscisic acid 123
ACC deaminase 180
actinorhizal symbiosis 135
ammonification 3, **40–1**
ammonium assimilation 61–2
ammonium toxicity 215–17
approach application 35
aspartate aminotransferase 106
autophagy 104
auxin 121; response factor 6, 122

B

bacteroid 11, 62, 133, 155–6, 179
bioelectrocatalysis 149
biological nitrogen fixation 3, 10–12, 23, 129–37
brassinolide 207

C

calcification 216
chlorosis 4, 25, 29, 31, 70, 220
critical concentration 30, 46, 215
cross-inoculation 189
cytokinin 122

D

denitrification 3, 42
denitrifier 42, 48
diazotrophs 129–32
dinitrogen tetroxide 203–4, 206
dinitrogenase reductase 12, 144, 145, 149, 177, 178

E

epigenetic regulation 9
ethylene 123

F

fertigation technique 35
fix genes 161
flavonoids 133–6
foliar spplication 35

G

GDH pathway 8, 199, 233
glutamate dehydrogenase 62
glutamate synthase 62
glutamine synthase 42, 228
glutamine synthetase 61
GS/GOGAT cycle 61, 64, 186, 188–9, 199, 233
guaiacol peroxidase 10

H

Haber–Bosch process 3, 143
heterocyst 133, 138, 147–8, 239
high-affinity transport system 44, 72
histone modification response 9

I

industrial nitrogen fixation 144

J

jasmonic acid 123

M

micorhizal association 76

N

necrosis 25
Nif genes 160
nitrate reductase 4, 45, 47–8
nitrate toxicity 220
nitrate transport 6, 45, 46, 60
nitrate transporter 6, 42, 43, 46, 121, 122, 123, 188, 189, 198
nitrate uptake 56–7, 73
nitric acid 3, 20, 23
nitrification 3, 6, **42**
nitrite reductase 48
nitrite toxicity 219
nitrobacter 42
nitrococcus 42
nitrogenase regulation 145–8
nitrogen autotrophy 98, 99
nitrogen bio-fertilizers 12, 171–81
nitrogen cycle **22–4, 40–1**, 238
nitrogenise enzyme complex 11, 143
nitrogen mineralization 2, 18
nitrogen pollution 39
nitrogen toxicity 213–14
nitrogen uptake efficiency 10
nitrogen use efficiency 6, 10, 33, 39
nitrogen utilization efficiency 10
nitrosamines 10
nitrosomonas 42
nitrosonium 207–8
nitrosothiols 208
nod genes 157–60
nodulation 11, 133–7, 156–60, 177–80
nodulin 11, 109, 158

P

peribacteroid 11, 156
peroxynitrite 207

persulfide 207
photosynthetic nitrogen-use efficiency 89
phytoelaxins 135
placement method 34
pnicogens 20
polyphenol oxidase 10
primary salinization 195
programmed cell death 206–7

R

reactive nitrogen species 9, 23, 203–8
remobilization 7, 91, 98–110
rhizobium 173

S

salinization 195–6
secondary salinization 195
senescence 97
S-nitrosylation 203, 207–8
soil compaction 33
specific ion stress 197
strigolactones 119, 205–7
symbiosome 156
symbiotic islands 157

T

T3 pili 135
transamination reaction 56, **63**

U

ubiquitin/proteosome pathway 103
urease 77, 106
ureides 28, **63–4**, 121